S216 Environmental Science
Science: Level 2

The
Uni

Block 3

Water and Life

Prepared for the Module Team by Rachael James, Stuart Bennett and Colin Neal
David Gowing and Hilary Denny

The Open University
Walton Hall, Milton Keynes
MK7 6AA
First published 2002. [Second edition 2005. Third edition 2008. Fourth edition 2011]

Edited and designed by The Open University.
Typeset by The Open University.
Printed and bound in the United Kingdom by Halstan Printing Group, Amersham.

ISBN 978 1 8487 3972 7
4.1

MIX
Paper from responsible sources
FSC® C016278

PART 1
WATER

Rachael James, Stuart Bennett and Colin Neal

A blue planet

1

Without water, there can be no life as we know it on Earth. Even on Earth itself, there are stark differences from region to region that relate to water. Where rainfall is plentiful in the tropics, there are lush forests teeming with all manner of organisms. Where there is little rain, there are deserts that are relatively sparse in life.

Box 1.1 Are we alone?

The presence of liquid water is probably the principal requirement for life on other planetary bodies. Although the evidence is tenuous, it's thought that liquid water may have been present on Mars, and also on some of Jupiter's Galilean satellites.

Mars today is a frozen desert. It's too cold for liquid water to exist on its surface and too cold to rain. The planet's atmosphere is also too thin to permit any significant amount of snowfall.

Even if some internal heat source warmed the planet up enough for ice to melt, it wouldn't yield liquid water. The martian atmosphere is so thin that even if the temperature rose above freezing the ice would change directly to water vapour.

But some scientists believe that there must have been water, and plenty of it, in Mars's past. Images of the martian surface seem to show evidence of dry river beds, flood plains, gullies on martian cliffs and sedimentary deposits (Figure 1.1), all of which can only form in the presence of water.

Figure 1.1 This image of Mars taken by the Mars Global Surveyor shows distinct, thick layers of rock within a crater-like depression, which may have formed in lakes or shallow seas. Such layered rock structures where there were once lakes are common on Earth.

However, if Mars had abundant liquid water in its past, then we would expect to find carbonate minerals in its rock record. Imaging equipment on board the Mars Global Surveyor spacecraft has found no evidence for carbonates on the surface of Mars to date.

If indeed there was once water on Mars, then where did it come from and where did it go? For Mars to be warm enough to rain, it would need a much thicker atmosphere than it has today. As yet, there is no clear explanation for how such an atmosphere could have formed. The answer to the question 'where did the water go?' is equally uncertain. Was it absorbed into the ground, where it remains today, frozen? Or did it dissipate into the martian atmosphere, where it was subsequently lost to space?

Thus the debate continues. The next step is to land an instrument on Mars that can characterize the martian surface, including its sediments (such an instrument is called a lander; Figure 1.2). It will only be then that we are able to better understand what is going on, and what has gone on, on Mars.

Figure 1.2 Simulation of *Beagle 2* on the martian surface. *Beagle 2* is a lander that is able to analyze and sample the martian surface. It is part of the Mars Express mission in the 2003 programme of the European Space Agency. (All Rights Reserved *Beagle 2*.)

Water is stored in many parts of the Earth, including the oceans, the atmosphere, lakes and ice-caps. These are called **reservoirs**. Different reservoirs differ enormously in size (Table 1.1). The total volume of water on Earth is about 1.39×10^9 km^3, with by far the largest proportion (97%) being stored as seawater in the oceans. The remaining 3% is either on (or in) the continents or in the atmosphere. The amount of water in the atmosphere, in the form of water vapour, is very small in comparison with other reservoirs, only around 0.001% of the total.

❍ Although the atmosphere is only a small reservoir of water, it is enormously important. Bearing in mind what you have learnt in Block 2, Part 1, can you suggest why?

● Water in the atmosphere is important because the exchanges of moisture and heat by precipitation and evaporation between the atmosphere and the Earth's surface, fundamentally affect Earth's climate.

Of the freshwater stored on the continents, around three-quarters is in the form of polar ice-caps and glaciers. Another 15% is located in ground over 1 km deep and is relatively inaccessible. The principal sources of water for human use are freshwater lakes, rivers, soil moisture, and relatively shallow groundwater. The usable portion of these sources is estimated to be only about 200 000 km^3 of water, less than 1% of all freshwater on Earth and only about 0.01% of all water on the planet.

Table 1.1 Location of water on Earth.

Location	Volume/10^3 km^3
atmosphere	1.3×10
ocean	1.34×10^6
land	
biological	1
rivers	2
soil	1.6×10
marshes, wetlands*	1.2×10
inland lakes (saline)	8.5×10
inland lakes (fresh)	9.1×10
groundwater (saline)	1.29×10^4
groundwater (fresh; less than 1 km deep)	4.00×10^3
groundwater (fresh; greater than 1 km deep)	6.00×10^3
ice-caps, glaciers	2.70×10^4
total water on Earth	1.39×10^6
total freshwater on Earth	3.71×10^4

*Marshes, wetlands, and water stored in flora and fauna are often mixed salt- and freshwater.

An individual water molecule doesn't remain in any one reservoir forever. Water is continually moving from one reservoir to another: water in the atmosphere will eventually fall as rain, and water in rivers will evaporate or reach the sea. This movement of water is known as the water or the hydrological cycle, and will be discussed in detail in Section 5. What is important is that while water is being continuously transformed, the volume of water on Earth is relatively constant over time and water can be considered a renewable resource.

Each year about 1×10^5 km^3 of water falls as precipitation (precipitation includes rain, snow, sleet and mist) on the continents. With a world population estimated at 5700 million, that means that we each receive on average about 50 000 litres of freshwater per day.

However, this global average hides the considerable variation in the distribution of freshwater in both space and time. For example, some places receive enormous quantities of water regularly; others are extremely dry and arid. Seasonal cycles of rainfall are the rule, not the exception, and the variability from one period to another can be large.

On 16 October 2000 the following news story was filed from Zurich, Switzerland.

Box 1.2

Alpine floods *17 dead and 27 missing*

More than two days of torrential rain high in the Alps bordering Italy, Switzerland and France have created a deluge of huge proportions. More than 60 cm of rain has fallen resulting in unrelenting floods and landslides in this mountainous region.

The village of Gondo, home to some 140 people, lies in a narrow valley between Brig in Switzerland and Domodossola in Italy, near the Simplon Pass. A huge mudslide swept through the village destroying the Stockalper tower (dated to 1670) and many houses and other buildings. Heavy rain in the mountains caused water to cascade through the crevasses, carrying along rocks and mud with enormous force. This mudslide was not entirely natural in that a dam built above the village to trap floodwaters collapsed. The dam was built after floods in 1993.

Heavy rains were causing problems elsewhere in Switzerland. The city of Brig which was also flooded in 1993 was trying to cope with floodwater from the rapidly rising river Rhone. Lake Maggiore reached its highest level for 160 years and flooded part of Locarno and in the Tessin and Bernese Oberland region the situation was critical with several caves flooded and rivers about to break their banks.

Figure 1.3 Photo of part of the village of Gondo showing flood damage.

An indication of the damage caused by the flood can be seen from Figure 1.3 which also shows the rocks and mud loosened and swept along by the floodwater. The damage was made worse by the failure of a dam above the village (supposedly designed to prevent such damage) but the real cause was the volume of water falling in a short time. Rainfall of 60 cm in a little over two days represents about 40 % of the annual rainfall for this area (Figure 1.4).

On 12 October 2000, an area of deep low pressure (relative to the surrounding air) was centred over the UK (Figure 1.5) and cold, arctic air was forced south over Spain and France. The white dots to the west of the Bay of Biscay represent squall clouds and indicate wind speeds of well over 30 knots. At the same time, a band of warm moist air originating over the Equator, west of Senegal was moving towards southern Europe. Three days later, the warm, moist air was driven above the denser cold air and the result was heavy rain (Figure 1.6). The front was relatively stationary for almost three days, being fed with cold air from the north and warm, moist air from the south, and the rain was continuous.

Figure 1.4 Map of alpine Europe showing the area worst hit by the floods (outlined in white).

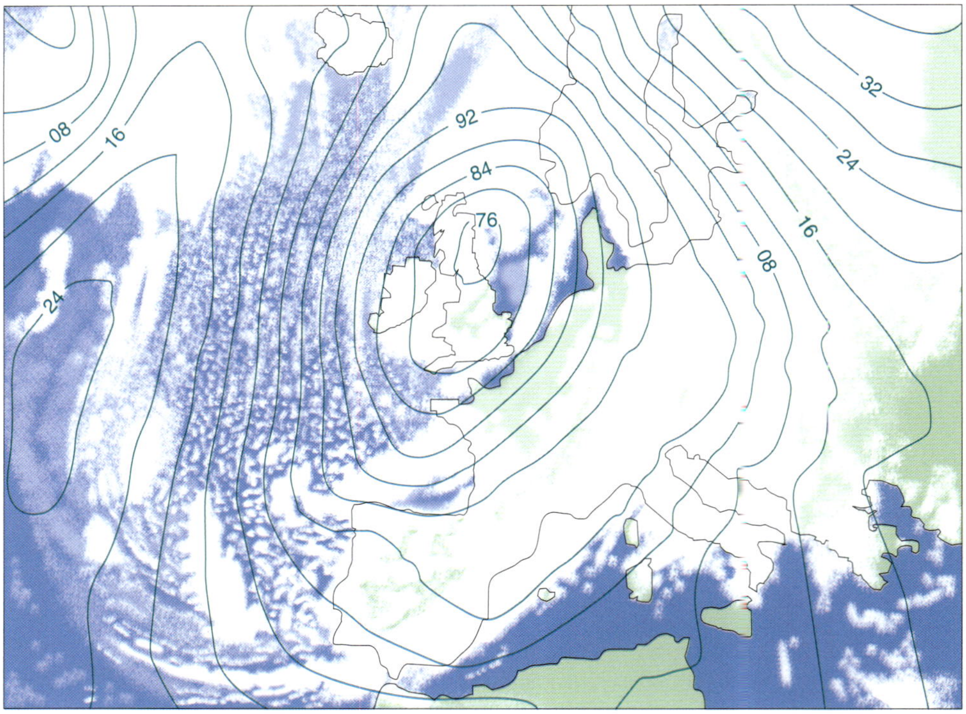

Figure 1.5 Cloud and pressure map for Europe on 12 October 2000 at 00.00 GMT. Note that, for Figure 1.5 and Figure 1.6, the isobars are labelled with the last two digits of the pressure in millibars. For example, the label 84 represents 984 mbar and the label 08 represents 1008 mbar.

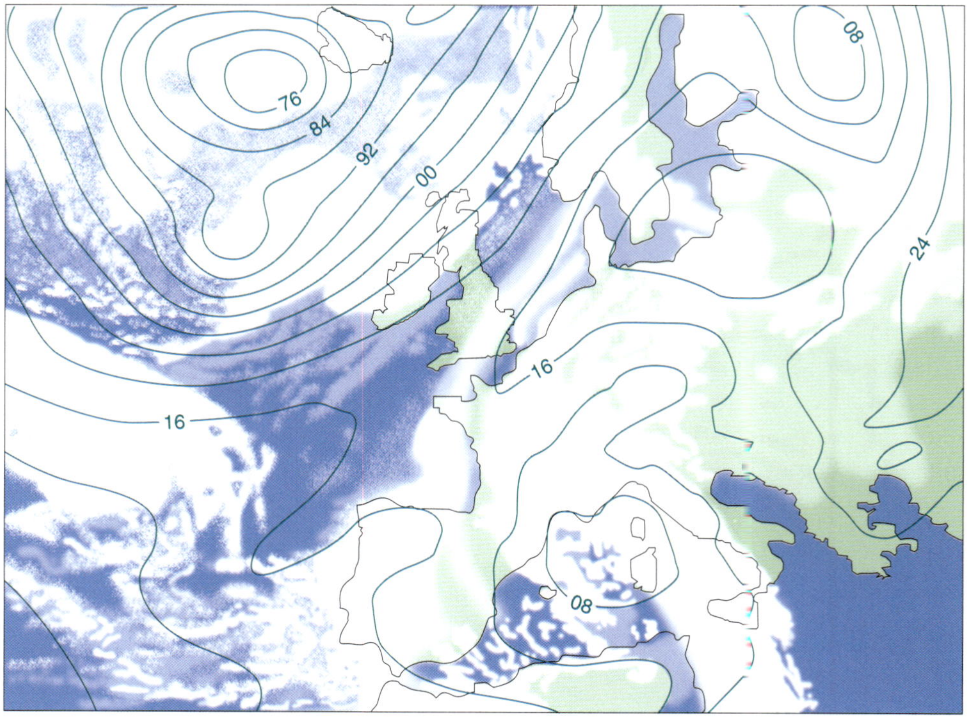

Figure 1.6 Cloud and pressure map for Europe on 15 October 2000 at 00.00 GMT.

Although this was not a large flood by world standards, when the cost of the damage can run into millions of dollars and thousands of lives lost, it does illustrate one of the reasons why it is important to study water. Knowing the limits that can be reached before flooding occurs is essential in terms of planning flood defences, sites for building and for the collection and distribution of drinking water.

1.1 Hydrology: the study of water

The study of water movement upon and beneath the ground and the physics and chemistry of water is called **hydrology**. You saw above that the study of hydrology can be crucial to flood management, but it is important in other ways too:

- *Water resource management.* Water is an important resource for drinking, industry, agriculture, recreation and so on. Demand for water quadrupled from the 1940s to the 1990s yet the stock of freshwater is unchanged.
- *Water quality.* The study of water chemistry is important to assess whether water is suitable for consumption or other purposes. The quality of freshwater on Earth is of major importance to human health, the economy, and the growth and type of organisms found in it.
- *Global climate* The hydrological cycle influences climate in a variety of ways. Exchange of heat during phase transitions from liquid to vapour and vice versa, respectively cool and warm the environment. Water vapour acts as a powerful greenhouse gas and clouds control climate by altering the Earth's radiation budget.

Here our focus will be mainly land water, water both *on* the land and *in* the land. This is not to say that the oceans are not important in the context of study of the environment. The oceans play a major role in climate, in changes of energy distribution on the Earth, in regulating the availability of carbon dioxide and in many other areas. Indeed, the study of the ocean, oceanography, is a whole endeavour in itself and we address this area later in the course. Our immediate concern is that small proportion of global water that is the freshwater of rivers and lakes, water in the ground but close to the surface, water in vegetation and water in the atmosphere. This is the water that is of major importance to life. In order to manage this resource effectively, we need to know how it moves, how much moves, how quickly it moves, what makes it move, and what effect the movement of water over and through the land has in moving materials from one area to another.

For example, try asking the simple question 'What happens to rain?'. A common (and valid) response may be that rain runs off the land into streams and rivers which flow into the ocean. You might be lucky and get a further response indicating that water evaporates from the oceans, forms clouds and rain falls, completing the cycle. The fact that only about one-third of the water from rain reaches the ocean via streams and rivers suggests that there is much more to the process. The disappearance of a puddle after a rainstorm is a common observation but again, ask a question 'Where has the water gone?'. 'It has dried up' would be a frequent answer but this is an answer that does not *explain* anything. As soon as we start probing even the most simple of observations, it becomes clear that the explanations of the observations are far from simple. Nature is like that.

1.2 Summary of Section 1

1 Reservoirs are parts of the Earth that store water.

2 Hydrology is the study of water within reservoirs, and the movement of water and interactions between them.

3 The study of hydrology is important because water is a natural resource that is essential for life on Earth. It is also important for our understanding of the Earth's climate.

What happens to rain?

2

Before we get involved in trying to find out what happens to the rain that falls on the surface of the Earth, we need to look at how much rain falls. You have seen why rain (and other precipitation) occurs and how it is measured using a raingauge. The basic system in the UK is relatively simple involving a sharp-rimmed funnel with a diameter of about 12.7 cm sitting in a copper can, such that the funnel rim is 305 mm above the ground. Remember, though, that it is not only rainfall that is measured by the raingauge. The gauge actually records the total precipitation and condensation from the atmosphere, which includes, in addition to rain itself, snow, sleet and hail and dew, mist, cloud water and frost. The usual measurement interval is 24 hours. So each day, at the same time (usually 09.00 GMT in the UK), someone must trek to the gauge, empty the contents into a calibrated cylinder and record the water depth. This use of a manual storage raingauge is labour intensive and is not ideally suited to remote, inaccessible areas.

2.1 Validity of rainfall data

Siting of a raingauge is every bit as important as its construction. A gauge sticking above ground level presents an obstruction to wind and creates turbulence, Figure 2.1. The result is that some rain droplets which would have fallen into the gauge are carried downwind. This loss can be relatively greater in temperate latitudes where raindrops tend to be smaller than in tropical regions.

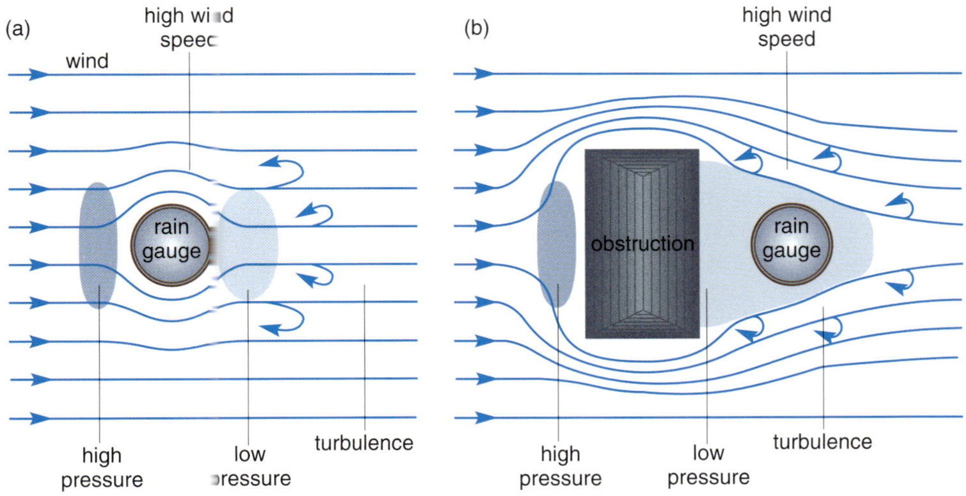

Figure 2.1 Problems with siting of raingauges. (a) The effect of wind creating a pressure gradient across the mouth of the gauge and increasing turbulence, and (b) adjacent obstacles affecting wind flow over the gauge.

The answer seems to be to set the gauge into the ground so that the rim is at ground level. This avoids wind turbulence but makes the gauge more prone to being blocked with snow, leaves or other detritus. We also need to be careful where the gauge is sited. If it is near an obstacle (a large rock or building) it may be protected from the full rainfall and the obstacle could in itself affect the local wind speed by creating eddies. Gauges should generally be separated from any obstacle by a lateral distance of at least four times the height of the obstacle. Wind speed often changes close to sloping ground, so again this has to be taken into account when siting the gauge. Gauges are usually sited on relatively level ground, as shown in Figure 2.2. To reduce the effect of turbulence in particularly windy regions, shields of various designs can be placed round the rim.

Figure 2.2 Modern raingauge sited in an arid area.

A further variation in the design of gauges is reflected in the data that are required. In some circumstances, it may be perfectly satisfactory to have someone trudging the countryside and recording the rainfall at each gauge once per week or even once per month. This would give perfectly good data for weekly, monthly or annual figures. In addition, it is often important to know the pattern of rainfall. For example, does light rain fall steadily over a long period of time or does the rain come in sudden, heavy showers?

Continuously recording gauges are essential where data are required relating to short-term events. A storm may last only an hour or two but 25 mm of rain over that period would be much more likely to cause flooding than would the same amount of rain over 24 hours. Many raingauges today are automatic in that they record the data which may be collected by infrequent visits to the gauge or they are linked to a recording station so that data are gathered without raingauge visits. The storage raingauge, by contrast, records the amount of accumulated water at daily or longer intervals. To measure the intensity of rainfall, a recording raingauge is used that either records small increments (under 0.5 mm) of rain or continuously records changing water levels in the gauge.

The more sophisticated a gauge, the more likely it is to malfunction. Mechanisms can jam, detritus can block the orifice, animals (and humans) can interfere with the operation and the collection of the data may not be performed with due diligence world-wide. These factors, coupled with the care needed in siting the raingauge and its inherent design compromise, mean that rainfall data have to be interpreted with a degree of caution.

So far, we have talked only about individual raingauges which give us a measure of rainfall at a particular point. What is often of interest is rainfall over a particular area, the so-called **areal rainfall**. To have a high density of raingauges would clearly be ideal, but there is a major cost in supply, maintenance and in data collection and processing. A very low density of gauges would result in the loss of detail in the data but there are situations where a low density can be reasonably satisfactory. In general, the greater the local areal variation in rainfall, the greater the density of raingauges needed. An estimate of the minimum density of raingauges needed for different environments comes from the World Meteorological Organization. This ranges from one raingauge per 10 000 km^2 for arid and polar regions through to about one gauge per 600 km^2 in temperate climes, one gauge to 250 km^2 for mountainous areas and a density as high as one gauge for 25 km^2 for small mountainous islands. The average density in the UK is one gauge to 60 km^2, in the USA one gauge to 1040 km^2, and in Saudi Arabia about one gauge to 8000 km^2. Even within these broad figures, siting of raingauges needs to accommodate local variations, taking in factors that are likely to influence rainfall, such as altitude, distance from the sea or proximity to high ground. The use that is to be made of the data is also a factor. For example, if flood information is important, then gauges should be sited in areas that are likely to contribute most to the water supply. There is always a tension between the resources available and the quality of the data obtainable. Siting of raingauges is inevitably a compromise. Rainfall data obtained from raingauges are increasingly being supplemented by data from weather radar and satellites (see Block 2, Part 1). Such methods have the major advantage of being able to address large areas with relative ease, once the system is in place.

Once precipitation data are collected, how are they to be interpreted? Figure 2.3 shows a map of part of Cumbria, UK, with the location of a series of raingauges. Rainfall amounts for each gauge over one month are shown in Table 2.1. An indication of the topography can be gained both from the map and from the photograph (Figure 2.4). The question is 'What is the average rainfall for the area covered by the map for the month?'.

- Which raingauge represents the 'average' rainfall over the area in Figure 2.3, or is there no such thing? Can we estimate the average rainfall simply by taking the mean of all gauges and use this figure to estimate the total volume of water that fell on the area during the month?

Figure 2.3 Raingauge locations in northwest Cumbria, UK. Map scale 1 : 200 000.

Figure 2.4 Photograph taken from British National Grid Reference NY207307 at 140°.

Table 2.1 Rainfall depths over one month recorded by each of the raingauges in Figure 2.3.

Raingauge identifier	Rainfall depth over one month/mm
1	48.6
2	61.1
3	55.4
4	63.4
5	82.9
6	120
7	117
8	131
9	59.7
10	43.9
11	98.6
12	101
13	48.0
14	61.3
15	52.6

- The mean rainfall figure from the 15 gauges is 76.3 mm (which does not correspond to the data for any one raingauge).

  The area covered by the map is 30 km × 30 km.

  So the volume of water over one month

  $= 76.3 \times 10^{-3}\,\text{m} \times 30 \times 10^{3}\,\text{m} \times 30 \times 10^{3}\,\text{m} = 6.87 \times 10^{7}\,\text{m}^{3}$.

Suppose that you were told that the total volume of water for the month falling on this area was, in fact, $6.12 \times 10^{7}\,\text{m}^{3}$. How would you account for this discrepancy?

Box 2.1 Rainfall measurement

The volume of water falling on to an area can be expressed in cubic metres (m^3) or, if the area is particularly large, km^3. Note that 1 km = 1000 m, but that 1 km^3 = 1000 m × 1000 m × 1000 m or $1 \times 10^{9}\,\text{m}^{3}$. You will also meet rainfall expressed in units of millimetres per hectare, mm ha^{-1}. The hectare is an area represented by a square of side 100 m. A volume of water represented by 1 mm ha^{-1} corresponds to a volume of 10 m^3. However, a more useful datum for the hydrologist is the depth of water per unit time, often represented as mm h^{-1}.

As you can see from the variation in the data from the raingauges, different parts of the catchment receive very different amounts of rain. Gauges in the low-lying areas, which represent the major area of the catchment, record lower rainfall than the higher areas facing the prevailing winds (which in this region are from the west and southwest). Data for the higher-altitude gauges are representative of only the smaller part of the area and should not therefore be weighted equally with the gauges in the lower areas. This problem is exacerbated by the greater area density of gauges in the higher areas of this catchment.

One method of estimating average rainfall over an area is to assign a weighting factor to each gauge. The factor is based on the proportion of the area that is nearest to each gauge. These areas are irregular polygons; the method of constructing them can be described quite easily, but it takes a bit of practice to master the technique.

The location of a simple array of just four raingauges is shown in Figure 2.5a with a superimposed grid. Rainfall data recorded by each gauge over a one month period are given in Table 2.2. Adjacent gauges are connected by a network of lines. A perpendicular line is then drawn at the mid-point of each line connecting two gauges (Figure 2.5b), and extensions of the perpendicular bisectors are used to draw polygons around each gauge (Figure 2.5c). The area of each polygon can be measured, and an estimate of rainfall for the whole area can be obtained.

Table 2.2 Rainfall depths over one month recorded by each of the raingauges in Figure 2.5.

Raingauge identifier	Rainfall depth over one month/mm
A	20.3
B	55.2
C	15.8
D	29.7

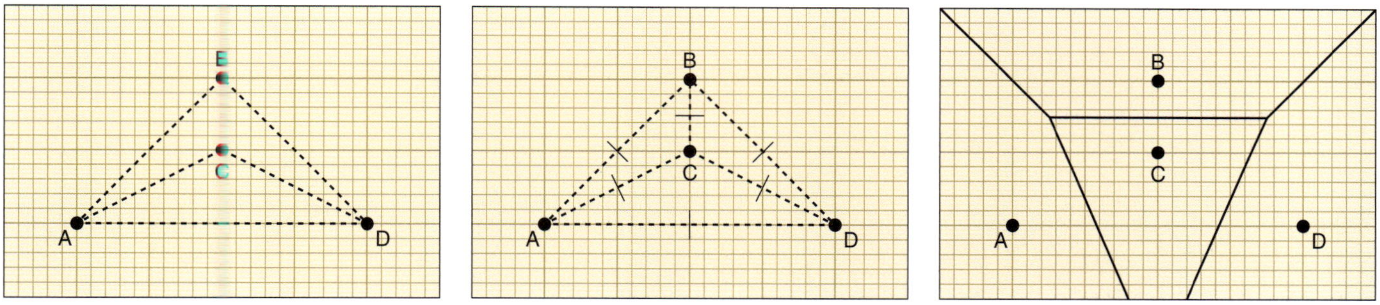

Figure 2.5 (a) Raingauge locations showing connecting lines; (b) perpendicular bisectors of connecting lines; (c) polygons around each gauge.

- ❍ The area shown in Figure 2.5 is divided into a total of 600 squares. Assign a proportion of these squares to each of the polygons A–D.

- ● By counting squares to the nearest complete square, or using basic geometry, the area covered by gauge A is 156 squares, B is 170 squares, C is 118 squares and D is 156 squares. Therefore, polygon A is:

  $$\frac{156}{600} = 0.26$$

  of the total area. In the same way, polygon B = 0.28, C = 0.20 and D = 0.26.

  The average rainfall over the area is therefore (0.26 × 20.3 mm) + (0.28 × 55.2 mm) + (0.20 × 15.8 mm) + (0.26 × 29.7 mm) = 31.6 mm.

This system of analysis, known as the **Theissen polygon method**, can be modified in various ways that take into account local topography and other conditions that might influence rainfall.

An alternative approach is to draw lines on the map connecting points of equal rainfall, known as **isohyets**, and calculating the areas between these lines. The difficult part of the **isohyet method** is to draw these lines, often from inadequate raingauge data. The method can take into account all the information about rainfall data available for the area and is regarded as being relatively reliable. Both these basic methods (and there are many variations on these and other methods) have become more accurate and easier to develop with the introduction of computational methods. For large areas that are not particularly hilly, radar methods are less time-intensive but generally can only provide data for grid squares upwards of 1 km^2 in area.

Question 2.1

Figure 2.6 shows polygons constructed around a series of five raingauges. You might like to try to construct the polygons yourself (you will need graph paper and a protractor). The annual rainfall recorded by each gauge is given in Table 2.3. Calculate the proportion of the total area covered by each gauge then calculate the average annual precipitation over this area.

Table 2.3 Rainfall depths over one year recorded by each of the raingauges in Figure 2.6.

Raingauge identifier	Rainfall/mm yr^{-1}
1	327
2	268
3	333
4	126
5	222

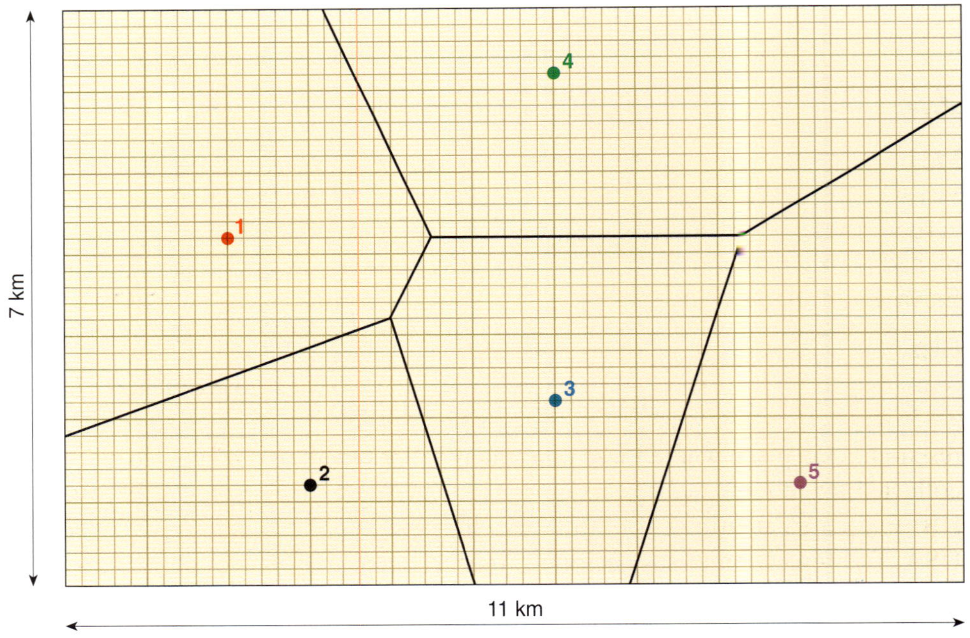

Figure 2.6 Raingauge location and polygon analysis of data in Table 2.3.

So far we have assumed that precipitation is liquid but snow can cause particular problems for the measurement of precipitation. We have seen that reducing turbulence around raingauges is important and for snow, this is even more important. Even the slightest turbulence results in a significant undercatch for snowflakes. Gauges with various wind shields have been designed but catches of snow are still unreliable. For rain in light winds the undercatch is usually only a few per cent, but for snow an undercatch of 50% is not unusual. In heavy snowfalls, the gauge itself can be blocked rendering any measurement meaningless.

Measurement of snow depth is another possible way of estimating frozen precipitation, although it is dependent on the snow not melting rapidly. Also any depth has to take account of unevenness of the snow pack caused by drifting. A further problem is in the conversion of a snow depth to an equivalent volume of water. The density of freshly fallen snow varies from about 50–200 kg m^{-3} (water has a density of about 1000 kg m^{-3}) but snowpack density generally increases on standing up to around 600 kg m^{-3}. Density data can be estimated using radioactive sources. A gamma source and detector is lowered through a vertical channel in the snowpack and changing density can be calculated from changes in the proportion of radiation that is absorbed at various depths.

Probably the best method of recording snowfall (as water equivalent) is to use a large flat bag filled with antifreeze. This is connected to a pressure detector and a continuous estimate of the weight of the snow falling on the bag can be monitored remotely.

Question 2.2

A raingauge without a wind deflector records a rainfall equivalent depth of 6.5 mm over a 24-hour period. However, all this precipitation fell as snow. Comment on the validity of the 6.5 mm measurement of rainfall.

2.2 What is in rainwater?

We have seen how to measure amounts of rainfall but there is more to rainfall than just water. Raindrops may contain small particles of solid materials such as dust or carbon from combustion processes and also dissolved materials, all of which are picked up from the moment that droplet is formed in a cloud through its journey to Earth. Even snow falling over the remotest parts of Antarctica is not pure water.

Clearly, knowing what is in rainwater and how this changes over the years is important in both a scientific sense and as a means of monitoring changes in our environment. Acquiring such data requires careful, painstaking, long-term research. One place where these (and many other types) of data have been collected for more than 30 years is the Hubbard Brook Experimental Forest, HBEF (located at latitude 43° 56′ N and longitude 71° 45′ W) in New Hampshire, USA. The forest comprises deciduous (beech, sugar maple, yellow birch, white ash, basswood, red maple, red oak, white elm) and coniferous (hemlock, red spruce, white pine) species over an altitude from 212 m to 1015 m. Winters are cold and summers warm. The mean air temperature is −9 °C in January rising to 19 °C in July. Average precipitation is 130 cm yr^{-1} of which about 30% is snow equivalent. We shall return to the HBEF a number of times as the work there represents a thoroughly researched, long-term set of data.

The data in Table 2.4 represent the major dissolved ions in precipitation reaching the HBEF. You can see that the forest receives a cocktail of chemicals but we need to ask from where do they come?

Table 2.4 Mean concentration of some dissolved ions in bulk precipitation at HBEF.

Ion	Concentration/mg l^{-1}
H^+	0.073
NH_4^+	0.22
Na^+	0.12
K^+	0.070
Mg^{2+}	0.040
Ca^{2+}	0.16
SO_4^{2-}	2.9
NO_3^-	1.5
Cl^-	0.48
PO_4^{3-}	0.0080
HCO_3^-	0.0060
pH*	4.14

* pH is dimensionless: it does not have units.

The origin of anions such as sulfate, SO_4^{2-}, nitrate, NO_3^- and hydrogen carbonate, HCO_3^-, is relatively easy to explain. All these ions in precipitation have their origins in non-metal oxides that are gaseous at normal temperatures. The reaction of sulfur dioxide SO_2, from burning fossil fuels and volcanic sources, with oxygen in the atmosphere gives sulfur trioxide, SO_3 (Equation 2.1) which dissolves in and reacts with water to give the sulfate anion, SO_4^{2-} (Equation 2.2). (Hydrogen ions are also produced and represent a source of acid rain.)

$$SO_2(g) + \tfrac{1}{2}O_2(g) = SO_3(g) \quad (2.1)$$

$$SO_3(g) + H_2O(l) = 2H^+(aq) + SO_4^{2-}(aq) \quad (2.2)$$

The nitrate anion NO_3^- can arise from nitrogen dioxide NO_2 produced by direct reaction of nitrogen and oxygen in the atmosphere (Equation 2.3) as a result of lightning or from nitrogen oxides (NO_x) from motor vehicle emissions. Subsequent reaction of NO_2 with water gives a mixture of nitrous (HNO_2) and nitric (HNO_3) acids, the latter ionizing to H^+ and the NO_3^- anion (Equation 2.4). Nitrogen oxides can also arise as a result of the decay of plant and animal material.

$$N_2(g) + 2O_2(g) = 2NO_2(g) \quad (2.3)$$

$$2NO_2(g) + H_2O(l) = HNO_2(aq) + H^+(aq) + NO_3^-(aq) \quad (2.4)$$

Atmospheric carbon dioxide CO_2 is the source of the hydrogen carbonate anion HCO_3^-:

$$CO_2(g) + H_2O(l) = H^+(aq) + HCO_3^-(aq) \quad (2.5)$$

Ions such as the magnesium cation Mg^{2+} or the sodium cation Na^+ do not have a related gaseous compound that can spend time in the atmosphere. The oxides of most metals are solids at normal temperatures. So how do these ions get into raindrops?

A clue to the origin of some of these ions can be gleaned from Figure 2.7, which shows the concentration of sodium cation Na^+ in rainwater over continental USA.

What is the significant feature of the distribution of Na^+?

The concentration of Na^+ is higher over coastal areas with generally lower levels inland. This suggests that the sodium cation might be derived from the ocean: after all, sodium is the most abundant metal ion dissolved in seawater (Table 2.5). Tiny droplets of water are blown from the surface of the ocean and form aerosols which can rise to heights of several kilometres in the atmosphere. The aerosols are incorporated into raindrops. However, even the higher coastal levels of Na^+ in rain are far lower than those in seawater itself (by a factor of 150 based on mass).

How could you test this view that sodium ions in rainwater are from the ocean?

If this is the case, and there is no disproportionate loss of material in the aerosol droplets or gain from non-oceanic sources, then the chemical composition of raindrops reaching the Earth should reflect the composition of the seawater.

Figure 2.8 shows a plot comprising a series of points each of which represents the concentration of chloride ions, Cl^-, for a particular sample of rainwater and the concentration of sodium ions in the same sample. The samples were taken from both coastal and inland locations.

Table 2.5 Concentration of the major cations and anions in seawater.

Cation	Concentration/g l^{-1}
Na^+	10.8
Mg^{2+}	1.31
Ca^{2+}	0.410
K^+	0.390
Sr^{2+}	0.0133
Cl^-	19.4
SO_4^{2-}	2.71
HCO_3^-	0.145
Br^-	0.067

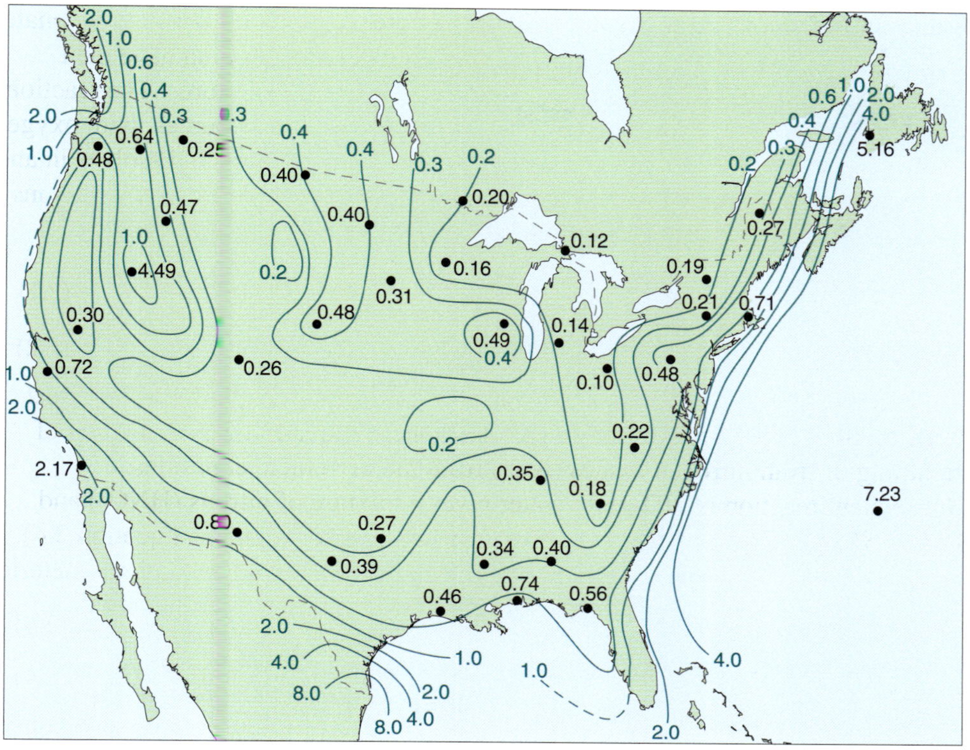

Figure 2.7 Concentration of the sodium cation in rainwater (mg l^{-1}) during the summer period over continental USA.

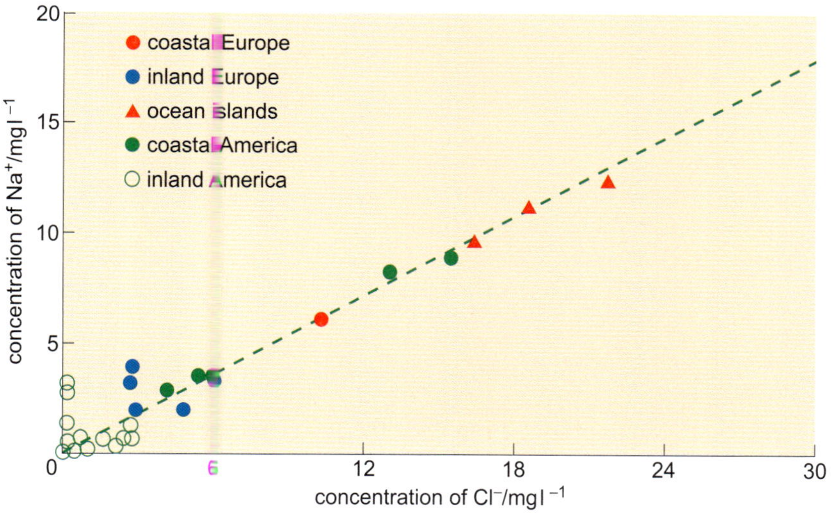

Figure 2.8 Concentration of sodium and chloride ions in samples of rainwater from coastal and inland locations. The dashed line represents the ratio of these two ions in seawater.

- What do the data in Figure 2.8 suggest about the origin of the sodium ions in these samples?

- The samples from coastal locations lie close to the dashed line and show a Na^+/Cl^- ratio similar to that in seawater. This suggests that these ions are derived from the ocean. The samples further away from the dashed line, have a Na^+/Cl^- ratio very different from the ratio in seawater. This indicates an input from non oceanic sources.

Question 2.3

Can you suggest why the concentration of sodium and chloride ions in rainwaters from ocean islands is greater than for rainwaters collected from coastal America and Europe?

Box 2.2 Concentration data

Ionic concentrations in precipitation and in stream water can be expressed as mass concentrations as above, but sometimes they appear as molar concentrations.

To change a molar concentration of the sulfate anion SO_4^{2-} of 30.0×10^{-6} mol l^{-1} to a mass concentration, we must calculate the relative ionic mass of the sulfate anion. This is done simply by adding together the relative atomic mass values of all the atoms in the ion, Table 2.6.

Table 2.6 Relative atomic and ionic mass data for the sulfate anion.

Atom	Relative atomic mass	Number of atoms	Contribution to relative ionic mass
S	32.1	1	32.1
O	16.0	4	64.0

Relative ionic mass of the sulfate anion is 96.1.

One mole of the sulfate anion has a mass of 96.1 g

So 30.0×10^{-6} mol SO_4^{2-} has a mass of $30.0 \times 10^{-6} \times 96.1$ g $= 2.88 \times 10^{-3}$ g

So the molar concentration of the sulfate anion of 30.0×10^{-6} mol l^{-1} is equivalent to 2.88×10^{-3} g l^{-1}.

Mass concentration is often expressed as milligrams per litre (mg l^{-1}). This is done to minimize the use of powers of ten in quoting concentrations. One gram is equivalent to 1000 milligrams (mg).

So the sulfate ion concentration is:

30.0×10^{-6} mol l^{-1}

or 2.88×10^{-3} g l^{-1}

or 2.88 mg l^{-1}

However, we need to take care here with the mass concentrations. The sulfate ion concentration of 2.88×10^{-3} g l^{-1} refers to mass of SO_4^{2-} in a volume of solution of one litre. If we wanted to refer to the concentration of the element sulfur (which comprises just the fraction of $\frac{3.21}{96.1}$ of the mass of sulfate), the concentration would be:

$$2.88 \times 10^{-3}\,\text{g l}^{-1} \times \frac{32.1}{96.1} = 9.62 \times 10^{-4}\,\text{g l}^{-1}$$

To distinguish clearly between these two situations we should write the former concentration based on the mass of sulfate as 2.88×10^{-3} g (SO_4^{2-}) l^{-1} and the one based on the mass of sulfur in the sulfate anion as 9.62×10^{-4} g(SO_4^{2-}–S) l^{-1}.

The use of moles in concentration data avoids this potential source of confusion.

Question 2.4

Express the dissolved ion concentration in mol l^{-1} for water that has concentrations of Na^+ of 1.32 g l^{-1}, Al^{3+} of 0.64 g l^{-1}, Cl^- of 2.05 g l^{-1} and SO_4^{2-} of 3.38 g SO_4^{2-} l^{-1}. The relative ionic mass of Na^+ is 23.0, Al^{3+} is 27.0, Cl^- is 35.5, and SO_4^{2-} is 96.1.

Do the precipitation data from Hubbard Brook (HBEF) in Table 2.4 suggest an oceanic origin of the sodium cation?

Na^+ ion concentration = 0.12 mg l^{-1}
Cl^- ion concentration = 0.48 mg l^{-1}

$$\text{The Cl}^-/\text{Na}^+ \text{ ratio is } \frac{0.48 \text{ mg l}^{-1}}{0.12 \text{ mg l}^{-1}} = 4.0$$

This figure is much higher than the Cl^-/Na^+ ratio of seawater (1.8; Table 2.5) suggesting that, *if* the sodium cation is solely marine-derived, then there must be an additional input of chloride ions in the rain (and possibly sodium ion input).

Another source of material in rain is dust. Winds continually move fine particles of dust (from weathering of rocks and soils) into the atmosphere which become incorporated in water droplets. Soluble materials are leached from the dust (and chemical reactions occur). At any one collection site, the dissolved material in precipitation can vary according to the direction of the wind, whether it passed over industrial areas, over agricultural land where perhaps fertilizer has been used or over areas that are undergoing active erosion. Of the principal ions dissolved in rainwater, Ca^{2+}, SO_4^{2-}, NO_3^- and NH_4^+ have a significant input from continental sources.

- ❍ Which of the ions listed above is most likely to be derived from weathering of rocks?

- ● Ca^{2+} is most likely to be derived from rock weathering.

  The other ions, SO_4^{2-}, NO_3^- and NH_4^+, are largely anthropogenic in origin; NO_3^- and NH_4^+ are found in fertilizers, while SO_4^{2-} is derived from combustion of fossil fuels.

2.3 Where does the water go?

So far, we have assumed that all the rain that falls actually reaches the ground. We have ground level (or something very close to it) as the major location of our raingauges. If the ground is covered with vegetation, only a proportion of the falling rain may actually reach the ground beneath, the throughfall. Rain may be intercepted by vegetation such as the leaves of the trees of a forest and retained (temporarily) on the leaves (and stems). Some of this water may eventually find its way to the ground beneath but some may evaporate directly back into the atmosphere. So the net rainfall under a forest canopy may be less (and sometimes considerably less) than the rainfall falling on the canopy. The water getting through to the forest floor is also unevenly distributed. If you have tried to shelter under a tree during rain, you will be well aware of this. Where the canopy of the tree is densest (usually near the trunk), throughfall is minimal but further out much more water drips through and runs off at a greater rate. Also the drop size of the water that reaches the ground via the leaf surfaces is usually larger than the drop size of the rain and this can have implications for soil erosion, particularly on bare patches of forest floor.

The deficit of water that would have otherwise have reached the ground surface, the so-called **interception loss**, I, is shown in Figure 2.9 and can be represented by

$$I = P - T - S \tag{2.6}$$

where, over a given time,

P, the **precipitation**, represents the depth of the water (often measured in mm) that would have reached the ground over a particular area had there been no vegetation cover. Although we have been discussing rain, precipitation is the appropriate term as it encompasses rain, snow, hail and mist all as sources of water.

T, the **throughfall**, is the depth of water that drips through the leaf cover and reaches the ground over the same area and

S, the **stemflow**, is the depth of water that reaches the ground by running down the stems or trunks of the vegetation. Stemflow is usually small relative to *I* and *T*, although with deciduous trees in winter it may exceed *I*.

Figure 2.9 The fate of rain falling on a forested area: P = precipitation; T = throughfall and S = stemflow.

In a forested area, precipitation *P* can be measured by putting raingauges above the forest canopy. The difficult task is to measure both *T* and *S*. Stemflow can be estimated by attaching ring collection devices (Figure 2.10) around the trunks of trees in a sample area of the forest. Individual raingauges on the ground beneath the forest canopy can give an estimate of throughfall, but locating them in the test area is not easy. As water dripping from the leaves of trees is likely to be greatest at the edge of the canopy of each tree, a raingauge here would register an above average indication for water reaching the ground in the test area. Similarly, there would be an area under the dense canopy where very little water drips through. An alternative method is to cover an area of ground with large impermeable sheets and collect the total throughfall over the area. This is obviously much less convenient but does avoid the problem of choosing the sites for the individual raingauges. To minimize variations an array of collectors can be used at different locations. Mobile collectors are particularly useful in these circumstances.

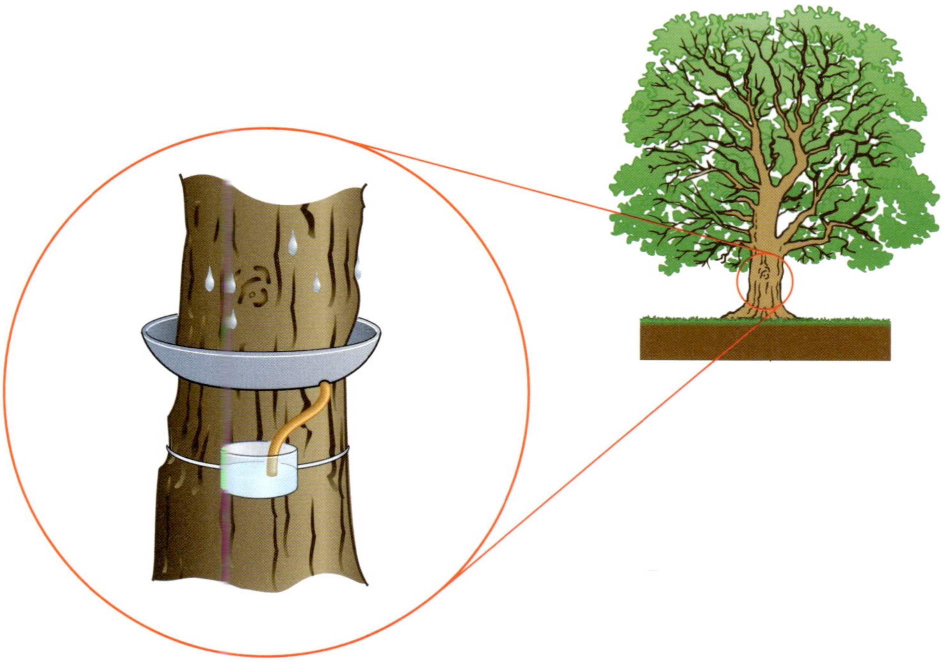

Figure 2.10 Estimation of stemflow.

Throughfall depends on canopy coverage, the type of leaves (smooth or rough, deciduous or evergreen), their orientation and the surface area of the leaves.

Unfortunately (for those who wish to use these data), rain is not continuous and the interpretation of event-based (rainstorms) data has to be done with care. Table 2.7 shows some data for throughfall, and precipitation, for an area of a temperate, evergreen forest for a series of ten rainstorms, that is, for each rainstorm, a value for total throughfall, *T*, and the total rainfall, *P*.

Table 2.7 Data for the total throughfall and total precipitation for temperate evergreen forest for a series of rainstorms.

Rainstorm	Throughfall, *T*/mm	Precipitation, *P*/mm
1	8.1	10.0
2	0.0	1.7
3	2.3	4.4
4	0.0	0.7
5	4.0	5.0
6	4.7	6.6
7	5.4	7.7
8	0.0	1.0
9	6.2	8.0
10	1.6	3.0

In practice, data from many more rainstorm events would be included in a valid analysis and, particularly with seasonal deciduous forests, account would need to be taken of the leaf variation during the year. However, some of the difficulties in interpretation can be seen from the data in Table 2.7.

- ❍ From a visual examination of the data, what seems to be the relationship between throughfall, *T*, and precipitation, *P*?

- ● As you might expect, throughfall increases as precipitation increases. However, for small amounts of precipitation, throughfall is zero.

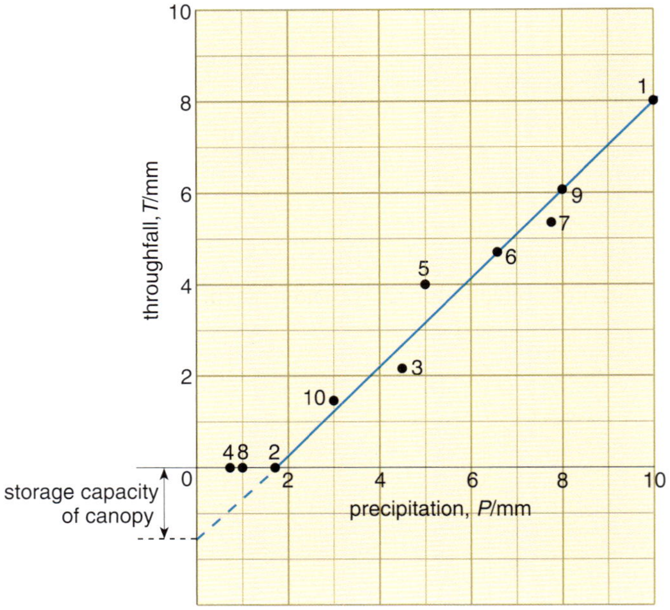

Figure 2.11 Plot of throughfall against precipitation for a series of ten rainstorms (1–10). The solid line represents the line of best fit through the data.

The data from Table 2.7 are shown in Figure 2.11, with throughfall plotted against precipitation. The data have a significant scatter but the approximately linear relationship between throughfall and gross precipitation is clear for all values of precipitation above 1.8 mm. The plot also shows the extension of the line of best fit to the throughfall axis. When $P = 0$, T has a value of −1.8 mm.

What does this negative intercept represent as a negative throughfall does not have a physical meaning? Think what happens when rain begins to fall on the canopy. At first, there is no throughfall as the rain is simply wetting the surface of the leaves. Only when the leaves have their (usually upper only) surfaces thoroughly wetted, does water begin to run off the leaves and fall to the forest floor beneath. With a dense canopy, it can be a considerable time before the lower leaves become drenched and throughfall begins. This negative intercept in Figure 2.11 then represents the surface **storage capacity** of the canopy. Comparative storage capacity data for different vegetation types are shown in Table 2.8.

Table 2.8 Data showing the variation in canopy storage with type of vegetation.

Vegetation	Surface storage capacity/mm
coniferous	
Douglas fir	1.2
Sitga spruce	1.8
Norway spruce	1.6
deciduous	
oak	0.9
hornbeam	1.0
grasses	1.2
heather	1.1
tropical forest	~ 2.5

These data vary significantly from one study to another but they do all indicate similar trends. You will notice that the table includes data for grasses and heather. These data are much more difficult to obtain (and more unreliable) owing to the difficulty of measuring throughfall for ground cover vegetation.

The greatest relative proportion of rainfall that is intercepted occurs in the early stages of a storm when the leaves are relatively dry. It follows then that there is likely to be a greater interception with discrete, short duration rainstorms. The leaves dry between the short rainstorms but this does not happen when the same total rainfall occurs in a single continuous event. So the amount of water reaching the ground *at any one spot* is not just a function of climate and weather patterns but depends critically on vegetation cover. However, as we shall see, the net amount of water sinking into the ground *overall* may be largely independent of ground cover.

Not only does a vegetation canopy change the distribution of water reaching the ground, the composition of the water is different too. The data in Table 2.9 represent a comparison of the major ions in rainwater above a forest canopy, with the ion composition of water that has come through the canopy by running off leaves and from flowing down stems (or trunks of trees). About 85–90% of water falling on the top of the canopy may reach the forest floor beneath, the rest evaporates from (and is absorbed by) the leaves and stems.

Table 2.9 Chemical composition of precipitation above and beneath (at ground level) a canopy of maple, birch and beech trees during the summer period.

Element	Precipitation above canopy/mg l^{-1}	Precipitation beneath canopy/mg l^{-1}
calcium (as Ca^{2+})	0.16	1.6
magnesium (as Mg^{2+})	0.03	0.45
potassium (as K^{+})	0.07	6.4
sodium (as Na^{+})	0.06	0.14
nitrogen (as NO_3^{-})	0.22	0.67
nitrogen (as NH_4^{+})	0.21	1.2
phosphorus (as PO_4^{3-})	0.0026	0.15
sulfur (as SO_4^{2-})	0.90	5.4
chlorine (as Cl^{-})	0.45	1.5
hydrogen (as H^{+})	0.087	0.010

Note that the number of collectors required when chemical data are being sought should be high. Pollutants and other materials can dissolve in water to different extents within forests and at the forest edges. Local variations can make the collection of valid data difficult. What is the most noticeable difference in the data in Table 2.9 between the composition of the above-canopy and ground-level water composition?

All the ions in Table 2.9 show an increase in concentration as the water passes through the canopy and down the stems with one exception. There is a marked decrease in the concentration of hydrogen ion. As we have seen, some ions are present in the incident rainfall. If some of this rain evaporates directly from the surface of the leaf, the dissolved materials in the water will remain on the leaf until they are dissolved by further rain flowing over the leaf and down to the forest floor. The canopy does not represent a clean, inert leaf surface. Leaves intercept aerosols, adsorb gases such as the oxides of nitrogen and sulfur and have dust from the atmosphere deposited on them. Forested areas downwind from arable farmland are likely to be dusted with nitrate, phosphate and potassium during and after fertilizer application. Rain falling on to the surface of a leaf will dissolve these materials and enrich the throughflow. It is also possible that some material may be leached from within the leaves themselves rather than just from the surface.

Less obvious is why the concentration of one of the ions in rain (the hydrogen ion) should actually decrease as the water passes through the canopy. Where are the hydrogen ions going (and what is replacing them)? Rather than just sitting on the surface of a leaf, ions can be adsorbed on the surface. In Figure 2.12a, hydrogen ions are replaced in the rainwater by sodium and magnesium ions from the leaf, and the water, depleted in hydrogen ions, passes through the canopy. Note that the ion replacement does not alter the overall neutral charge of the water. Each hydrogen ion in water is replaced by one sodium ion, but two hydrogen ions are replaced by a single magnesium ion. The molar concentration of charge of the incoming and outgoing water is unchanged for ion exchange processes.

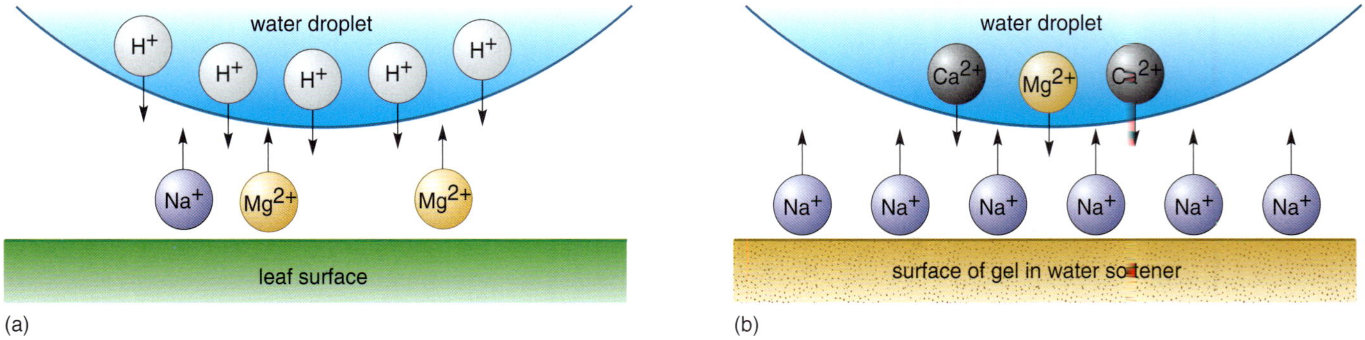

Figure 2.12 Exchange of ions (a) on the surface of a leaf and (b) in a domestic water-softening unit.

This process of ion exchange is similar to that which occurs in domestic water softeners. However, here the calcium and magnesium ions that are in hard water are replaced by sodium ions adsorbed onto a gel. Salt (sodium chloride) used to regenerate water softeners simply removes the calcium and magnesium ions from the gel and restores the sodium ions, ready for more water softening, Figure 2.12b.

Already we have seen that ten to fifteen per cent of water falling on the canopy evaporates without ever reaching the forest floor. Evaporation is an important process not just for forested areas. Evaporation ultimately results in all precipitation returning to the atmosphere and forms the focus of the next stage of our examination of the fate of rainwater.

Question 2.5

Data for the concentration of the nitrate ion, NO_3^-, in precipitation above a forest canopy and in throughfall is shown below. Data A are from a forest close to an urban area and Data B are from a forest more than 50 km away from significant population centres. How would you account for the differences in the two sets of data?

	Data A/mg l^{-1}		Data B/mg l^{-1}	
ion	precipitation	throughfall	precipitation	throughfall
NO_3^-	5.8	10.2	2.5	6.3

2.4 Evaporation

Evaporation is the process by which liquid water is converted to a gas. Water molecules are removed from the surface of the liquid and enter the gas phase. Imagine some liquid water and dry air is put into a closed container. Initially, water molecules from the surface of the liquid water enter the gas phase. This is evaporation. However, as molecules accumulate in the gas phase, the chance of a molecule returning to the liquid surface increases. Eventually the rate of evaporation of molecules from the liquid surface is the same as the rate at which water molecules condense back to the liquid as shown in Figure 2.13. The net rate of evaporation is said to be zero. Although there is no net change in the amount of water in the liquid and gas phase, evaporation and condensation are going on.

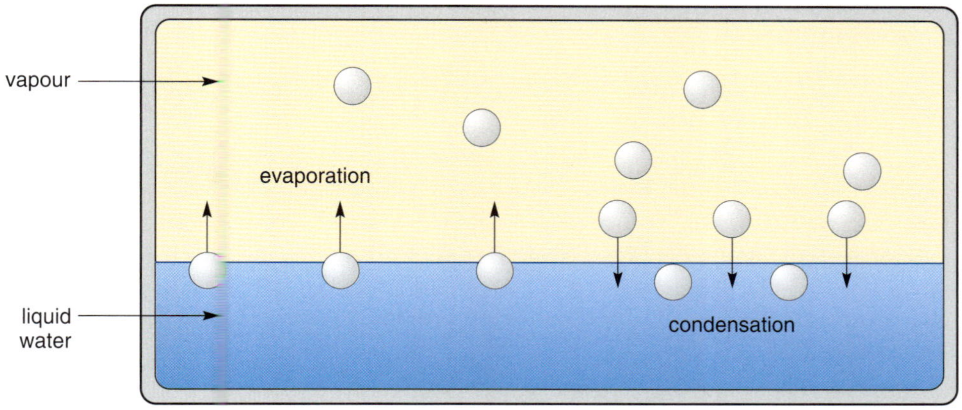

Figure 2.13 Evaporation and condensation in a closed container.

However, we are not usually dealing with a closed container. A puddle left by a rain shower eventually dries up. It may take a long time on a cold, damp day but the puddle will disappear more rapidly in warm, dry weather. What we are seeing here is an evaporation rate that is greater than a condensation rate. The air above the puddle will be on the move and replaced with air that contains less water.

Evaporation from liquid water can take place at any temperature under normal atmospheric pressure. It occurs only at the surface of the liquid. With boiling, the conversion of liquid water to gaseous water takes place throughout the bulk of the liquid and occurs at a fixed temperature for a given pressure. The boiling temperature of water under normal atmospheric pressure is 100 °C but this is significantly lower at high altitude where the pressure is lower.

The energy requirement to convert water from the liquid to the gas phase is considerable. To convert one kilogram of liquid water at 100 °C to the gas phase at the same temperature requires an energy input of about 2.4×10^6 J (joule). This represents the latent heat of vaporization of water. Its value varies a little with temperature being $2.47 \times 10^6\ \text{J kg}^{-1}$ at 10 °C and $2.44 \times 10^6\ \text{J kg}^{-1}$ at 25 °C. To raise the temperature of one kilogram of liquid water from 0 °C to 100 °C requires an input of just 4.18×10^5 J, just about one sixth of the energy required to convert the water from a liquid to a gas. The large quantities of energy involved with evaporation have a major effect on the Earth's energy budget. Energy changes in evaporation processes represent over 70% of the radiation energy from the Sun reaching the Earth.

- ❍ The average annual rainfall over the UK is 0.874 m and the land surface area is $2.45 \times 10^5\,km^2$. How much energy would be needed to convert this quantity of water into gaseous water at 10 °C?

- ● The volume of annual rainfall over the UK is $0.874\,m \times 2.45 \times 10^5\,km^2$

  $= 0.874 \times 2.45 \times 10^5 \times 10^6\,m^3 = 2.14 \times 10^{11}\,m^3$

  $(1\,km^2 = 1\,km \times km = 1 \times 10^3\,m \times 1 \times 10^3\,m = 1 \times 10^6\,m^2)$

  As one cubic metre of water has a mass of $1 \times 10^3\,kg$, the mass of this water is:

  $2.14 \times 10^{11} \times 10^3\,kg = 2.14 \times 10^{14}\,kg$

  At 10 °C, the latent heat of vaporization of water is $2.47 \times 10^6\,J\,kg^{-1}$.

  So the energy required to convert the annual rain falling on the UK back into gaseous water is:

  $2.14 \times 10^{14} \times 2.47 \times 10^6\,J = 5.29 \times 10^{20}\,J$ or $5.29 \times 10^{14}\,MJ$ $(1\,MJ = 1 \times 10^6\,J)$.

As about one-third of rainfall finds its way from the land into the ocean, annual land evaporation in the UK consumes about 3.53×10^{14} MJ. By way of comparison, the annual energy consumption of all homes in the UK amounts to about 4×10^{11} MJ: the difference is a *factor* of about 1×10^3, or one thousand.

Question 2.6

Assume that a domestic garden of area 0.25 ha loses water at the rate of $0.47\,l\,m^{-2}\,d^{-1}$ by evaporation. Calculate the energy required for vaporization if the temperature is 25 °C.

2.4.1 Estimating evaporation

Some of the radiant energy reaching the surface of the Earth is reflected back into space and is not available for evaporation. Of the net radiant energy (absorbed minus reflected), R, some is used to heat the soil, vegetation or surface water, G, some is used for heating the air above the land surface, H, and only a tiny proportion goes towards the growth of plants. In the following equation, E is the latent heat of vaporization of water:

$$R = H + G + E \qquad (2.7)$$

The relative values of these is different for particular regions. In arid areas H and G are much greater than E, but for tropical rainforests E dominates.

The amount of radiation reaching the surface of the Earth at a particular point varies with latitude, season, cloud cover and time of day. The proportion of that radiation that is reflected back into space, the albedo, also depends critically on the nature of the ground surface. Fresh snow is a better reflector than is bare soil but the albedo varies with particular types of vegetation (Table 2.10).

The effect of solar elevation (the angle between the sun and the ground surface) on the albedo for different vegetation is shown in Figure 2.14. Radiation is reflected more efficiently when the elevation is small.

Table 2.10 Albedo value ranges for different surfaces.

Surface	Albedo
forest (deciduous)	0.16–0.22
forest (coniferous)	0.05–0.14
grass and grain crops	0.12–0.28
water	0.08–0.14
snow	0.35–0.85
soil and rock	0.05–0.45

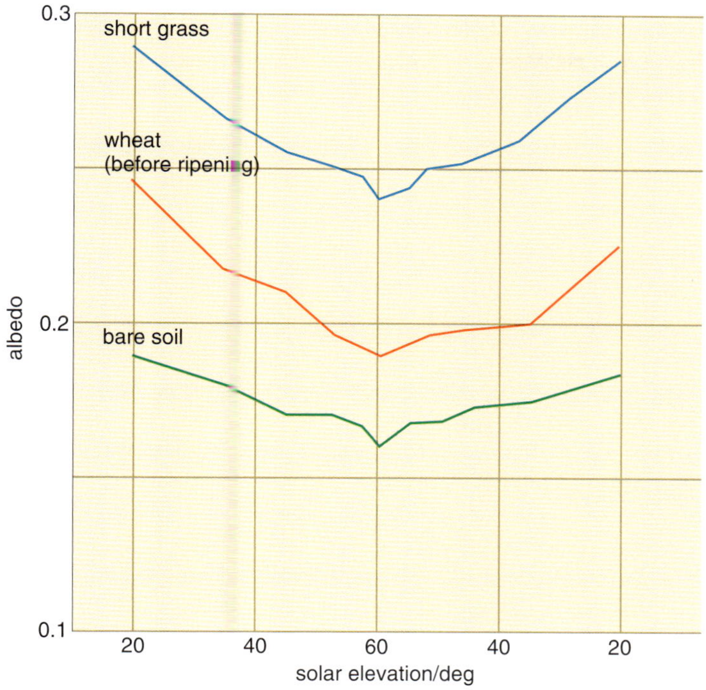

Figure 2.14 Effect on albedo of solar elevation for short grass, wheat (before ripening) and bare soil.

Accurate measurement of energy flows (fluxes) is relatively easy to achieve in the laboratory but much more difficult in the field where control of the variables is more difficult. In addition, air movement has a major effect on evaporation rates and this is a factor that can be excised in laboratory estimates. Nevertheless, there are some ingenious methods devised for field estimation of evaporation. These include a pan of water sunk into the ground surface, a container of soil that can be weighed at regular intervals (**lysimeter**) and even small-scale, labour-intensive methods which might involve the containment of individual leaves of a plant.

Open water provides the simplest situation for evaporation. Here the rate of evaporation is not limited by the availability of water. Rates of evaporation can be high from the surface of open water and represent a major loss of water from reservoirs (in the UK, typically 0.8 m depth yr^{-1}). The surface area of the water body is significant (as is the temperature of the water). Imagine dry air moving over land then reaching the edge of a lake. Initially the evaporation rate will be high but as the humidity of the air rises the rate of evaporation will decrease until, with a large lake, the *net* evaporation rate will reach zero. Air travelling over oceans has essentially constant humidity. For small lakes a continuous high rate of evaporation is achieved as the humid air close to the water surface quickly diffuses away. Water depth can have an influence on evaporation rates. The water temperature of shallow lakes closely parallels that of the air, so evaporation rates are much higher when the air is warm in the summer than in the winter. For large, deep lakes, water temperature lags behind air temperature. (In temperate climates, the ocean is always more agreeable for bathing at the end of the summer than it is at the beginning of the summer.) So for deep lakes, evaporation rates may fall to a minimum in the early spring and actually be higher in the winter. Rates of evaporation are higher for freshwater than salt (saline) water. Seawater contains about 35 g of dissolved salts per kilogram of water and has an evaporation rate about three per cent less than that of pure water. A simplistic way of rationalizing this is to view the dissolved material occupying a proportion of the surface of the seawater and thereby reducing the effective area from which water molecules can evaporate.

Evaporation from bare soil (in contrast from open water) can be limited by the supply of water. Water is evaporated from any surface water film and from water filling the spaces between grains of soil near the surface. Evaporation rates, initially high after rainfall, decrease with time and may reach zero if there are no mechanisms available (such as plant roots) to bring deeper lying water to the surface. Capillary attraction can bring water to the surface and it is more effective with small-grained soils, where the spaces between the grains are small, than it is with coarse-textured soils. However, much of the soil is not bare, being covered with agricultural or natural vegetation.

2.4.2 Evapotranspiration

Water evaporated from the surface of vegetation comes not only from evaporation of water from the wet surface of the vegetation, but also from water that is lost *from* the plant *through* the leaves as water vapour. The term **evapotranspiration** is used to describe both of these processes together. The actual process of evaporation is the same no matter what the source of the water. Movement of water through a plant from intake at the root to its removal through the leaves is known as **transpiration**. Most of this water plays no chemical part in the growth of the plant: only a small proportion (less than one percent) is used to manufacture plant tissue.

Briefly, water is drawn into a plant rootlet from the soil moisture and moves through the plant to its leaves. The water is passed as vapour through small pores in the leaf known as stomata. These are mainly on the underside of the leaf's surface.

Each stoma (plural stomata) comprises two curved cells which control the size of the opening into the interior of the leaf (Figure 2.15). A fully open stoma allows water to be lost rapidly. In times of water shortage, the stomata tend to close to limit water loss but this also limits the intake of gases (carbon dioxide), which can restrict plant growth. You will investigate the movement of water through plants in more detail in Part 2 of this block.

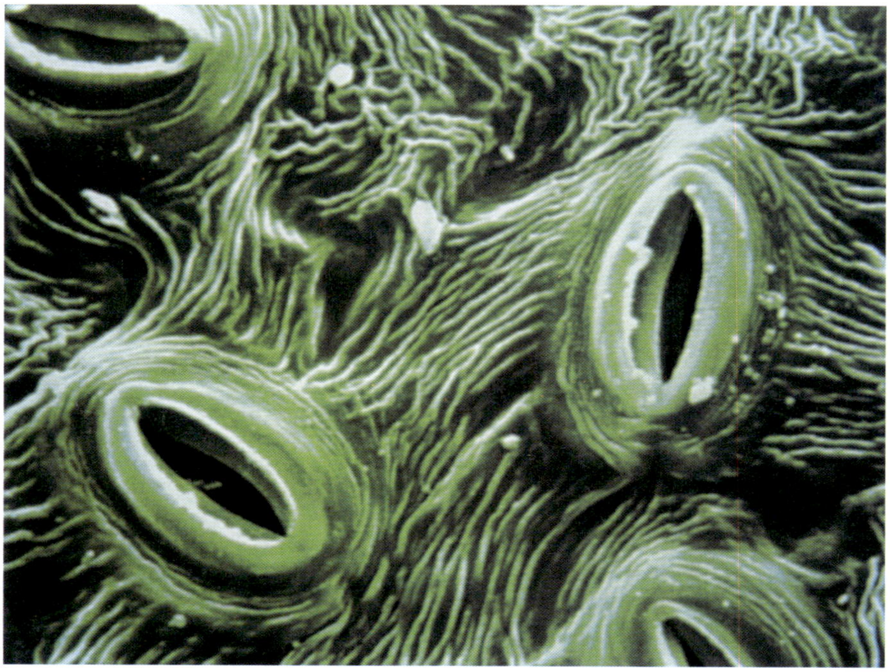

Figure 2.15 Stomata on the underside of a leaf.

2.4.3 Modelling evapotranspiration

Estimating evapotranspiration from vegetation-covered ground is particularly difficult. Evapotranspiration has several components that vary in magnitude. Type of vegetation, size of vegetation, soil water levels, density of ground cover, energy available, air humidity and wind speed are all influential and largely independent variables which can vary in importance over even short periods of time.

There have been many attempts to estimate evapotranspiration empirically but the translation of experiments from the laboratory to the field can be notoriously difficult. There are also several theoretical and semi-empirical models which produce results that may be appropriate for some situations but are less effective for others. In general, the more complex the model (and the more demanding of data) the better the match with field observations, but over-complexity makes a model inconvenient to use.

The concept of **potential evapotranspiration** has been developed in order to model evapotranspiration. Potential evapotranspiration is determined by atmospheric demand, that is, it assumes that the evaporative process is not limited by the supply of water. Thus the calculated rate of potential evapotranspiration must normally be reduced to yield an estimate of **actual evapotranspiration**. The difference between the potential and actual rates of evapotranspiration depends on a number of factors, including soil moisture content, vegetation type, rainfall, and air temperature (Figure 2.16). For example, a dry wind on a sunny day blowing over an arid area would result in very little actual evapotranspiration but the potential evapotranspiration would be very high.

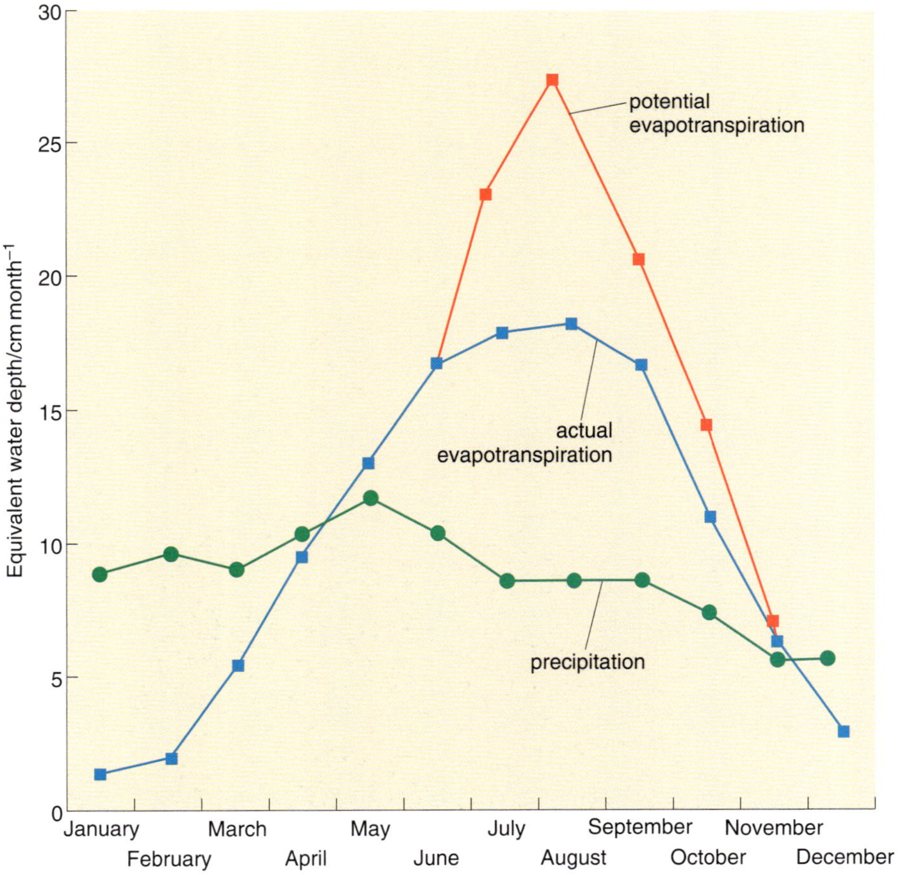

Figure 2.16 Diagrams showing potential and actual evapotranspiration and precipitation over a 12-month period.

The most widely adopted model for estimating potential evapotranspiration is the Penman–Monteith model. The original form of the model was developed in 1948 by H. L. Penman alone. It represented evaporation from an open water surface as a function of (i) the 'drying power' of the air (i.e. its humidity and the wind speed) and (ii) the net radiation energy available for evaporation, and was related to a vegetation-covered surface by use of a coefficient, f, where f for grass varies from 0.8 in summer to 0.6 in winter. This model was tolerably accurate, but a missing factor was the resistance to evaporation from the vegetation.

There are two principle resistances from vegetation that discourage evaporation; the **stomatal resistance** (which controls transpiration) and the **aerodynamic resistance** (which controls the rate at which water vapour can move away from the leaf canopy into the atmosphere). A 'rough' surface will cause much more turbulence in moving air than will a smooth surface. Wind blowing over a freshly mown lawn will be less effective in causing evaporation than will wind of the same speed blowing across a forest with trees of different heights. This effect is the aerodynamic resistance, and it depends solely on the physical properties of the vegetation cover. Water vapour diffuses through the stomata into the atmosphere partly in response to meteorological variables (available energy, humidity and windspeed), and partly depends on the vegetation type (the size and number of stomata). The stomatal resistance is of major importance in the evaporative process because it is usually an order of magnitude greater than aerodynamic resistance (Table 2.11). In terms of modelling evapotranspiration, it is the resistance of the canopy as a whole, rather then each individual leaf, which is important. The resistance of the canopy (called the **canopy resistance**) is generally lower than stomatal resistance of an individual leaf. Table 2.11 shows typical values of aerodynamic, stomatal and canopy resistances.

Table 2.11 Typical values of aerodynamic, stomatal, and canopy resistance for different land uses. Units are in s m^{-1}, the inverse of the units for speed because we are dealing with *resistance* to speed.

Land cover	Aerodynamic resistance	Stomatal resistance	Canopy resistance
open water	125	0	0
grass	50–70	100–400	40–70
arable	30–60	100–500	50–100
heather	20–80	200–600	60–100
forest	5–10	200–700	80–150

Incorporation of the aerodynamic and canopy resistance terms into the Penman model was proposed by J. L. Monteith in 1965. The result is the Penman–Monteith formulation for the potential evapotranspiration rate, E:

$$E = \frac{1}{\lambda}\left[\frac{\Delta R + c_p \rho D / r_a}{\Delta + \gamma(1 + r_s / r_a)}\right] \qquad (2.8)$$

This equation looks complicated, but do not be alarmed! You have met the majority of the terms already. Respectively, r_a and r_s are the aerodynamic and surface resistance (which incorporates stomatal and canopy resistance). The terms λ, ρ, and c_p are concerned with the physical properties of water: λ is the latent heat of vaporization, ρ is the density and c_p is the specific heat capacity. R is the net radiant energy (Equation 2.7). Remember from Block 2, Part 1, that the ability of air to carry water increases with temperature. Also recall that vapour pressure is a common measure of humidity. With this in mind we can define Δ and D as follows: Δ is the rate of change of the saturation vapour pressure with temperature, and D is the vapour pressure deficit (which is a measure of humidity) and γ is simply a constant. You do not need to know the units of these individual variables, nor do you need to remember this equation; the important thing, is that the Penman–Monteith model expresses evapotranspiration in terms of humidity, available radiant energy, and vegetation resistance.

Estimates for the mean annual potential evapotranspiration (measured like rainfall as water depth) over Europe are shown in Figure 2.17. The influence of energy availability is evident with an increase in evapotranspiration from the cooler climates of the north to the Mediterranean warmth. The west–east effect is interesting too. The west coasts are likely to have higher winds than the inland areas, increasing evapotranspiration. However, the air having travelled over the ocean is moist which reduces its potential for evapotranspiration.

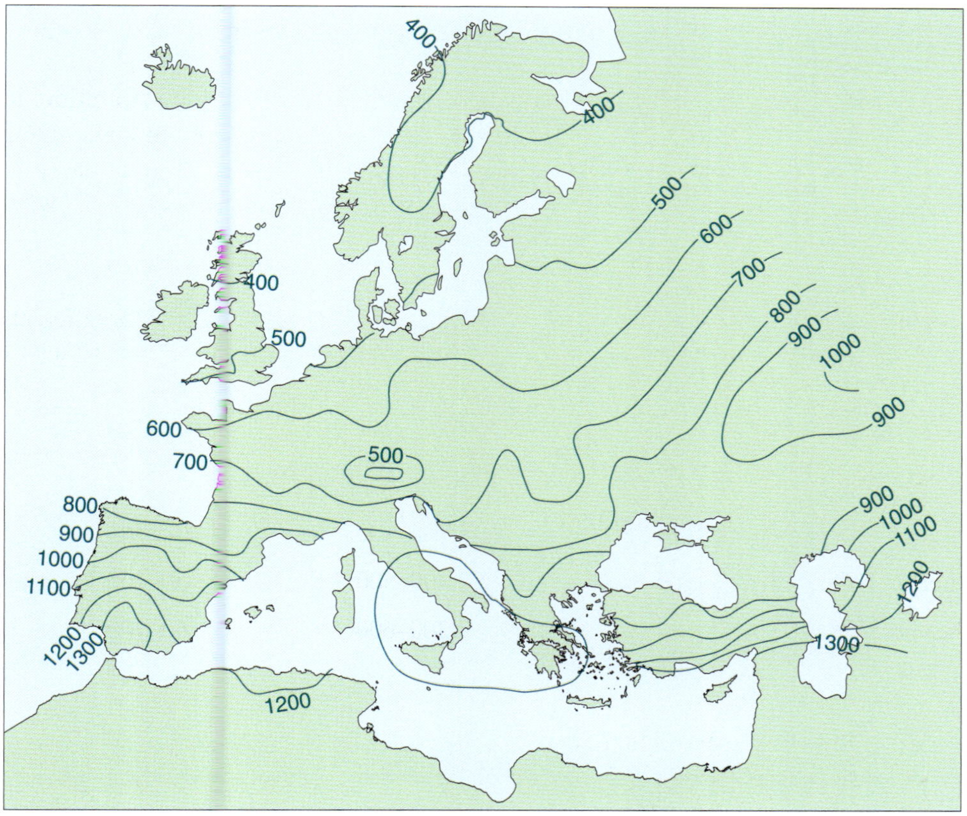

Figure 2.17 Penman–Monteith estimates for mean annual potential evapotranspiration (equivalent water depth/mm) over Europe.

One criticism of the Penman–Monteith model is that it does not consider the effects of **advection**; which can be regarded as the horizontal movement of energy in the atmosphere in contrast to the vertical movement through convection. The surface from which evapotranspiration occurs and the airmasses above the surface are unlikely to be homogeneous, so the link between humidity, available radiant energy, and vegetation resistance may be broken when air with different characteristics is imported from another area. If an airmass moves

Figure 2.18 Map showing (a) the River Sprint in south Cumbria and (b) the river catchment (outlined in green).

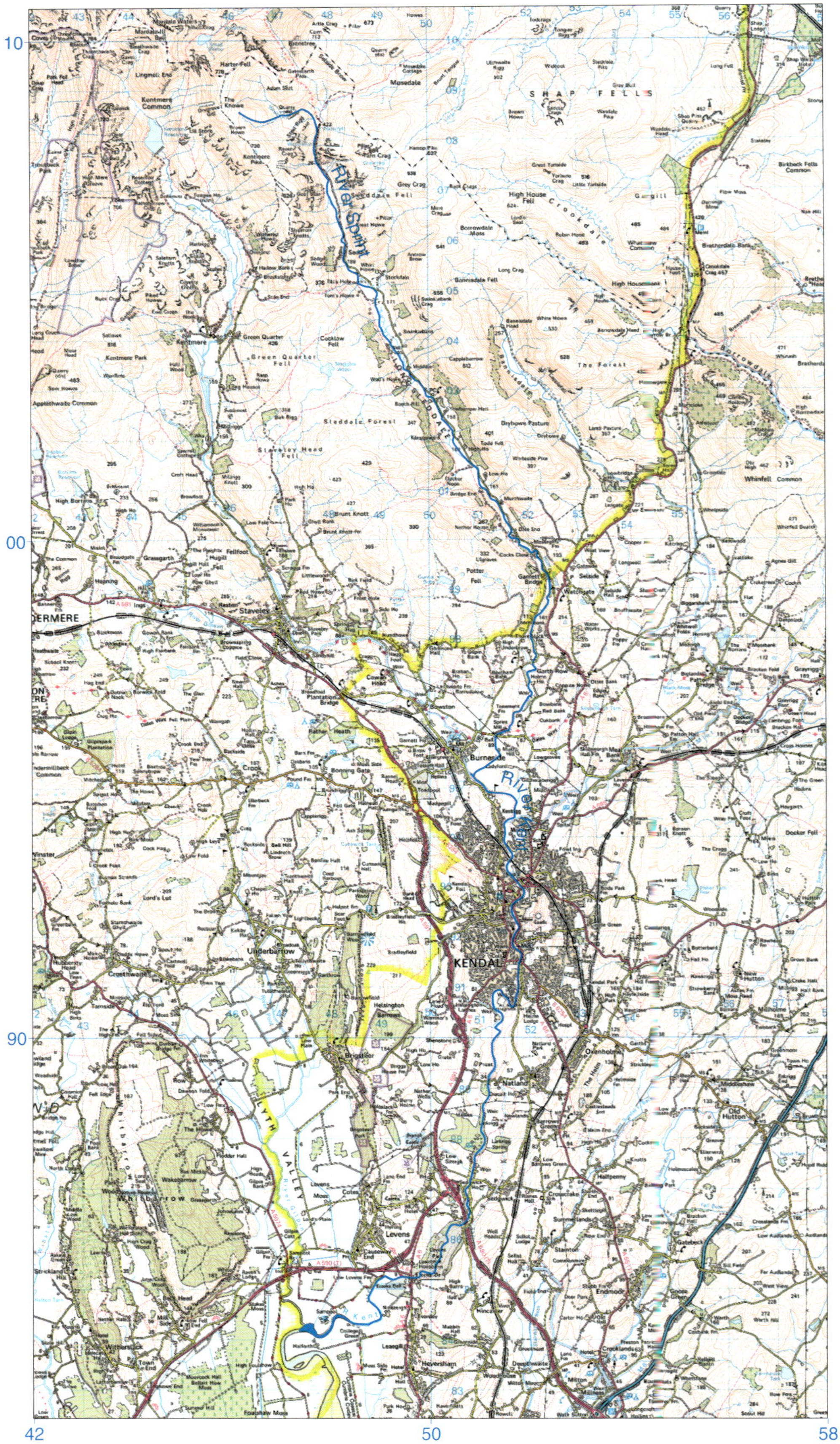

(a)

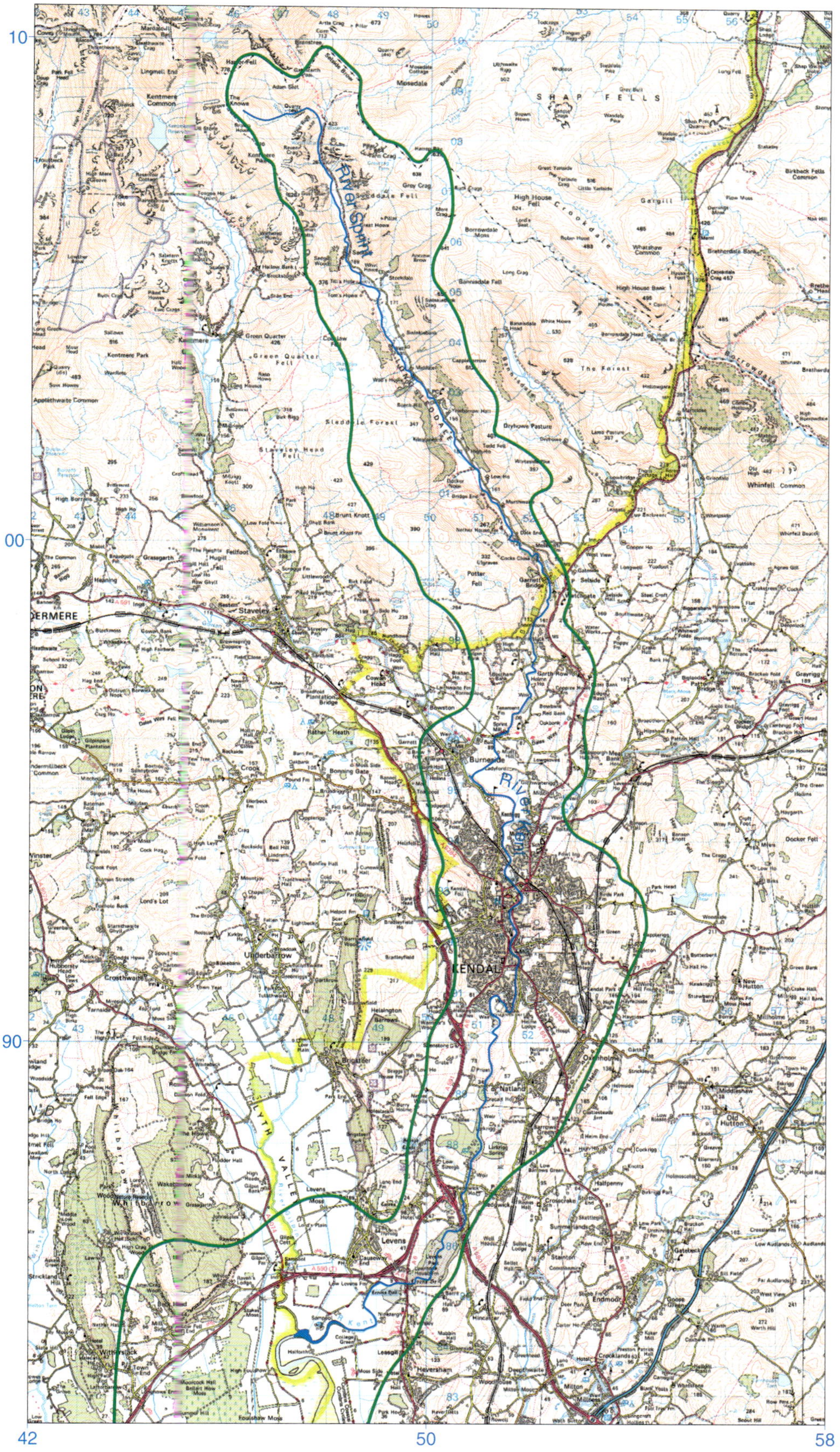
10
00
90
42
50
58
SHAP FELLS
River Sprint
River Kent
KENDAL
Staveley
Burneside
Levens

(b)

across the boundary between two surfaces of different wetness, then a gradient in the vapour pressure deficit is produced, leading to 'advectional enhancement' if the air moves from a drier to a wetter surface (i.e. the vapour pressure deficit increases) and to advectional depression when the movement is in the opposite sense and vapour pressure deficit decreases. These advective processes can have a significant effect on evapotranspiration rates.

Vegetation cover plays a major role in the redistribution of rainfall. Interception can prevent some rainfall from reaching the ground and dense forest canopies can influence where water reaches the ground, shielding some areas and increasing the amount of water reaching the ground in others. Also vegetation can remove water from beneath the ground from significant depths through deep root systems of large trees. Now we need to take a look at the water that actually reaches the ground that does not directly evaporate back into the atmosphere.

2.5 Runoff

Although a most obvious result of a rain storm is water flowing over land and increased flow in rivers, this accounts for only about one-third of the rain that falls. This is **runoff**, which may also be variously referred to as **streamflow** or **discharge**, and it represents the second main mechanism by which precipitation falling on the land is removed. To examine runoff in more detail it is convenient to think in terms of a **catchment**. Typically this might be an extended valley which encloses a river system. Rain falling over the system will tend to flow under gravity ultimately to the main river of the system. If rain does not flow towards the main river, then it is not falling within the catchment.

Using Figure 2.18a (which shows part of south Cumbria in the UK), sketch the catchment area associated with the River Sprint and compare your catchment boundary with that in Figure 2.18b. (You do not need to cover the whole catchment of the River Kent into which the River Sprint flows.)

However, not all rain flows over the surface of the ground to the streams and rivers. Water that percolates into the ground may not reach the main outflow of our area if the underlying rocks or soils (as you saw in the Block 2, Part 2) provide a route for water to flow laterally away from the catchment. Figure 2.19 shows just such a situation where a layer of impervious rock slopes away from the catchment area. It is important to distinguish between impermeable catchments, where the hydrological catchment is consistent, with the topographical catchment and permeable catchments, where this may not necessarily be the case.

Our 'ideal' catchment area is one where all the precipitation, P is removed either by evaporative processes or appears in the outflow river(s) of the catchment. In this case the topographical catchment coincides with the hydraulic catchment. The only occasion when a mismatch might occur would be if there were a change in the quantity of water within the ground. So what routes are available for rain that falls over the catchment to reach the catchment (or basin) exit?

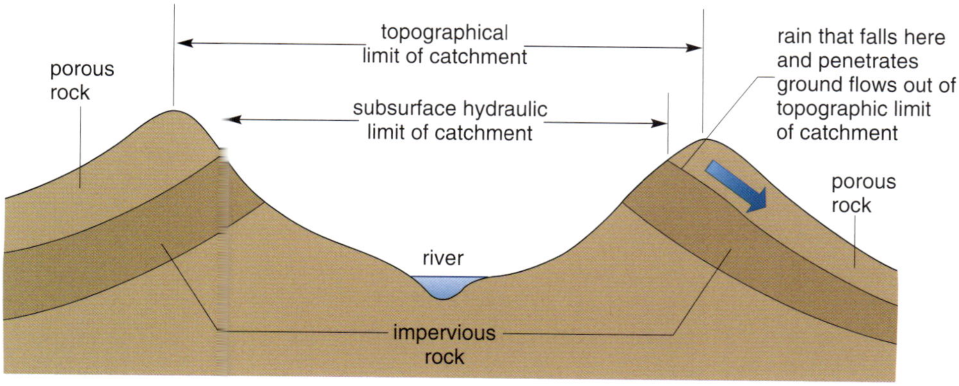

Figure 2.19 Cross-section of a catchment showing underlying rocks.

Probably the most obvious route is for rain that falls directly into drainage channels. This water is known as **channel precipitation,** Q_p and simply flows along the drainage system ultimately to the catchment outlet. Water also reaches the drainage channels by flowing over the land, the appropriately termed **overland flow,** Q_o. Depending on the nature of the ground, the slope and the vegetation cover, this may be as a thin layer of flowing water or, more commonly, as tiny trickles or rivulets. This leaves the water that penetrates the ground surface. Some of this percolates through to where the ground is saturated and flows through this region as **groundwater flow,** Q_g which we shall examine in some detail in the next section. The remaining water that penetrates the ground surface flows through the upper unsaturated layers, **throughflow,** Q_t. These four main routes for water moving through the catchment are shown as a flow chart in Figure 2.20 and are indicated in a cross-sectional diagram of part of a catchment in Figure 2.21, where P_c is the precipitation falling within the catchment.

Box 2.3 Units of runoff

Runoff (which is also commonly referred to as streamflow or discharge) is usually expressed as volume per unit of time. The **cumec**, which is one cubic metre per second ($m^3\ s^{-1}$), and cumecs per square kilometre ($m^3\ s^{-1}\ km^{-2}$) are commonly used units. Runoff may also be expressed as a depth equivalent over a catchment; millimetres per day or month or year. This is a particularly useful unit for comparing precipitation and runoff rates, since precipitation is invariably expressed in this way.

Runoff, Q ($m^3\ s^{-1}$), at any point along the course of a river is given by the product of its cross-sectional area, A (m^2), and average speed, v ($m\ s^{-1}$):

$$Q = Av$$

The relative importance of the four components to the total runoff, Q_p, Q_o, Q_g and Q_t, is a function of the topography, soils, geology and vegetation of the particular catchment area. However, the relative size of these components also changes with time. Precipitation is not constant: we are not operating in a steady-state situation, and it is the changes that occur over time that make the study of

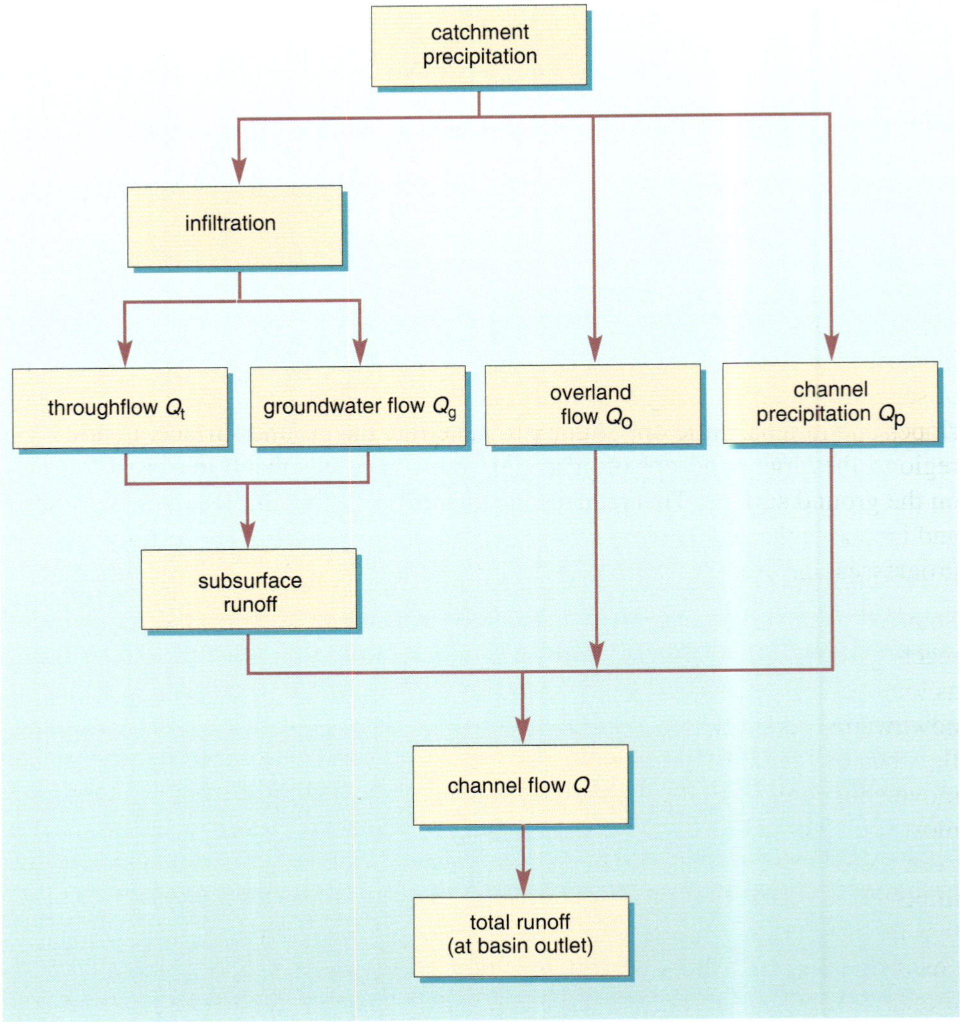

Figure 2.20 Main routes for precipitation falling in a catchment.

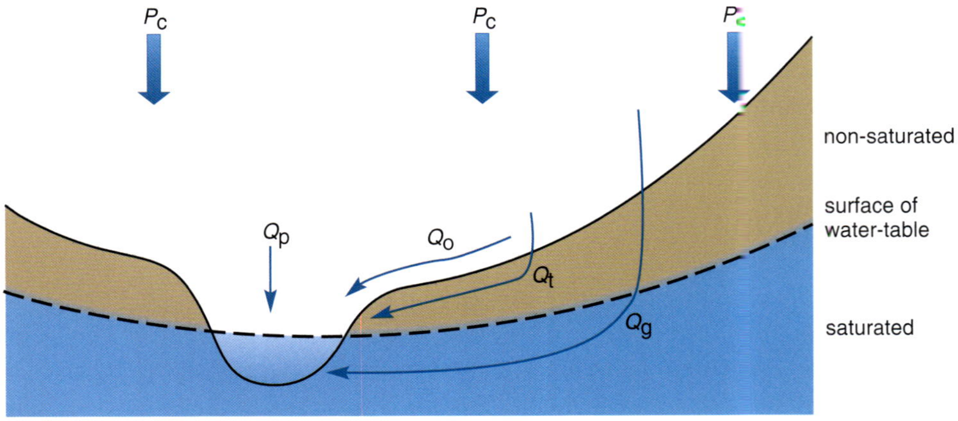

Figure 2.21 Section of a catchment, showing main routes of water movement through the system.

hydrology difficult (and interesting). Variations occur during an individual rain storm, over different parts of the catchment and over seasons and even years.

The relative proportion Q_p of rain falling directly into drainage channels generally represents a minor (around 2%) component of total runoff simply because channels represent a small part of the surface area of most catchments. For catchments that have significant areas of lakes (northern Minnesota in the USA) or swamp (Okavango region of Botswana), channel precipitation is a more significant contributor and may represent 20% or more of catchments. These figures do not stay constant for a particular area. During a rainstorm, particularly where the topography is relatively flat, the area occupied by channels may increase significantly, and the contribution of Q_p could temporarily rise to 50% or so. Overland flow is a major contributor to total runoff where there are steep slopes and rain has little opportunity to penetrate the ground surface. In arid regions, the large raindrops of infrequent rainstorms can result in a crust forming on the ground surface. This reduces the quantity of water infiltrating the ground and increases the relative importance of Q_o. However, as the rainstorm progresses, the crust can break down and infiltration increases.

Some of the infiltrated water flows just below the ground surface and eventually reaches the stream channels. During a downpour, this throughflow Q_t increases as long as it is easier for the water to move laterally rather than vertically downwards. As the amount of water increases, this laterally moving, saturated flow can be 'perched' on an unsaturated layer. A higher lateral rate of water movement compared with vertical movement is not an uncommon situation. In most soils the surface layers tend to be less compacted than deeper down. This is even more marked where there is a thin, conductive soil layer overlying impermeable rock or where there are soil strata of differing permeability. Water that flows through gaps and fissure in the soil or rock reaches the stream channels very much more quickly than does water that has worked its way through the bulk material and this distinction between **quick throughflow**, $\boldsymbol{Q_{qt}}$, and **delayed throughflow**, $\boldsymbol{Q_{dt}}$, is indicated in Figure 2.22. Also shown is a route connecting quick throughflow to surface runoff. This represents throughflow that reappears at the surface and becomes overland flow before it reaches the stream channel.

The relative immediacy with which some rivers show an increased flow rate after the onset of rain indicates a rapid route for rainwater to the river. This **quickflow** is usually a combination of overland flow, Q_o and quick throughflow, Q_{qt}. Water that takes a longer route and contributes to the flow of rivers in non-rainy periods is know as **baseflow** and is often a combination of delayed throughflow, Q_{dt} and groundwater flow, Q_g.

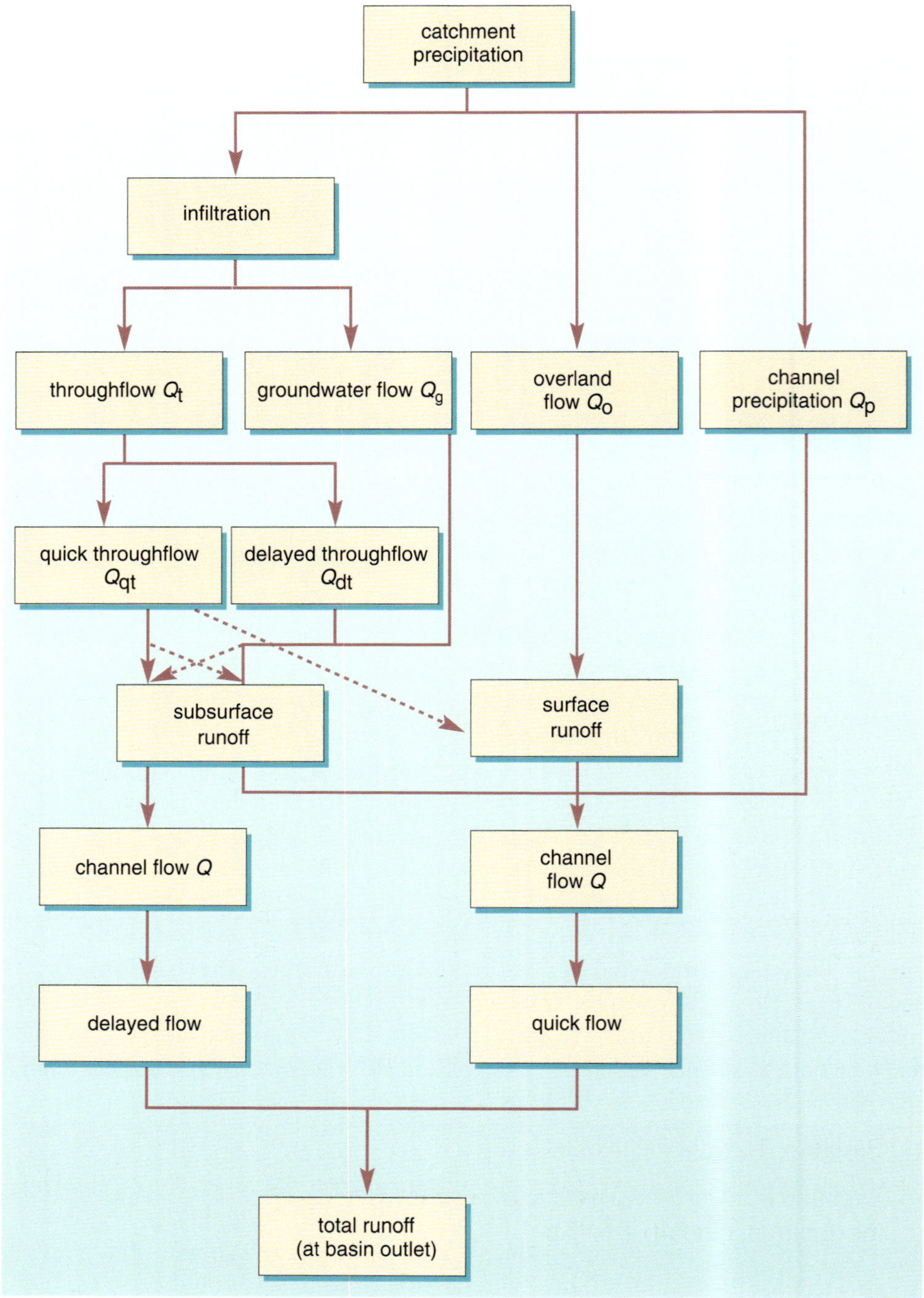

Figure 2.22 Flow through a catchment, incorporating quick throughflow and delayed throughflow. Dashed lines indicate cross-over flows.

2.5.1 Hydrographs

A plot of river discharge at a particular cross section of river against time is know as a **hydrograph**. The hydrograph for a stream in Figure 2.23 also shows rainfall in a part of the river catchment over the same time-scale. The superimposition of rainfall data on the hydrograph is a common and useful addition but note the difference in scales and units. The rainfall data here are measured in mm over specific time intervals.

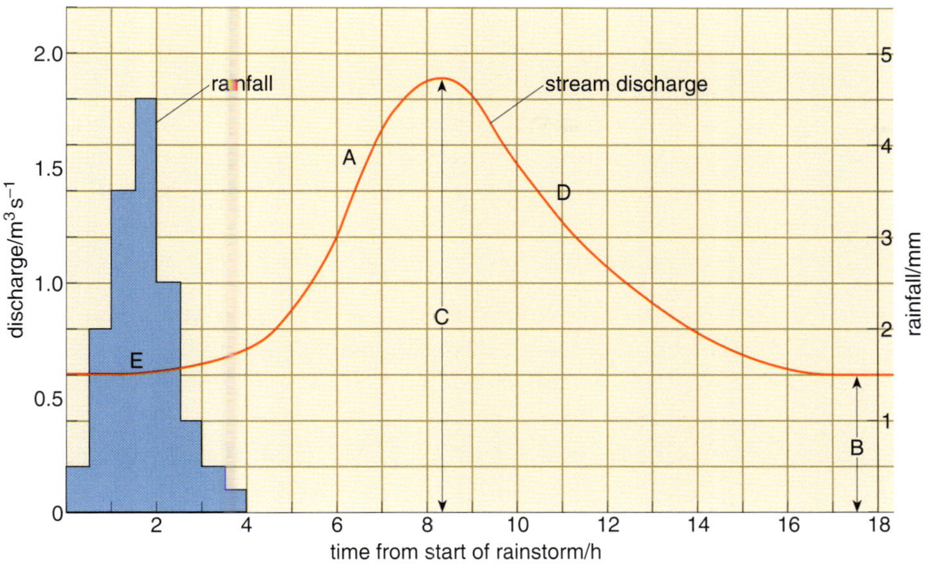

Figure 2.23 River hydrograph also showing rainfall data.

The **rising limb** of the hydrograph, A, represents the period where river flow rate is increasing and the flow rate decreases during the **recession limb**, D. **Peak discharge**, C is the maximum flow rate and the baseflow, B represents the normal flow rate in the absence of recent rain. In this case, baseflow may be estimated by a horizontal line from point E to its intersection with the recession limb. This is a somewhat simplistic way of estimating baseflow which is adequate for our argument here, but you should be aware that better estimates can be obtained (by more complex methods). The **lag time** is the time between the maximum rainfall rate and the time of peak discharge.

- ❍ From Figure 2.23, estimate values for peak discharge, baseflow and lag time. Also estimate the volume of rain that fell in the catchment, assuming that the rainfall data represent a weighted average over the catchment which has an area of 22.4 km^2.

- ● The peak discharge is the maximum of the river discharge C and has a value of 1.9 $m^3\ s^{-1}$.

  Baseflow, B, has a value of 0.6 $m^3\ s^{-1}$.

  Lag time is given by 8.2 h (time of river discharge peak) −1.75 h (time of maximum in rainfall) = 6.45 h

To calculate the rainfall, we need to find the total depth of rain represented by the barchart part of Figure 2.23. In the first 0.5 h of the rainstorm 0.5 mm of rain fell. The next 0.5 h showed 2.0 mm of rain. We can construct a table showing all the rainfall time intervals (Table 2.12).

Table 2.12 Rainfall over time intervals from Figure 2.23.

Time interval/h	Rainfall/mm
0.0–0.5	0.5
0.5–1.0	2.0
1.0–1.5	3.5
1.5–2.0	4.5
2.0–2.5	2.5
2.5–3.0	1.0
3.0–3.5	0.5
3.5–4.0	0.25
total	14.75

The volume of rain falling on the catchment during this event is given by the product of its surface area and the total depth of rainfall:

$$\text{volume} = 22.4 \times 10^6\ m^2 \times 14.75 \times 10^{-3}\ m = 3.3 \times 10^5\ m^3$$

Lag time is dependent on a number of factors including the slope of the land, intensity and duration of the rain, underlying soil and rock and land use. Trees

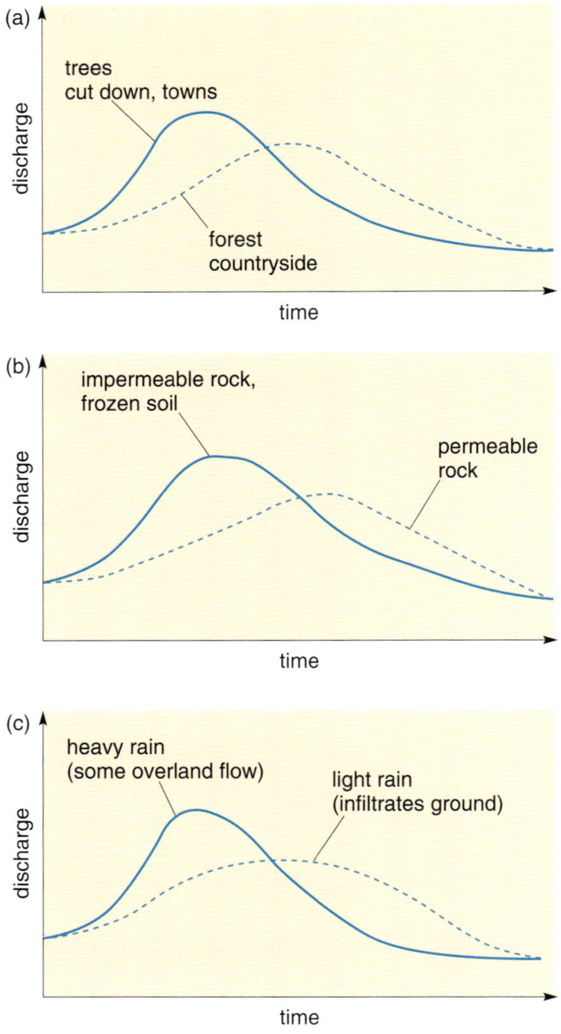

Figure 2.24 The effect of (a) changes of land use, (b) underlying rock type and (c) rain intensity on river hydrographs responding to similar rainstorms.

and vegetation tend to increase lag time by intercepting rain and delaying its route to the ground. If trees are removed and bare soil exposed or even if light crops are planted, water can run over the land more quickly with a shortening of lag time. The same effect is observed when hard, impermeable surfaces such as roads and buildings replace vegetation. As is evident from Figure 2.24a, not only is lag time reduced but peak discharge is increased, increasing the probability of flood.

If the ground is relatively impermeable, which can occur when the soil is frozen or heavily crusted or there is underlying impermeable rock, lag time is reduced and peak discharge is increased (Figure 2.24b) and the same effect is observed with heavy rain (Figure 2.24c). There is time for light rain to infiltrate the ground but overland flow can occur rapidly with heavy rain.

❍ Relate the river hydrograph data to the rainfall data in Figure 2.25.

● The river showed relatively little response from the initial rainstorm but peak discharge increased significantly after the second storm. The suggestion is that most of the rain in the initial storm infiltrated the ground and reached the river by a slow route. The semi-waterlogged ground resulted in a greater proportion of overland flow in the second storm, indicated by the higher peak discharge and shorter lag time. By the third storm, the ground was waterlogged. Rainwater took the rapid overland route to the river, with a consequent even higher peak discharge and shorter lag time. Although none of these three storms in isolation would have contributed to flooding, the short time interval between them and consequent changes to the ability of the ground to absorb water greatly increases the probability of flood.

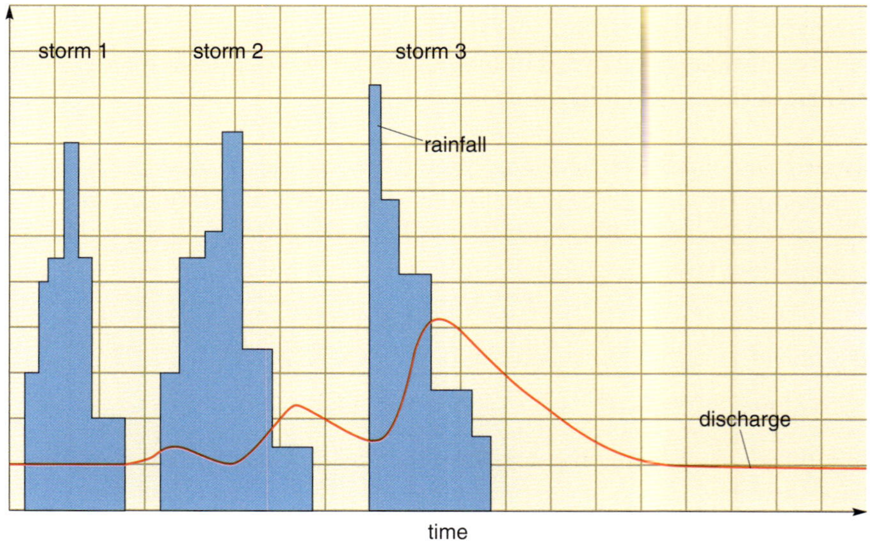

Figure 2.25 Rainfall and river hydrograph data for a series of three storms in a catchment.

Question 2.7

Describe how a hydrograph (a) for a river catchment subject to a total of 16 mm of rain falling over 4 hours would differ from a hydrograph (b) for the same catchment subject to the same amount of rainfall over 24 hours.

One of the important things that hydrologists need to know is the amount of runoff that will occur from a given rainfall event. For example, structures such as storm sewers that carry water are designed on the basis of the peak discharge. The simplest way to make this prediction is to assume that if it rains long enough (such that rain falling in the most distant parts of the catchment reaches the point of the discharge while the rain is still falling), the peak discharge (Q_P) from the drainage basin will be the average rainfall multiplied by the drainage basin area, reduced by a factor to account for infiltration:

$$Q_P = C \times I \times A$$

Where C is a runoff coefficient (which is dimensionless), I is the average rate of rainfall (m s^{-1}) and A is the surface area of the catchment (m^2). Values of C vary depending on land use due to differing rates of infiltration. Typical values of C are given in Table 2.13. Urbanized areas have a high percentage of impervious surface and a greater percentage of runoff. For each land use a range of the value of C is given. The lower number is used for storms of low intensity; storms of greater intensity will have proportionally more runoff and a higher value of C. We will explore the reason for this in a following section.

Table 2.13 Runoff coefficients, C, for calculation of peak discharge.

Principal land use	C
business district	0.70–0.95
residential suburb	0.25–0.40
parkland	0.10–0.25
lawn, sandy soil	
flat	0.05–0.10
steeply sloping	0.15–0.20
lawn, clay-rich soil	
flat	0.13–0.17
steeply sloping	0.25–0.35

2.5.2 Flood flows

Most river floods result from excessively heavy and/or excessively prolonged rainfall. In cold winter areas, where snowfall accumulates, substantial flooding can occur during the melt season in spring and early summer, particularly if melt rates are high. Floods can be intensified or ameliorated by a number of factors, including the area, slope and altitude of the catchment, soil type, geology, vegetation cover, the nature of the drainage basin network, and the shape of the channel itself.

For flood prediction purposes, it is useful to know how often a particular discharge is likely to occur. Data going back over many years are used to plot discharge against the average time interval before a similar discharge occurs again, the **return period**. Such a frequency curve is shown in Figure 2.26 for the Mississippi River near St Louis in the USA.

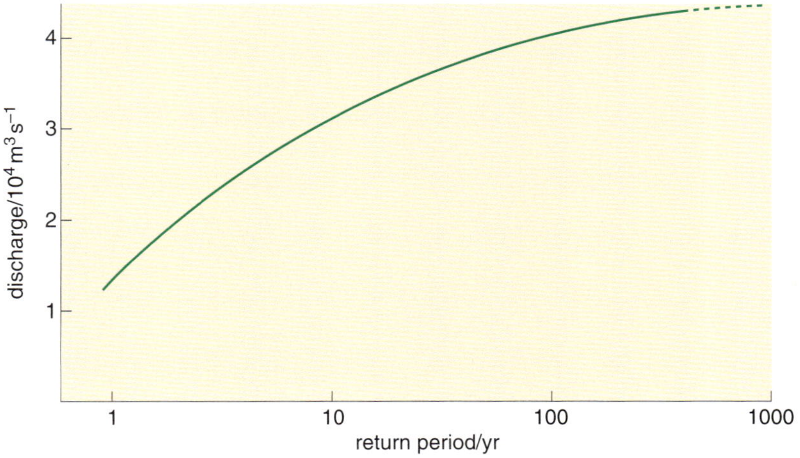

Figure 2.26 Frequency curve for Mississippi River near St Louis, USA. Note that the scale on the horizontal axis is not linear; each division corresponds to a factor of ten increase in the return period.

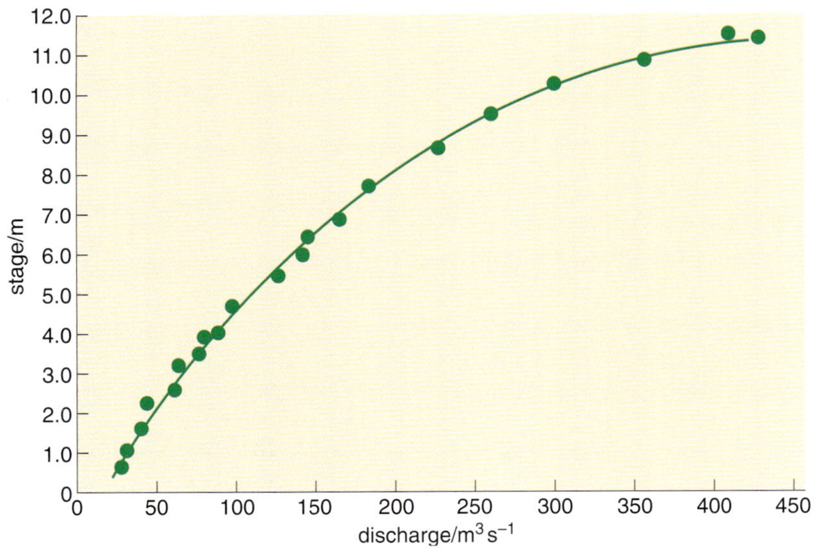

Figure 2.27 Typical stage–discharge rating curve for a river.

❍ From Figure 2.26 what is the return period of a discharge of 3.0×10^4 m^3 s^{-1}?

● The return period is approximately 10 years.

If the maximum discharge that could be carried by the river channel at this point were 2.8×10^4 m^3 s^{-1}, there would be serious risk of flooding on average every decade. Yet building in such situations has occurred despite the hydrological evidence pointing to the consequences.

While measurements of discharge are useful, it is the *height* of the water in the river that is usually more important for flood warning purposes; is the river going to burst its banks? It is also much quicker and simpler to measure height than it is to measure discharge because height can be measured automatically by means of a well (called a **stilling well**) connected to the river. Data from the stilling well can be recorded in digital form and automatically transmitted for processing and monitoring purposes.

River height (usually referred to as the river **stage**, that is, the elevation of the water surface above a fixed point) can be empirically related to discharge by a **rating curve**, Figure 2.27. This is obtained by simultaneously measuring the discharge of a river and its stage at a particular location, and then repeating the measurements for a number of different stage heights. Providing the river channel does not scour during flooding, the rating curve should not vary significantly over time.

As you might expect, frequency curves differ for different catchments. Two frequency curves (which plot stage against return period) are shown in Figure 2.28.

❍ Assign each curve to one of the following catchments, A or B.

Catchment A: Located on impermeable bedrock, soil coverage is thin and the catchment is in an upland area with a steep-sided river valley.

Catchment B: Thick covering of permeable soils underlain by unconsolidated sand. The catchment is low-lying, with only a shallow gradient from its boundary to the river valley.

● The stage for any given return period is much higher in curve 2 than in curve 1. Curve 2 corresponds to catchment A where the steep-sided valley and impermeable bedrock would result in a rapid rise in river level. The wide valley and porous soils of catchment B would result in a smaller river height rise for a given discharge.

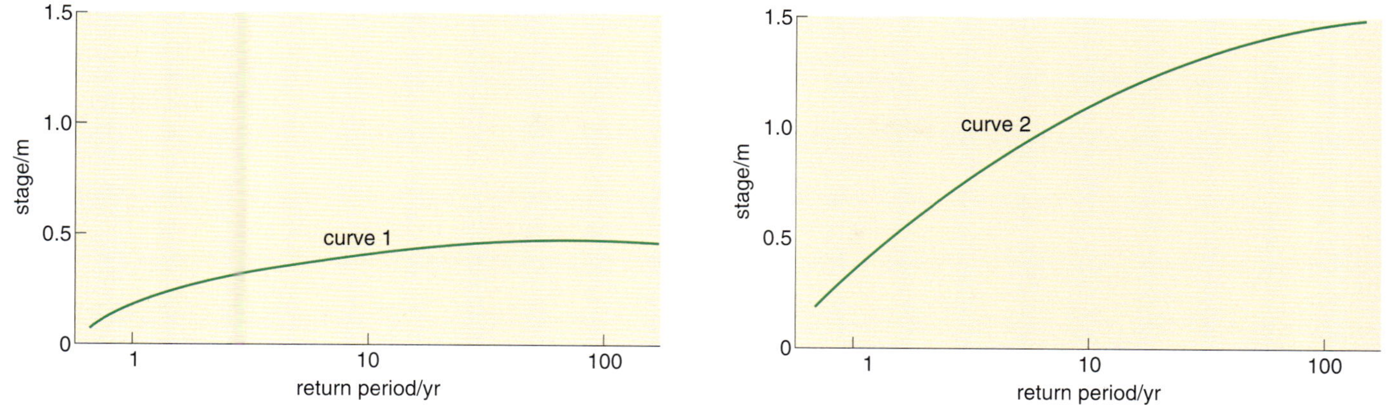

Figure 2.28 Stage (or river height) versus return period for two river catchments.

2.5.3 Horton and Hewlett hypotheses

The complexity associated with the routes that precipitation can take to reach stream channels is increased because, even for the same area, the relative importance of each route changes with time. It was in an attempt to bring some structure to these event-based changes that the **Horton hypothesis** was developed in the 1930s. The basic idea is relatively simple.

Rain falling on the ground surface is divided initially into two components: one part penetrates the surface and becomes throughflow or, if it penetrates more deeply into a saturated region, groundwater flow. The other part is overland flow, the water that does not penetrate the ground. The factor that determines whether rain falling on the ground penetrates the ground or not is the infiltration capacity of the soil: this is the maximum rate at which rain can be absorbed by a particular piece of ground. It is important to realize that the infiltration capability of the ground varies with time, the ground is subject to change particularly during rainstorms. Consider a period of time, t, during a storm where rain falls at a rate, i, which is greater than the rate, f, at which it can be absorbed by the ground (Figure 2.29). There is an **excess precipitation, P_e**, which will flow over the ground surface and is identified as overland flow, Q_o.

$$P_e = (i - f)t = Q_o \qquad (2.9)$$

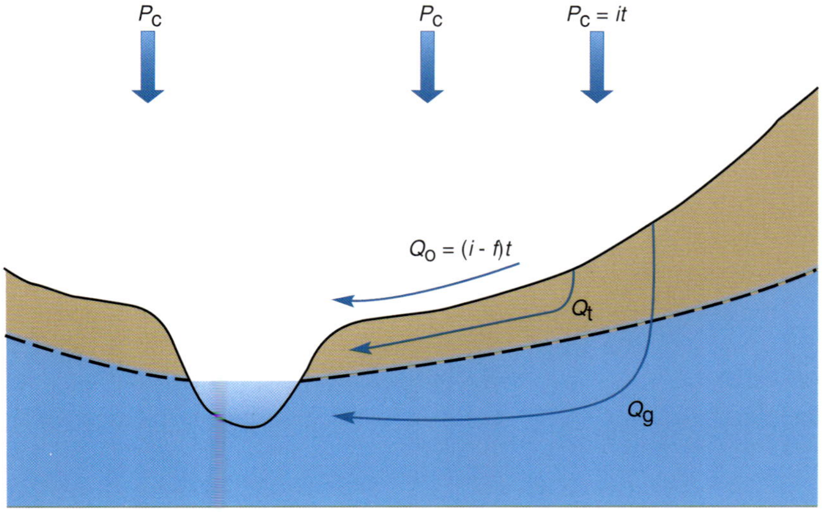

Figure 2.29 Illustration of the Horton hypothesis.

At the onset of a rainstorm, the infiltration capacity of the ground, *f*, can be quite high but usually decreases rapidly. The raindrops cause soil to swell which closes up any small cracks and passageways and, if the rain is heavy, the upper layer of the soil is compacted. After this initial, quite rapid decline in *f*, its value steadies and diminishes only very slowly during the rest of the storm. Recovery of the value of *f* occurs rapidly once the rain has ceased. This basic idea has a sound experimental basis but does not take into account changes deep within the ground. If, for example, there is an underlying layer where percolation downwards is rapid, *f* will not decrease to such a low value early in the storm.

One can imagine three situations. With light rain, all water falling on the surface infiltrates and there is no overland flow. Moderate rain results in no overland flow until the (initially high) infiltration capacity of the soil falls and only then will there be overland flow. High intensity rain rapidly results in a precipitation excess and there is overland flow for almost the whole of the duration of the storm. Water reaches the ground at a rate that is greater than it can be absorbed by the ground.

This approach of Horton was extended in the 1960s by Hewlett who adopted a somewhat different starting point. The idea behind the **Hewlett hypothesis** is that all precipitation initially penetrates the soil surface, even in heavy storms. The **water-table** is a surface beneath which the ground is saturated with water. Where the water-table is near the surface the additional water from heavy rain causes the table to rise. It can happen that the water-table rises to the ground surface and any additional water will not penetrate the saturated ground (Figure 2.30). The infiltration capacity of the soil is zero and all precipitation falling on the surface is excess precipitation translating as overland flow.

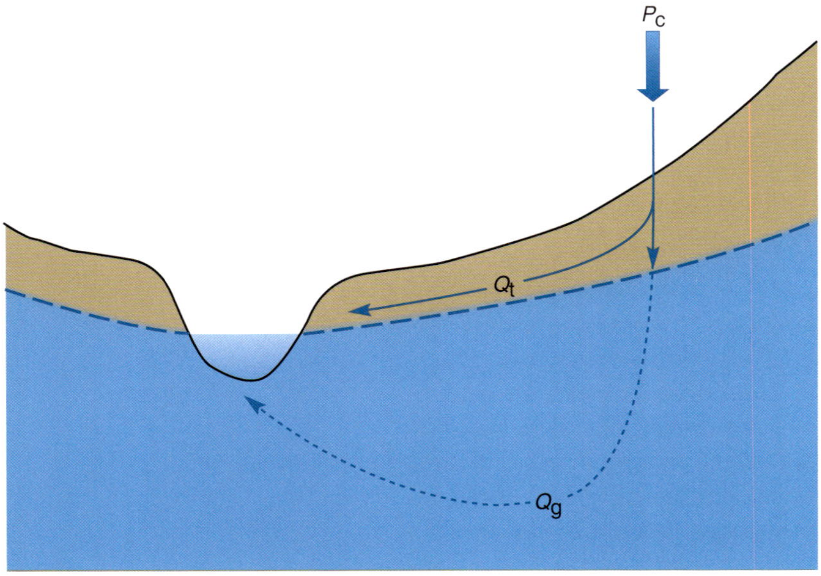

(a)

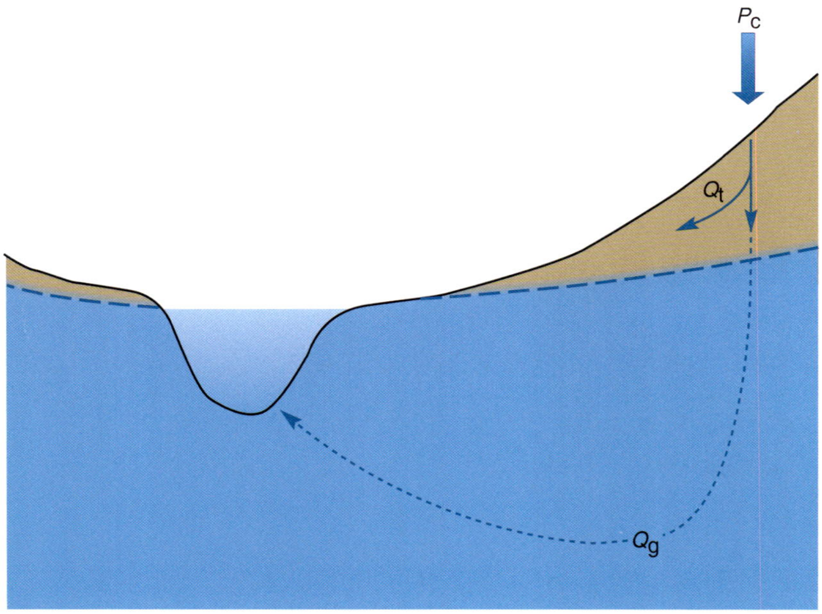

(b)

Figure 2.30 (a) Initial stage of a rainstorm and (b) a later stage according to the Hewlett hypothesis.

Since they were developed, the Hewlett and Horton hypotheses have to some extent been reinterpreted, yet they remain two of the single most important conceptual advances in the history of hydrology.

2.6 Summary of Section 2

1 Collection and interpretation of precipitation data is fraught with difficulties. Some of these are related to the design and positioning of raingauges and how the collected precipitation is recorded and monitored. Snow presents particular difficulties and there are alternative means of estimating snowfall. Average rainfall over an area where there are several raingauges needs to be estimated using methods such as the Theissen polygon approach.

2 Rainwater is not pure water and it already contains a cocktail of dissolved ions before it reaches the surface of the Earth. Some of the ions arise from the dissolution and reaction of gaseous oxides in the atmosphere and others as a result of leaching from airborne dust or incorporation of oceanic aerosols. Where rain falls through vegetation before reaching the ground there is the opportunity for incorporation and exchange of ions adsorbed on leaf surfaces.

3 Rainwater returns to the atmosphere by a variety of routes, the major direct one being through evapotranspiration. This process can be modelled by bringing together the factors of available energy, the drying power of the air and the nature of the vegetation as is done with the Penman–Monteith approach. Rain reaching the ground may run over the surface to river channels, penetrate the ground and progress either as throughflow or as groundwater flow. The factors determining the route include the nature of the ground surface and underlying soil and rock, the size and distribution of raindrops, the duration of the storm and the existing water content of the ground. A variety of explanations has been proposed in an attempt to model the routes and those advanced by Horton and by Hewlett have considerable merit.

Question 2.8

Construct Theissen polygons for the raingauge locations that feature in Figure 2.3. Calculate a mean rainfall for the land area (exclude ocean below the high water mark) using your polygons and the data in Table 2.1. The total map area shown in Figure 2.3 is $30 \times 30\,km^2$.

Learning outcomes for Section 2

When you have completed this section, you should be able to:

2.1 Calculate mean rainfall over an area from data from irregularly distributed raingauges.

2.2 Identify difficulties in the collection and processing of rainfall data and identify ways of overcoming these difficulties.

2.3 Account for the origin of dissolved ions in rainwater and cite evidence in support.

2.4 Convert between mass and molar concentration data for ions in water.

2.5 Outline methods of estimating throughflow, stemflow, and surface storage capacity in vegetated areas.

2.6 Account for the changes in ion concentrations when rain passes through a forest canopy.

2.7 Estimate the energy required for the vaporization of water.

2.8 Outline the major factors in evapotranspiration in the field and link them to the main ideas of the Penman–Monteith model.

2.9 Account for the components of runoff and indicate the factors that affect the relative proportions of these components.

2.10 Interpret hydrographs linking rainstorm and river flow rate with time and estimate flood frequencies from discharge and stage data.

2.11 Outline briefly and identify differences between the Horton and Hewlett hypotheses.

Groundwater

3

> And these waters, falling on these mountains through the ground and cracks, always descend and do not stop until they find some region blocked by stones or rock very close set and condensed.
>
> *Discours Admirables,* Bernard Palissy (*c.* 1510–1590)

3.1 Introduction

If runoff were the only source of water entering a river then the river would dry up a relatively short time after rain stops. This very rarely happens, which means that there must be some other source of water going into the river. To understand, we need to follow the path of raindrops which infiltrate the ground surface. Water that passes through the surface will move down through the underlying soil and rock until it reaches a level where all the pore spaces are full of water. This level is called the water-table and it separates the **unsaturated zone** from the **saturated zone** (Figure 3.1). Water in the saturated zone is called **groundwater**.

Immediately above the water-table, water is drawn up into pore spaces by capillary forces into a zone called the **capillary fringe**. What do we mean by capillary forces? Imagine a sugar cube held so that its lower surface is in contact with coffee. The liquid can be seen to be drawn up into the pore spaces in the sugar lump; this is because the water molecules in the coffee are more strongly attracted to the sugar molecules than they are to each other. The attractive force between the water and the sugar molecules is called **surface tension**; the larger the surface area of the solid that is exposed to the fluid, the greater this force will be.

Thus the thickness of the capillary fringe is controlled by the grain size of the rocks. In rocks with fine pores which have a high surface area and therefore a high surface tension, it can be over 10 metres thick. In coarse-grained rocks it can be no more than a few centimetres thick.

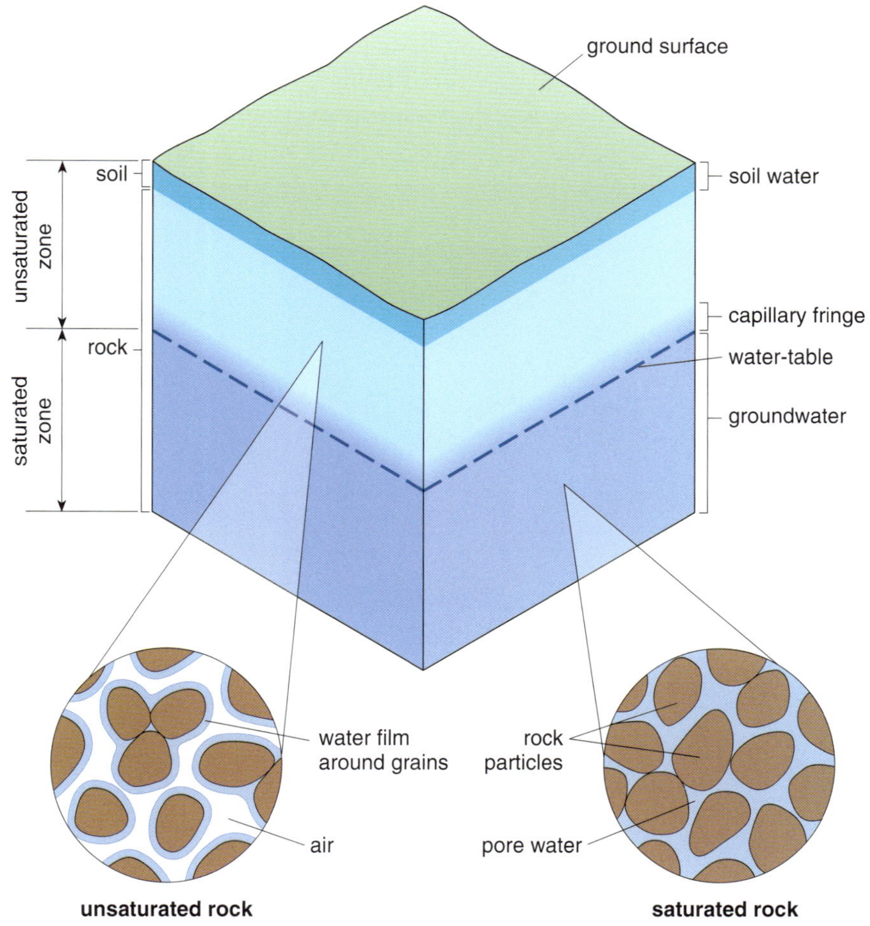

Figure 3.1 Profile of subsurface water. Below hills the unsaturated zone can be tens of metres thick. Most groundwater flows in the upper 200–300 metres of the saturated zone.

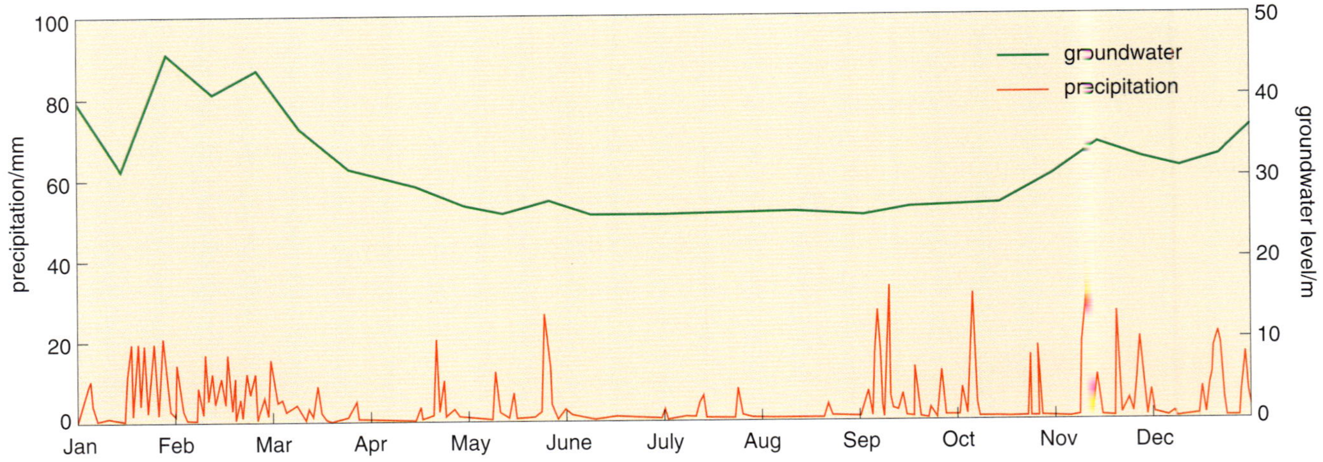

Figure 3.2 Variation of the groundwater level measured in at Stoneycombe Quarry in the Teign catchment, Devon, UK. Precipitation data for the same period are also shown.

The zones shown in Figure 3.1 are not horizontal. For example, the water-table slopes towards places where water leaves the ground — the river valleys — and away from where water is coming in — the hill sides and high ground. It is thus higher beneath hills than beneath valleys, following the landscape in a subdued way. Furthermore, the depth of each zone beneath the ground surface at any point is not fixed. The height of the water-table will fluctuate with precipitation; it increases in elevation following wet periods and decreases in elevation following dry periods (Figure 3.2).

- Figure 3.2 shows that highest rainfall is in autumn, but highest groundwater levels are in February. What is the reason for this offset?
- As you saw in the Block 2, Part 2, soils lose water over the dry summer months. Autumn rains must first replenish the soil water until it reaches its field capacity, before they can contribute to groundwater.

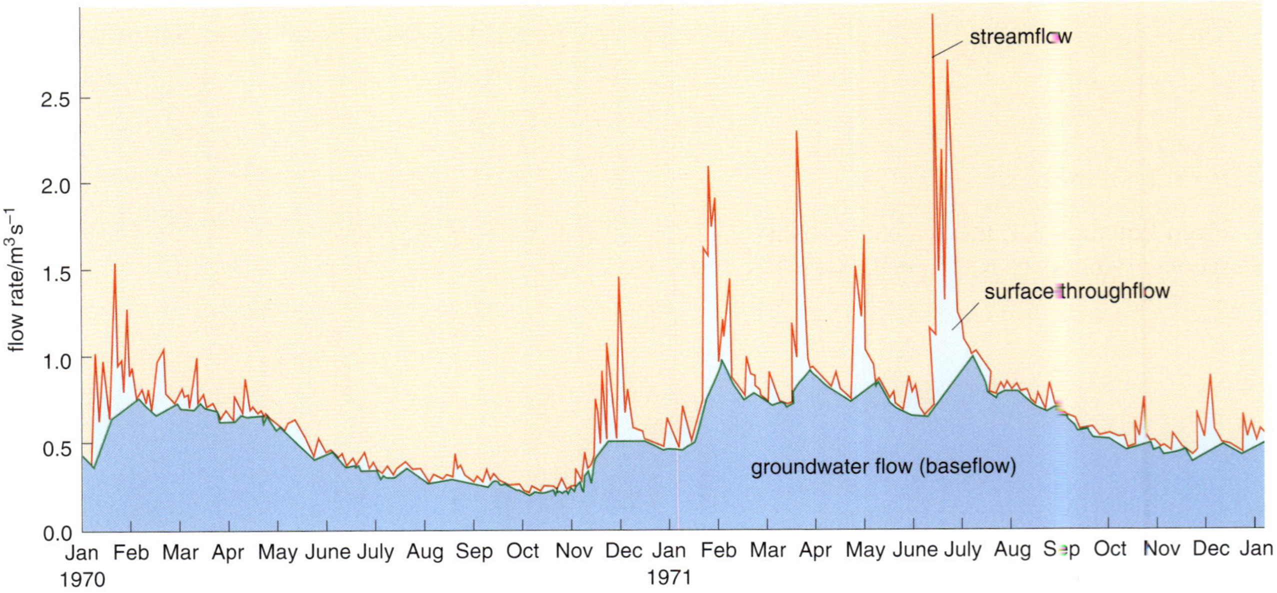

Figure 3.3 A hydrograph of the River Pang in Berkshire, UK showing the baseflow component. The river drains a chalk catchment partly covered by clay. Baseflow from the chalk provides almost the entire flow in dry periods.

After temporary storage in the ground, groundwater drains from springs into streams and rivers. Maximum discharges occur at the end of the winter when groundwater levels are high following seasonally high precipitation. They steadily decline throughout the summer into the autumn because the rate of evapotranspiration is higher than the rate of precipitation (Section 2.5.2). The contribution that groundwater makes to the flow of rivers is called baseflow and it is responsible for maintaining the flow of rivers during extended periods of dry weather, when surface throughflow virtually ceases (Figure 3.3).

3.2 Storing water in the ground

The rocks of the upper part of the Earth's crust contain many holes. Some are caverns, but most of them are tiny pores — such as the spaces between grains of sand in a sandstone — or networks of fine cracks. Groundwater is stored within these pores. Some rocks are more porous than others, and more importantly, the pores in some rocks are either large or join up so that water can flow through them easily.

3.2.1 Porosity

The amount of groundwater stored in a saturated material depends on its **porosity**. Total porosity is defined as the percentage of the total rock volume that is void space:

$$\text{porosity} = \frac{V_p}{V_{tot}} \times 100 \qquad (3.1)$$

where V_p is the volume of void space and V_{tot} is the total volume of rock. **Primary porosity** is due to voids created by grains as the rock formed (Figure 3.4a), while **secondary porosity** is due to voids or fractures created in the rock after its formation (Figure 3.4b).

- ❍ The volume of void space in a limestone due to (a) voids created by individual grains as the rock formed, and (b) fracturing after the rock was laid down is, respectively, $5\,\text{cm}^3$ and $15\,\text{cm}^3$. The total volume of the rock is $100\,\text{cm}^3$. What are the primary porosity, the secondary porosity, and the total porosity of the rock?

- ● From Equation 3.1, the primary porosity is $(5\,\text{cm}^3/100\,\text{cm}^3) \times 100 = 5\%$, the secondary porosity is $(15\,\text{cm}^3/100\,\text{cm}^3) \times 100 = 15\%$, and the total porosity is $(15\,\text{cm}^3 + 5\,\text{cm}^3)/100\,\text{cm}^3 \times 100 = 20\%$.

(a) (b)

Figure 3.4 (a) Electron micrographs of a sandstone showing the void spaces between the grains (primary porosity). The porosity is about 30%. (b) Photograph of an outcrop of limestone showing spaces resulting from bedding planes, joints, fractures and fissures (secondary porosity). Joints, fractures and fissures are all created after the deposit is laid down.

The principal factors that control porosity are grain size and shape, the uniformity of the grain sizes, the extent of cementing of the grains and the amount of fracturing. Figure 3.5 shows some typical kinds of porosity associated with various rocks.

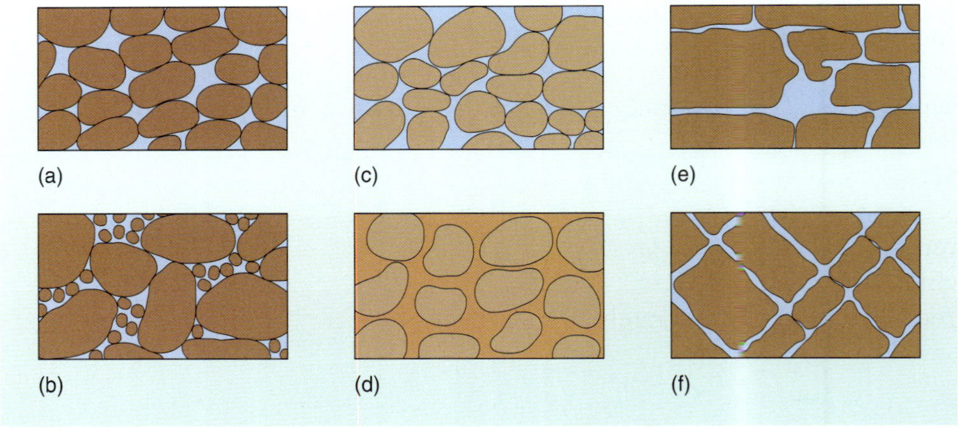

Figure 3.5 Types of porosity: (a) a well-sorted sedimentary deposit having high porosity; (b) a poorly-sorted sedimentary deposit having low porosity; (c) a well-sorted sedimentary deposit consisting of pebbles that are themselves porous, so that the deposit as a whole has very high porosity; (d) a well-sorted sedimentary deposit whose porosity has been diminished by cementing; (e) a rock rendered porous because it has partially dissolved; (f) a rock rendered porous because of fracturing. Dark brown rocks have low porosity; light brown rocks are relatively porous. Water is present in the areas coloured blue.

Porosity can range from zero or near zero to more than 60% (Table 3.1). Recently deposited sediments (**unconsolidated** — they have not been compacted or cemented) with rounded grains of uniform size are the most porous. Highly compacted (**consolidated**) rocks (such as shales), and dense crystalline rocks are usually less porous. Note that for recently deposited sediments, in general, the smaller the particle size, the higher the porosity.

Table 3.1 Porosities of different materials. The grain size classification of the unconsolidated sediments observes the International System.

Material	Grain size/mm	Porosity/%
unconsolidated sediments		
gravel	>2.0	25–35
sand	2.0–0.02	30–40
silt	0.02–0.002	40–50
clay	<0.002	45–60
consolidated sedimentary rocks		
limestone	variable	0.1–30*
sandstone	2.0–0.02	5–30
shale	<0.02	5–15
crystalline rocks		
granite	large (often cm size)	0.0001–1
basalt	small (usually <1mm)	0.001–1

* Higher values are associated with secondary porosity due to fracturing and dissolution.

3.2.2 Permeability

Although porosity determines how much water a saturated rock can hold, not all of this water is able to move freely from one place to another. The ease with which water can flow through a porous rock is measured by its **permeability**. Water flows most easily through rocks with large pores, or pores that join up. Such rocks are permeable; some sandstones and gravels are good examples of permeable rocks. In other rocks, water can hardly flow at all. Clay has a lot of very small pores which impart a high porosity, but the water in the pore spaces is virtually immobile because it is tightly held by capillary forces. Pumice is full of good-size holes but they rarely link up. Thus rocks such as clays and pumice can have high porosity, but low permeability.

Permeability and porosity decrease at depth as rocks compress. At some depth, rocks are no longer able to store water, and this marks the 'floor' of water storage in the Earth's crust. Geologists have found water deeper than 10 km, but the greatest depth at which water actively circulates is usually about a kilometre.

3.2.3 Aquifers

Layers of rock that are porous and permeable enough to store water and let it flow through them easily are called **aquifers**. The principal aquifers of the UK are found in the southeast of England. The most important are a type of limestone called 'the Chalk' (discussed in the Block 2, Part 2) other limestones and sandstones (Figure 3.6). Aquifers do occur in other rocks but these are generally older and therefore much harder and more compact, so they are regarded as of secondary importance in terms of water storage.

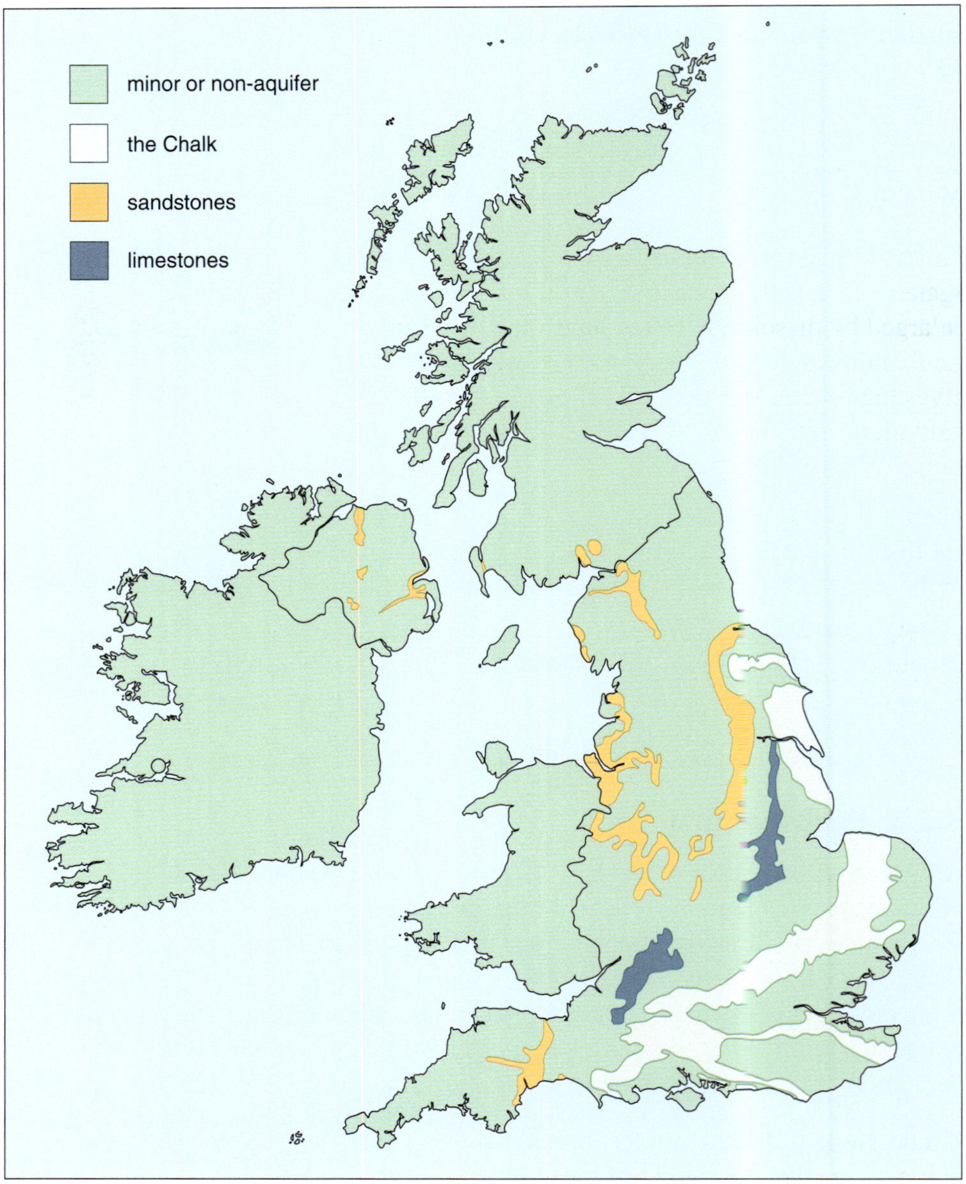

Figure 3.6 Distribution of the principal aquifers in the British Isles.

The Chalk, which is a soft, white limestone, underlies much of eastern and southern England. You may recall from the Block 2, Part 2 that it is a unique rock because it consists of minute calcareous shells and shell fragments of tiny marine animals. These impart a high porosity to the rock, but are so fine grained that the water contained in its pore spaces is virtually immobile, being held by surface tension. The Chalk owes its prominent role as an aquifer to a network of fine cracks that impart a high permeability (Figure 3.7). Individual boreholes in the Chalk can yield more than 10 million litres of water per day (Ml d^{-1}), which is sufficient to provide for the needs of about 70 000 people.

The sandstones are mainly brownish-red sandstones that originated in a desert environment (Figure 3.8). They are principally found in a series of sedimentary basins in western England and on the eastern and western flanks of the Pennines. The poorly cemented quartz grains in sandstone impart a porosity of about 30%. Groundwater can flow around the quartz grains but the presence of fractures enhances the permeability considerably. The sandstones are very permeable and yield a high proportion of the water that they store. Yields from boreholes are as much as 5–10 Ml d^{-1}.

Figure 3.7 An outcrop of the Chalk showing the interconnected network of fractures that gives the aquifer its high permeability.

Limestones are relatively hard with small pores, but again an extensive fracture network, with fractures enlarged by dissolution of the limestone by weakly acidic rainwater (as discussed in the Earth Block), gives high permeabilities. The largest yields from individual wells in the UK are provided by the Lincolnshire limestone — up to 30 Ml d^{-1}.

There are two types of aquifer, unconfined and confined, distinguished on the basis of the permeability of the surrounding rock in relation to the position of the water-table (Figure 3.9).

1 **Unconfined aquifers** are underlain by impermeable rock, but above this floor, the rock is permeable and rain soaks down to recharge the aquifer. The aquifer fills up with water until water reaches the surface of the land in one or more places, flowing out as springs or seepages. The top of the saturated rock is marked by the water-table. A thin impermeable layer sometimes occurs locally in an unconfined aquifer which stops or slows the downward flow of surface water. The permeable rock above this impermeable layer is a **perched aquifer**.

2 **Confined aquifers** are separated from the ground surface by an impermeable layer, and are underlain by impermeable rock. Thus water is trapped within this layer, and it presses on the confining layers above and below it. This means that at any point in a confined aquifer, the water pressure is larger than atmospheric pressure.

Figure 3.8 An outcrop of sandstone. Water flows through the sandstone between the individual quartz grains which make up the rock, and more rapidly along fractures.

If a borehole is drilled into a confined aquifer, water will rise up until the weight of the column of water balances the water pressure in the aquifer. If a series of such holes were drilled and the level of water was found in all of them, we could imagine a surface made by joining all the individual levels. This imaginary surface is called the **potentiometric surface** (Figure 3.9). It passes through the uppermost confining layer somewhere above the aquifer. Water in confined aquifers is called **artesian water** and a well that penetrates a confined aquifer is called an **artesian well**.

Note that the same aquifer can be confined in one area and unconfined elsewhere because an aquifer must be recharged somewhere.

- ❍ Water flows freely from the artesian well labelled (B) in Figure 3.9, but water from the artesian well labelled (A) must be pumped to the ground surface. With reference to Figure 3.9, can you suggest a reason for this?
- ● If potentiometric surface lies above the ground surface, then water will flow from the well. If the potentiometric surface lies below the ground surface, water must be pumped to the ground surface.

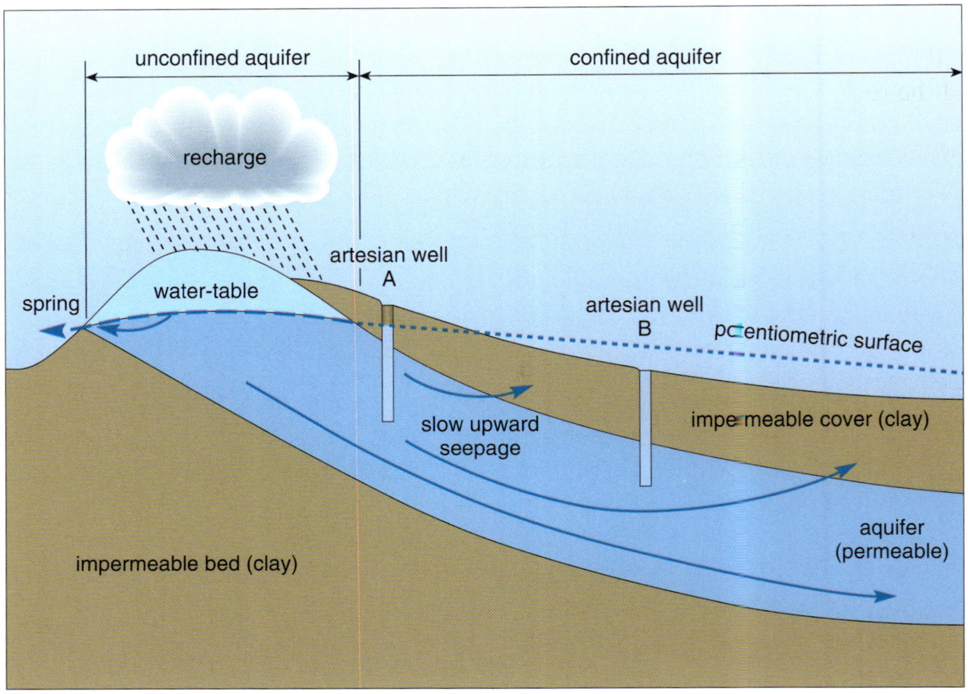

Figure 3.9 The relationship between unconfined and confined aquifers. Also shown are the water-table, the potentiometric surface, artesian wells, and areas of groundwater recharge and discharge.

Box 3.1 Water in the desert

The place where water leaves the ground may be a long way from where recharge occurs. An aquifer can therefore carry water from a wetter area to a drier one, even a desert. Then water may flow from the ground and form an **oasis** (Figure 3.10). For example, at Kufra in the Libyan Desert, water emerges from an aquifer called the Nubian Sandstone to form small lakes. The presence of such oases makes human travel and nomadic existence possible in arid lands.

Figure 3.10 An oasis in the desert.

3.2.4 Water storage in aquifers

In the case of an unconfined aquifer, water storage is mainly provided by the pore space. Changes in water storage are usually reflected directly by variations of groundwater level. However, the amount of water that can be recovered by pumping or drainage is usually less than the volume stored in the pore spaces. This is because some water is always retained around the individual grains that make up a rock by surface tension. This is known as **specific retention**. The **specific yield** is the maximum amount of water that can be recovered. Figure 3.11 illustrates the relationship between specific yield and porosity, which is expressed by the equation:

$$\text{specific yield} = \text{porosity} - \text{specific retention} \qquad (3.2)$$

All three terms in Equation 3.2 are expressed as *percentages* of the total volume of the rock.

- ❍ Why does the specific retention decrease with increasing grain size in unconsolidated sediments?
- ● Because a few large particles would have a smaller total surface area than a lot of small particles occupying the same volume, and a smaller surface area retains less water by surface tension.

The specific yield is greater for coarse sands than for gravels because porosity decreases more rapidly than specific retention with increasing grain size.

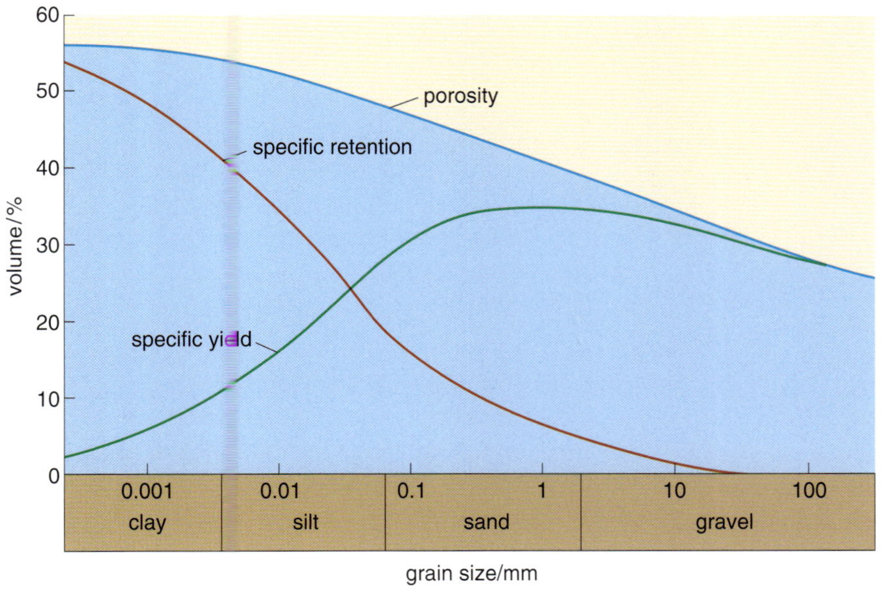

Figure 3.11 The relationship between specific yield, specific retention and porosity for unconsolidated sediments. Note that the scale on the grain size axis is not linear; each division corresponds to a factor of ten increase in the grain size.

In a confined aquifer, water can also be released from or taken into storage as a result of decompression or compression of both artesian water and the aquifer skeleton. However, the amount of water stored in this way is insignificant relative to the amount held in the pore spaces.

Box 3.2 Monitoring changes in groundwater storage

Many communities rely upon groundwater for their water supply, and have done so for thousands of years. Sometimes the names of settlements — names that in Britain end in '-well' or in the Middle East begin with 'Bir' or 'Beer' — testify to the nature of their water supply. In Britain today, groundwater supplies over 50% of the total public supply in the southeast; this figure is lower elsewhere, reflecting the distribution of aquifers and conditions for the development of this resource (Figure 3.6).

Pumping water from unconfined aquifers lowers the water-table. This may reduce or stop the natural 'overflow' from the aquifer causing springs or small streams to dry up. In Libya, pumping from wells around the Kufra Oasis has reduced the natural outflow. In south-east England, some streams on the Chalk flow less than they used to do; in part this is a result of abstraction.

But some changes are blamed unfairly on abstraction. In the 1850s, when there was still little pumping from the chalk aquifers in the UK, water levels were nearly as low as they are now; low rainfall was the cause then, and it is part of the cause of Europe's recent droughts.

In order to help maintain groundwater supplies, groundwater levels in aquifers are usually monitored on a regular basis. In the UK, some of the longest records come from private wells supplying water to large estates.

Chilgrove House is located upon the Chalk rocks of the South Downs in West Sussex, UK. A well one metre in diameter and 41.15 m deep was excavated in 1836, and later deepened to 62.03 m in 1934. Groundwater levels in the well have been measured since 1836 to the present day upon an almost continual basis, the longest period for any well in the UK. Measurements were started by the then owner of the property, Mr W. Leyland-Woods, and upon his death were continued by his children. Upon their death, the house came into the ownership of Colonel C. F. W. Dimond, who continued the measurements until 1952. Since 1952, measurements have been made by the local water authority, and latterly, by government institutions. The only break in this long period was for the thirteen months ending in April 1943, when the property had been requisitioned by the army during World War II.

Examples from the Chilgrove well hydrograph are shown in Figure 3.12. It shows no sign of human activity, such as large-scale groundwater abstraction. There are many natural fluctuations, however.

❍ Describe and account for the mean monthly variation in groundwater level over the period of the record, as denoted by the blue line in Figure 3.12.

● High water levels occur during winter months (November to April) due to natural replenishment by rainfall (and low evapotranspiration) and subsequent infiltration, and low levels occur during the summer months when there is little infiltration since evapotranspiration is generally sufficient to take up all the rainfall.

Thus, a summer drought will not increase the summer fall in groundwater levels, for there is little or no infiltration whatever the rainfall. A winter drought is another matter. Consider the hydrographs from 1975 and 1976. During the winter months there was virtually no replenishment, and groundwater levels fell continuously reaching minimum recorded levels in September.

During the period 1854 to late 1859, groundwater levels did not rise above the long-term monthly mean. By way of contrast, from 1912 to 1915 they seldom fall below the monthly means. Such long-term irregular, or **secular**, variations in groundwater levels are related to secular variations in rainfall. These are superimposed on the seasonal variations.

There are other interesting features in the hydrograph. Double peaks during the winter months, indicating breaks in infiltration, are not unusual. These breaks are often due, as in early 1987, to a lack of rainfall; January of that year, with rainfall about 30% of the monthly mean, was the third driest this century for the UK as a whole. Rather more occasionally, when the precipitation is in the form of snow rather than rain, infiltration may be delayed for as long as the snow lies. Infiltration starts again with the thaw. As the hydrograph for late January and February 1982 shows, the result is a fall followed by a recovery to produce the second peak.

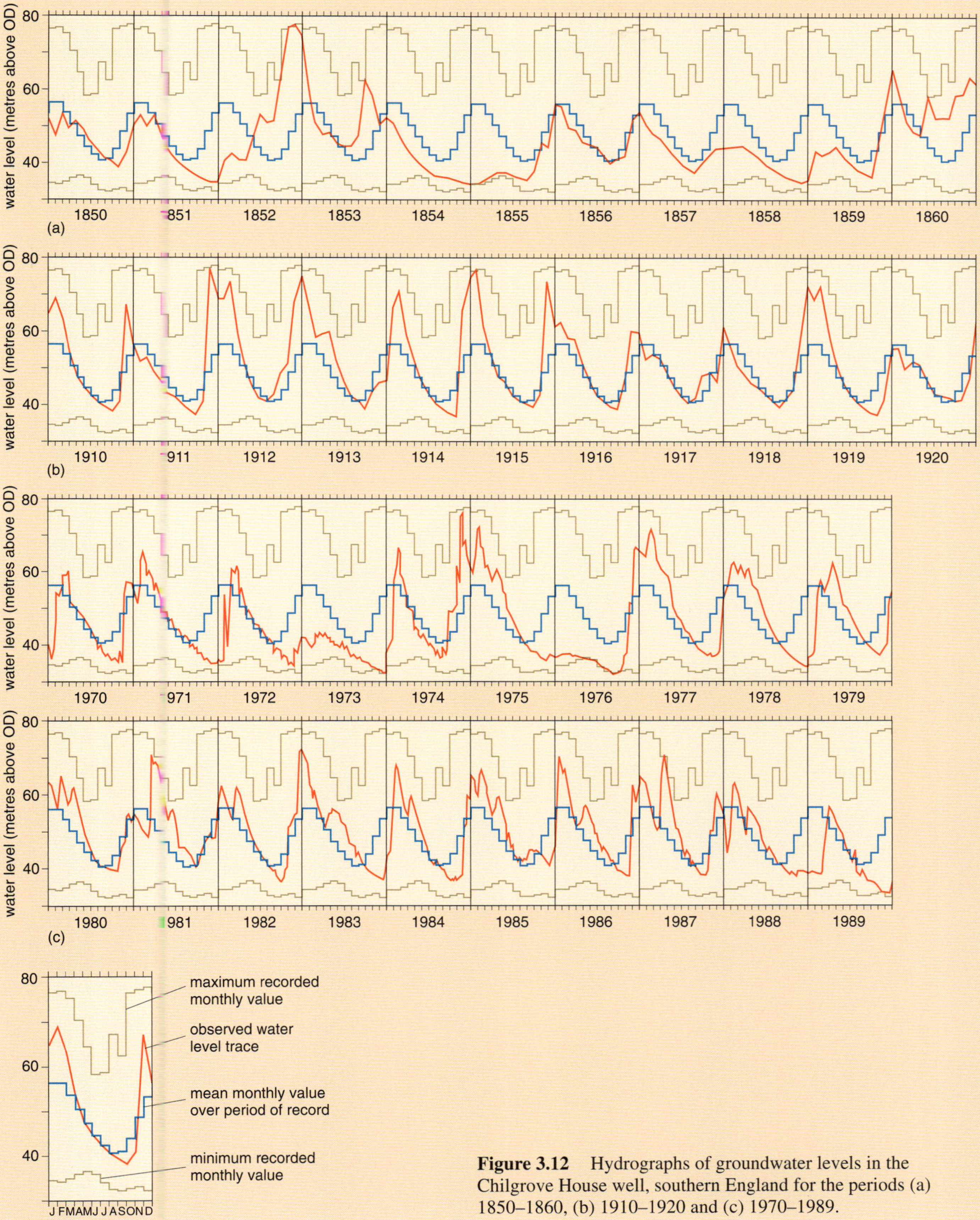

Figure 3.12 Hydrographs of groundwater levels in the Chilgrove House well, southern England for the periods (a) 1850–1860, (b) 1910–1920 and (c) 1970–1989.

3.3 Groundwater movement

Groundwater that is able to flow is in more or less continuous motion from areas of recharge to areas of discharge. Groundwater moves in response to pressure gradients, from an area of high pressure to an area of low pressure. Near the water-table, this means that groundwater usually flows 'downhill', i.e. from a higher level to a lower level, just as it would on the surface.

Maps of the water-table for an unconfined aquifer and of the potentiometric surface of a confined aquifer can be constructed from measurements of water-level elevations in wells. Sampling points of equal elevation can be joined by contour lines. Since groundwater levels can change with time, all water elevation readings need to be made within a short period of time. Data are usually expressed as height above some common datum; in the UK, Ordnance Datum (OD), which is roughly equivalent to mean sea-level, is the common datum for this purpose.

When constructing a map of the water-table, you should first plot the locations of any surface water features, such as lakes, springs, rivers and ponds. These can interact with the water-table. It is also useful to consult a map showing surface topography as the water-table roughly mimics this. The locations of the wells are plotted, and the water level elevations are noted. Lines of equal elevation can then be drawn. Interpolation of contours between data points is strongly influenced by the surface topography and surface water features. For example, water-table contours cannot be higher than the surface topography. The depth from the surface to the water-table will typically be greater beneath hills than beneath valleys. If a natural lake is present, the water-table lies at the lake surface. Water-table contours form a V shape, pointing upstream when they cross a stream that gains water from groundwater reserves beneath the water-table (Figure 3.13). Water-table contours bend downstream when they cross a stream that loses water to groundwater.

In general, the potentiometric surface of a confined aquifer is not influenced by surface topography and surface water features. Because there is no

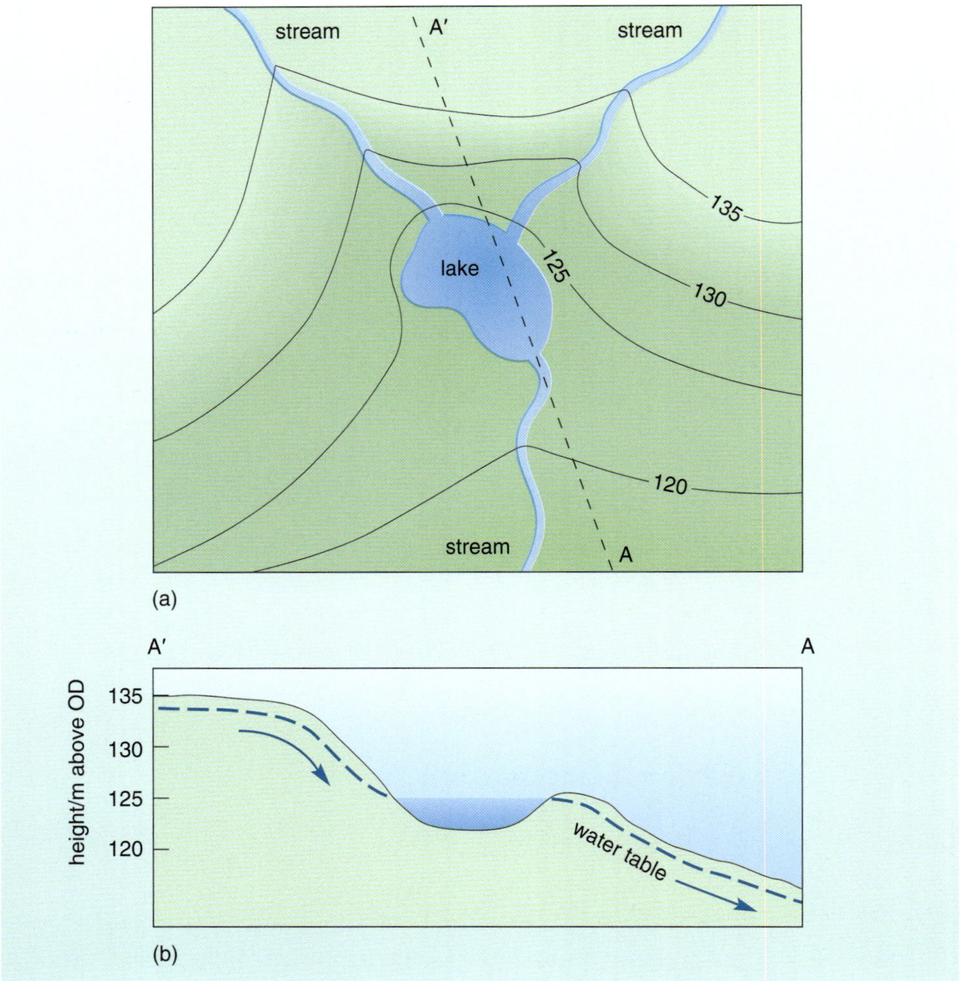

Figure 3.13 A water-table map in an area with a natural lake. (a) Two streams gaining water from groundwater reserves drain into the lake, and one stream gaining water from groundwater reserves drains from the lake. Water-table contours are given in metres above Ordnance Datum. (b) A cross-section through the lake from A to A′ is also shown; arrows show the direction of groundwater flow to the rivers.

hydraulic connection between a river and a confined aquifer beneath it, potentiometric surface contours are not influenced by the presence of the river. As shown in Figure 3.9, the potentiometric surface can even be above the land surface.

Just as with a topographical contoured map, or with isobars on a weather map, in areas where the water-table or potentiometric surface has a shallow gradient, the groundwater contours are spaced well apart. If the gradient is steep, the groundwater contours will be closer together. Groundwater will flow in the general direction that the water-table or potentiometric surface is sloping.

Question 3.1

Figure 3.14 is a contour map of water table elevations in the River Otter catchment, Devon, UK. Note that only a limited number of measurements are available, so contours are not given for the whole area of the map.

(a) Plot arrows on the map showing the general direction of groundwater movement.

(b) Is the speed of groundwater flow likely to be greater on the western or the eastern side of the river?

(c) Is the tributary in the southwest section of the river gaining, or losing, water from (or to) groundwater reserves?

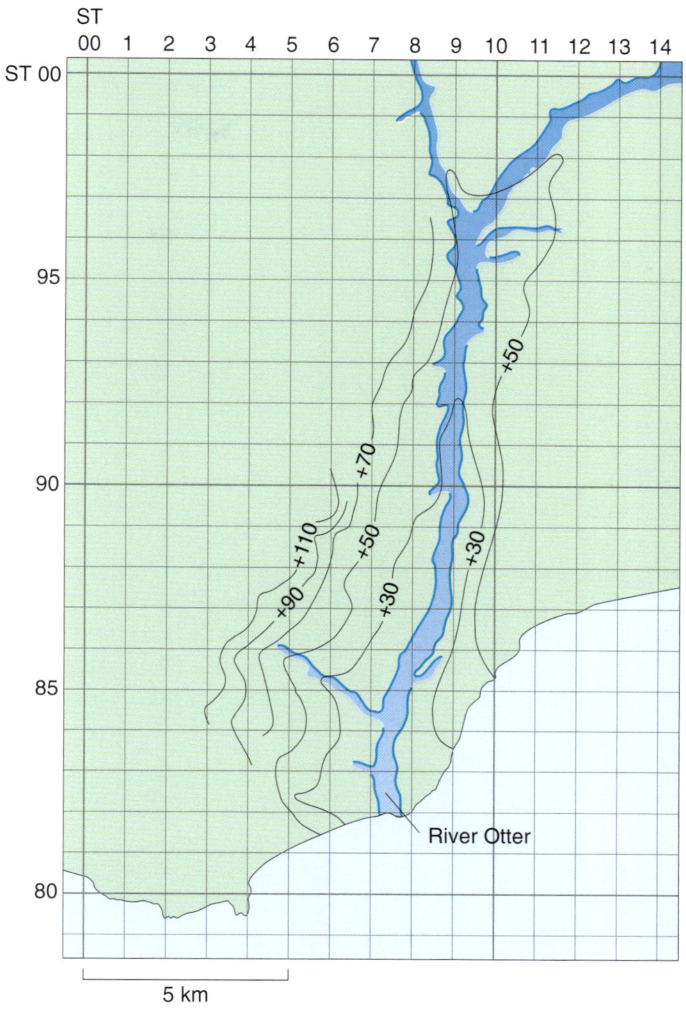

Figure 3.14 Contour map of water-table elevation (in metres above Ordnance Datum) in the catchment of the River Otter, Devon, UK.

3.3.1 Darcy's law

As you have seen in the Block 2, Part 2, the speed of water movement (v) through a saturated soil can be determined using Darcy's law:

$$v = K\frac{h}{l} \tag{3.3}$$

where K is the hydraulic conductivity and h/l is the hydraulic gradient. Darcy's law can also be used to calculate the speed of groundwater movement; in this case, the hydraulic gradient is the slope of the water-table (Figure 3.15). The value of K is dependent on the permeability of the rock, and also on the density and viscosity of the fluid flowing through it:

$$K = \rho g\frac{k}{\nu} \tag{3.4}$$

where ρ is the fluid density, ν is the fluid viscosity, g is the acceleration due to gravity, and k is the permeability of the rock. The units of hydraulic conductivity are metres per second ($m\,s^{-1}$) or metres per day ($m\,d^{-1}$). The larger the value of K, the more easily water flows through the rock. K is large for coarse materials (such as gravel), and small for fine materials (such as clays), see Table 3.2.

- ❍ Is the value of K smaller or larger for water than for a fluid of greater viscosity (such as treacle), assuming that the values of ρ and k are the same for both fluids?
- ● K is inversely proportional to fluid viscosity (Equation 3.4), so it is greater for the fluid with lower viscosity — in this case, water.

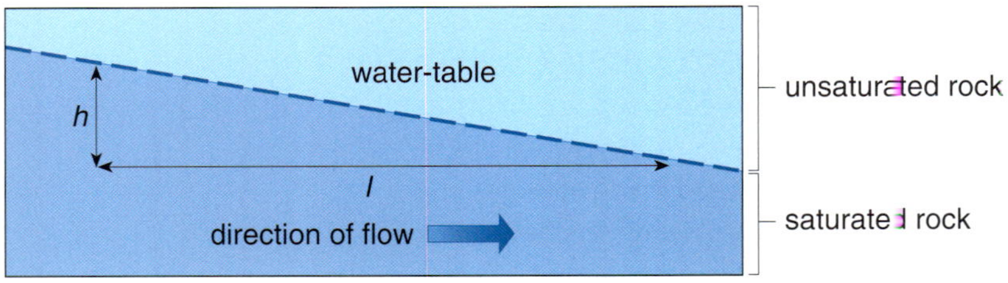

Figure 3.15 The flow of water through a permeable rock below the water table; h is the height (or head) of water over a horizontal distance l, so that h/l is the hydraulic gradient.

Rocks can be divided into two categories — permeable and impermeable — on the basis of their hydraulic conductivity. Rocks regarded as permeable have hydraulic conductivities of 1 m d^{-1} or more. Rocks with hydraulic conductivities of less than 10^{-3} m d^{-1} are usually regarded as impermeable.

Table 3.2 Hydraulic conductivites of different materials. For an individual rock type, higher values are associated with secondary permeability due to fracturing.

Material	Grain size/mm	Hydraulic conductivity/m d^{-1}
unconsolidated sediments		
gravel	> 2.0	5×10^2 to 1×10^4
sand	2.0–0.02	1 to 5×10^2
silt	0.02–0.002	10^{-2} to 1
clay	< 0.002	less than 10^{-2}
consolidated sedimentary rocks		
limestone	variable	10^{-5} to 10
sandstone	2.0–0.02	10^{-4} to 10
shale	< 0.02	5×10^{-8} to 5×10^{-6}
crystalline rocks		
granite	large (often cm size)	3×10^{-4} to 1
basalt	small (usually < 1 mm)	3×10^{-4} to 3

There are, however, a number of limitations in the application of Darcy's law.

(1) Darcy's law assumes that the rock unit is **homogeneous**; that is, it has the same properties at all locations. For a sandstone, this would indicate that the grain size distribution, porosity and degree of cementation are variable only within small limits. A limestone would have the same amount of cracks at all locations. In reality, rock units tend to be **heterogeneous**. For example, the formation of cracks in limestones is typically concentrated along preferred fractures or bedding planes.

At any point in the rock, the permeability will usually also vary with direction. For example, it is easier for water to flow between grains in the horizontal direction (parallel to the bedding) in the rock shown in Figure 3.16, than in the vertical direction. Equality of properties in all directions at a point is **isotropy**; variations in properties with direction is **anisotropy**. Most sedimentary rocks are anisotropic with respect to permeability because they contain grains that are not spherical but are elongated in one direction or shortened in another (Figure 3.16).

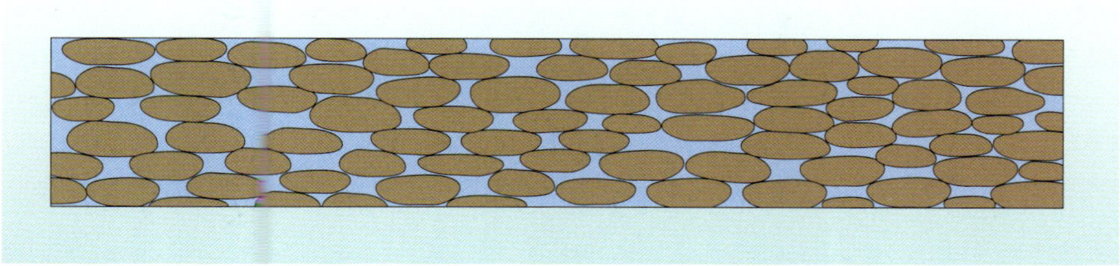

Figure 3.16 Anisotropy. Within a single rock type, elongate grains tend to be deposited with their longer axes horizontal. Fluids can pass more easily parallel to the long axis than perpendicular to it. Thus this texture causes permeability anisotropy — even though the rock may be homogeneous.

(2) Darcy's law takes no account of **tortuosity** — the fact that water must flow around the grains that make up the rock. This means that the actual flow of a water particle through a given length, *l*, will be longer than *l* as the fluid has to travel around the grains of the rock (Figure 3.17).

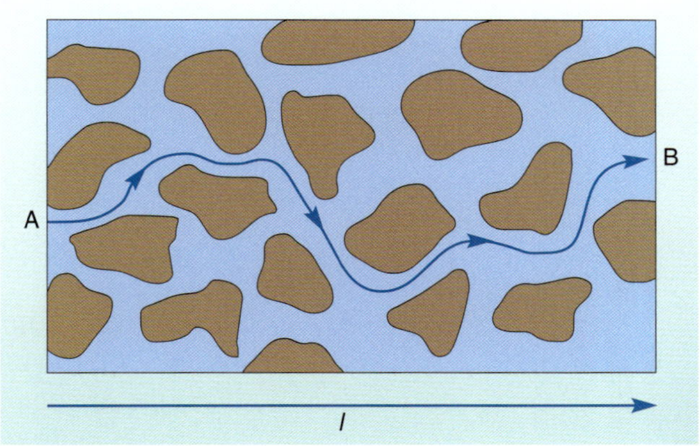

Figure 3.17 Flow paths in a porous rock. Because the fluid must travel around the rock particles, its pathlength from point A to point B is longer than *l*.

(3) Darcy's law applies only to very slowly moving groundwaters. In slowly moving fluids, fluid flow is **laminar** and molecules of water follow smooth lines, called **streamlines** (Figure 3.18). As the speed of the fluid increases, flow becomes **turbulent** and the water molecules no longer move along parallel streamlines. Turbulence can modify the relationship between hydraulic head and fluid speed, so Darcy's law breaks down.

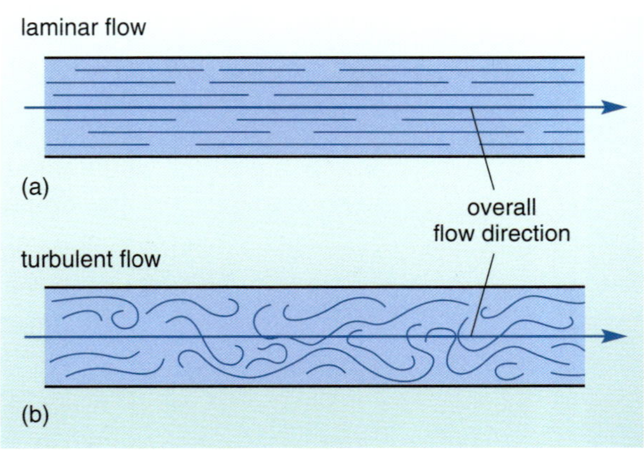

Figure 3.18 Flow paths of molecules of water in (a) laminar flow, and (b) turbulent flow.

3.3.2 Rate of flow of water through an aquifer

Darcy's law can also be used to calculate the rate of flow of water through an aquifer. Consider the case of the confined aquifer shown in Figure 3.19. The volume of water moving through the aquifer in Figure 3.19 per unit of time (the flow rate, Q) is given by:

$$Q = v \times b \times w = v \times A \tag{3.5}$$

where v is the speed of movement of the fluid and A is the cross-section area ($b \times w$) of the aquifer. Q is normally expressed in $m^3\,s^{-1}$. You already know from Darcy's law that v is equal to $K \times (h/l)$ (Equation 3.3), so:

$$Q = K \times \frac{h}{l} \times A \tag{3.6}$$

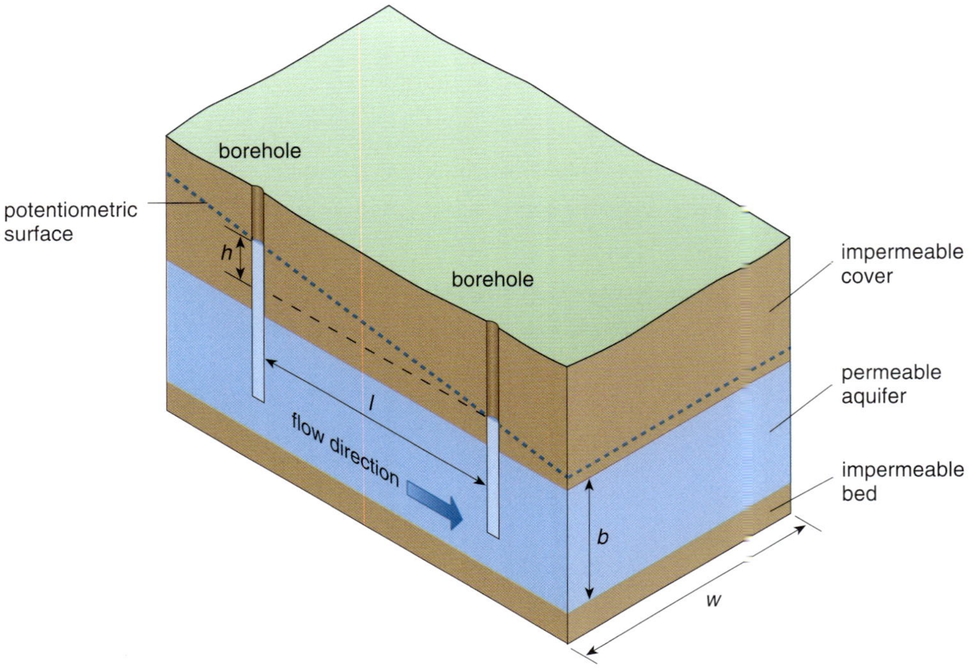

Figure 3.19 Flow through a confined aquifer. In a confined aquifer the saturated thickness, b, is constant. The flow rate, Q, through the aquifer is equal to $K \times (h/l) \times b \times w$, where K is the hydraulic conductivity.

Equation 3.6 shows that the effectiveness of a rock as an aquifer depends not only on its hydraulic conductivity but on its cross-sectional area perpendicular to the flow direction — i.e. on its thickness, b, and width, w. In particular, in carrying water to a well it is the product of hydraulic conductivity, K, and thickness, b, that is important. This product, $K \times b$, which is frequently given the symbol T, is called the **transmissivity** of the aquifer. Transmissivity has units of hydraulic conductivity multiplied by thickness; it is usually expressed in $m^2\,d^{-1}$. Important UK aquifers such as parts of the Chalk have transmissivities in excess of $1000\,m^2\,d^{-1}$.

Question 3.2

The dimensions of the aquifer shown in Figure 3.19 are $w = 8$ m, $b = 3$ m and $l = 10$ m. Water moves through the aquifer at a speed of 5×10^{-2} m d^{-1}; the hydraulic conductivity of the aquifer, K, is 1.2 m d^{-1}.

(a) Is the aquifer regarded as permeable or impermeable?

(b) What is the rate of fluid flow through the aquifer?

(c) What is the value of h, the hydraulic head?

(d) What is the transmissivity of the aquifer?

In an unconfined aquifer the slope of the water-table is a measure of the hydraulic gradient. Flow is occurring only through the saturated part of the aquifer. In this case, the transmissivity is the product of hydraulic conductivity, K, and b, where b is the *saturated* aquifer thickness. A complication is that because the water-table is sloping, b is not constant; in order to apply Darcy's law, we need to take an average value. A further complication is that a sloping water-table also results in fluid flow in the vertical direction as well as in the horizontal direction.

Flow to a well

When water is pumped from a borehole, the water level (or potentiometric surface) is lowered in the surrounding area. A hydraulic gradient is created in the aquifer which allows water to flow towards the borehole. The difference between the original water level and the water level during pumping is the **drawdown**, which is equivalent to the head of water necessary to produce a flow through the aquifer to the borehole — the greater the yield from the borehole, the greater the drawdown. The drawdown decreases with increasing distance from the borehole until a point is reached where the water level is unaffected. The surface of the pumping level is in the form of an inverted cone and is referred to as a cone of depression (Figure 3.20).

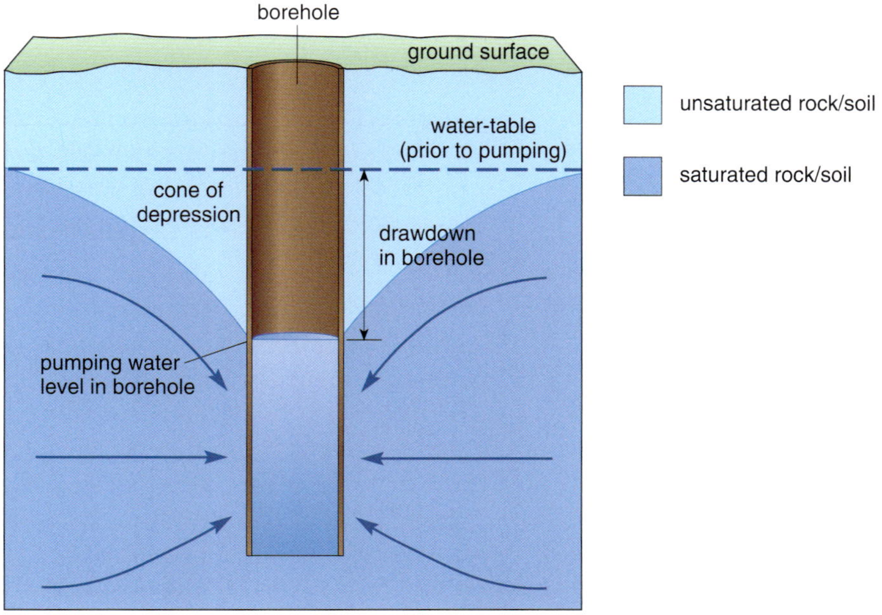

Figure 3.20 The drawdown of the water-table (or potentiometric surface) around a pumping borehole to form a cone of depression. The shape and extent of the cone of depression depends upon the rate of extraction, the duration of extraction, and the hydraulic properties of the aquifer.

Water flows into a borehole from all directions in response to pumping and, as the steepness of the water table (or potentiometric surface) increases towards the borehole, the speed of the fluid increases as it converges towards the borehole.

3.3.3 Exploring an aquifer

Three types of investigation are required in order to characterize an aquifer; geological investigations define the framework in which the groundwater occurs, hydrological investigations quantify the input and output of water to and from the groundwater catchment, and hydraulic investigations show how water moves within the aquifer.

Inputs and outputs of water to and from catchments are considered in Section 5. Here we take a brief look at some of the techniques used for assessing the geological and hydraulic properties of an aquifer.

Geological measurements

Basic information on the rocks that constitute the structure of the aquifer can be obtained from geological maps. Geological maps are available for all but the most remote parts of the world, though the detail and reliability can be variable. Hydrogeological maps, which combine information on basic geology with data on the hydraulic behaviour of the rocks and their usefulness for water supply, are now available in many countries. A section derived from a hydrogeological map showing the general relations of the rocks and the saturated zones of the principal aquifers in the southwest of the UK is shown in Figure 3.21.

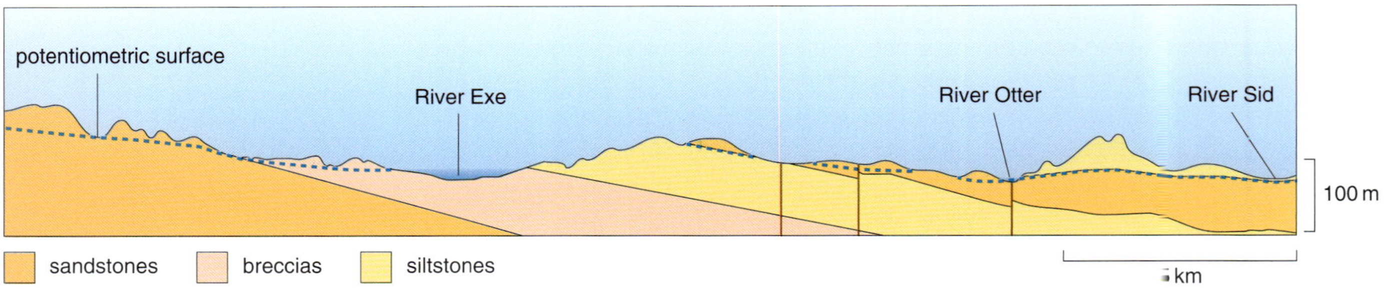

Figure 3.21 Cross-section showing the relationship between rock-type and aquifers in the southwest of the UK. The most important aquifers in this region are located within the sandstones and breccias. These aquifers are commonly confined by siltstones.

Mapping provides information about the ground surface and some idea of subsurface conditions can be inferred from this, but the only way to find out for certain what lies beneath the ground surface is to drill a borehole. This provides information on only one place, so a network of boreholes is generally required.

The technique most commonly used to study the distribution of groundwater is resistivity surveying. The electrical resistance of a material is a measure of its ability to resist electrical conduction — the flow of charged particles. Most common minerals have high electrical resistivity and are good insulators. However, if the rock contains water, which in turn contains dissolved ions, then electrical currents can flow by the movement of the charged ions in solution. The resistivity of rocks is therefore significantly reduced by the presence of pore water. The more porous the rock, the lower its resistivity. Resistivity surveys detect variations in the electrical conductivity of rocks underground, and may reveal bodies of rock with unusually high conductivity.

Question 3.3

Figure 3.22 shows the relationship between the electrical resistivity and rock type for a well located within a confined aquifer in the upper Brazos River Basin, Texas. Which of the different rock types has (a) the highest porosity, and (b) the lowest porosity, and how can you tell?

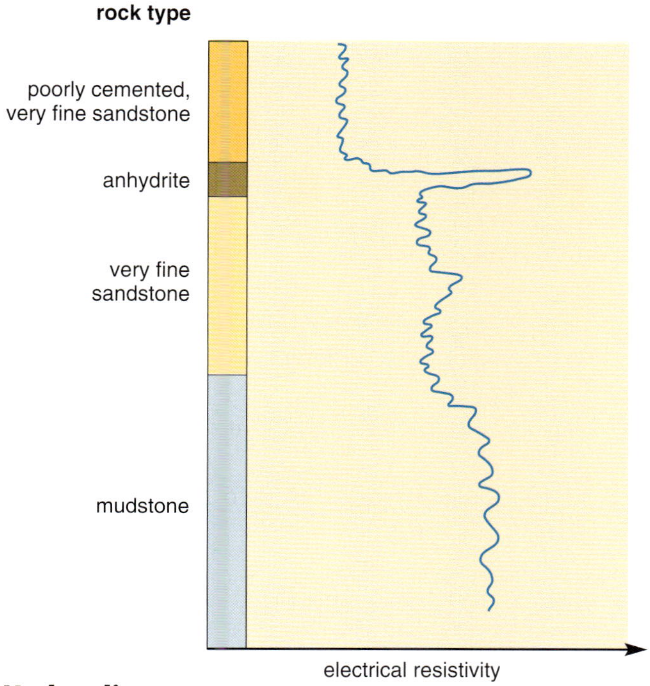

Figure 3.22 Profile of electrical resistivity and rock type in a well located in the upper Brazos River Basin, Texas, USA.

Hydraulic measurements

The most important hydraulic properties of the rocks are porosity, which controls how much water is held in storage; the specific storage and specific yield, which control how much of the water can be removed, and the transmissivity, which governs how readily the water can move through the rock to wells and natural outlets.

Porosity can be measured in the laboratory by weighing a known volume of rock first dried, and then saturated with pore water. The difference between the saturated and dry masses gives the pore space volume:

$$\text{pore space volume} = \frac{\text{mass rock saturated} - \text{mass rock dry}}{\text{density of pore water}} \quad (3.7)$$

The density of the pore water is usually close to $1000\,\text{kg m}^{-3}$. The porosity of the rock can be calculated from Equation 3.1.

❍ A rock of dry mass 1000 g occupies a volume of $250\,\text{cm}^3$. The saturated mass of the rock is 1125 g and the density of the pore fluid is $1000\,\text{kg m}^{-3}$ ($1\,\text{g cm}^{-3}$). What is the porosity of the rock?

● From Equation 3.7, pore space volume = $(1125\,\text{g} - 1000\,\text{g})/\ 1\,\text{g cm}^{-3}$ = $125\,\text{cm}^3$. From Equation 3.1, porosity = (pore space volume/ total volume of rock) × 100. Thus, porosity = $(125\,\text{cm}^3/250\,\text{cm}^3) \times 100 = 50\%$.

Box 3.3 Water dowsing

Some people claim to be able to locate an aquifer by walking over the surface of the ground until they observe a response with a forked stick, bent rods, a pendulum or some other apparatus, which is usually held in front of them with both hands. The origin of the practice is not clear, but the earliest sign of its usage dates from a 4500–5000 year-old grave inscription in Brittany, France.

There are various theories as to what causes a dowsing tool to move: electromagnetic or other subtle geological forces, suggestion from others or from geophysical explanations, ESP or other paranormal phenomena. Some scientific studies, such as the 'Scheunen' experiment conducted in Germany in 1987–1988, have concluded that the success of self-proclaimed dowsers was no better than would be expected by chance, while other studies report success rates as high as 96%.

Figure 3.23 Using dowsing rods to locate an aquifer.

There may or may not be something in these studies and theories. However, as you have seen, simply locating the presence of groundwater is only a small part of the groundwater story.

Laboratory instruments for measuring permeability are called permeameters. In essence a permeameter is simply a device for holding a sample so that a fluid can be passed through it and Darcy's law applied to determine the hydraulic conductivity. Holding a sample of rock so that fluid passes only through the rock and not through crevices between the rock and sample holder is difficult, particularly for hard rock samples. Permeabilities are usually determined in two directions in case the rock is anisotropic.

One drawback of laboratory tests is that samples are inevitably small, so larger scale features of the aquifer, such as cracks and fissures, may be overlooked. The only way to test more representative volumes is to move outside and conduct field measurements.

One of the most effective and frequently used methods of measuring aquifer properties is the field pumping test. As you saw in the previous section, when water is pumped from a well, a cone of depression is formed in the potentiometric surface (Figure 3.20). The steepness of the cone depends on the hydraulic gradient, which in turns depends on the pumping rate and on the transmissivity and specific retention of the aquifer. The greater the specific retention, the larger the cone must be for any given quantity of water extracted. Given these relationships, it follows that (in the absence of complicating factors) if we pump water from a well and observe the way the cone of depression expands, then we should be able to deduce the transmissivity and specific retention of the aquifer. Many models exist for the interpretation of well tests. These work well for confined, homogenous aquifers, but are less reliable for unconfined aquifers which are heterogeneous.

3.3.4 How long does water remain underground?

If we wanted to know how long a droplet of water had been underground, it would be useful to be able to label it with date, time, and place and introduce it to an aquifer and await its discharge.

Labels cannot be fixed to water droplets, but substances called **tracers** can be added to water and tracked as the water moves on. Ideally, the substance used for a tracer must move as part of the water, at the same rate, and not become filtered or separated from it; it must not be naturally present in the water; it must be readily detectable, easy to handle, and non-toxic.

One approach involves investigating levels of chlorofluorocarbons (CFCs) in groundwater.

Chlorofluorocarbons are synthetic compounds which are produced for a range of industrial and domestic purposes, such as aerosol sprays and refrigerants. Concentrations of the chlorofluorocarbons in the atmosphere have been increasing steadily over the past 50 years, as shown in Figure 3.24.

CFC concentrations in groundwater can therefore be used to estimate the age of waters less than about 50 years old. CFC concentrations in water at its time of infiltration are directly related to the atmospheric concentration at that time, so measured CFC concentrations in groundwater can be compared to the atmospheric concentration curve shown in Figure 3.24 to obtain the groundwater age. The sensitivity of this method depends on the rate of change of atmospheric CFC concentration with time. As a result of regulations regarding the use of CFCs (because they contribute to the destruction of the ozone layer), their production is declining, so the ability to date very young water will diminish with time.

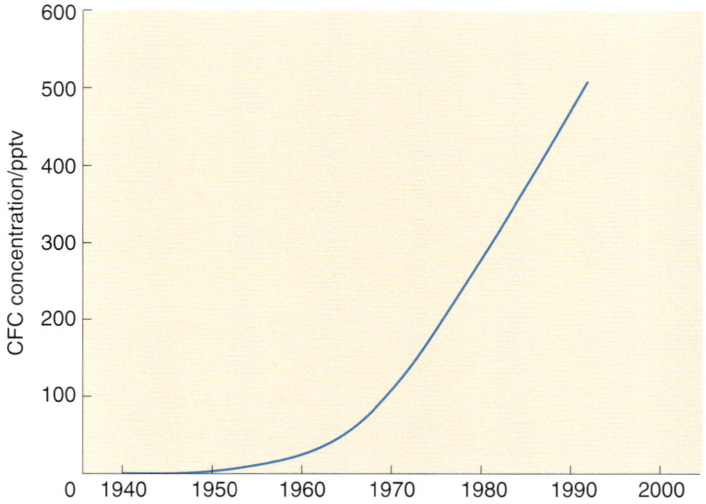

Figure 3.24 Atmospheric concentrations of CFCs from 1940 to 1990. Concentrations are expressed in parts per 10^{12} by volume (pptv).

- The concentration of CFCs measured in a groundwater sample in 1994 is equivalent to a concentration of 300 pptv in the atmosphere. What was the apparent age of the groundwater sample?

- A concentration of 300 pptv approximately corresponds to the year 1980 according to Figure 3.24. Thus the groundwater sample was about 14 years old.

Figure 3.25 shows a CFC profile measured in the Sturgeon Falls Aquifer, USA. Concentrations decrease almost linearly with depth; the calculated groundwater ages are in good agreement with other dating methods. The apparent age of 36 years for the sample from 19.1 m appears too young. This has been attributed to contamination during sampling.

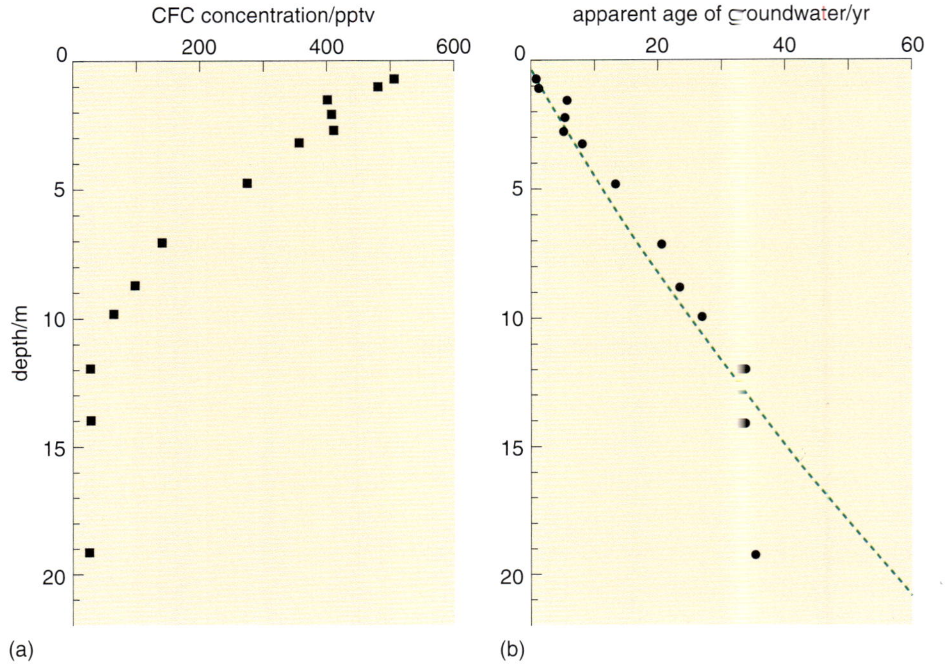

Figure 3.25 Profiles of (a) CFC concentration (in pptv) and (b) apparent age of groundwater in the Sturgeon Falls aquifer. The broken line represents results of a groundwater flow model calibrated using other dating methods.

In addition to problems with contamination, other drawbacks of CFC dating include; (1) adsorption of the CFC to aquifer materials, (2) degradation of the CFC where oxygen levels are low, and (3) time-lag associated with the diffusion of CFCs through the unsaturated zone. This time-lag increases with the depth of the unsaturated zone, and can be difficult to account for.

Groundwater is actually a mixture of waters from many sources of different ages. The age of a particular sample is the average age of all the constituent components. In general the age of groundwater increases with depth, and with distance from the area of aquifer recharge (Figure 3.26).

Another useful approach to groundwater dating involves investigating levels of radioactive isotopes in a fluid. If you need to refresh your understanding of isotopes and other aspects of the atomic structure of matter, read Box 3.4.

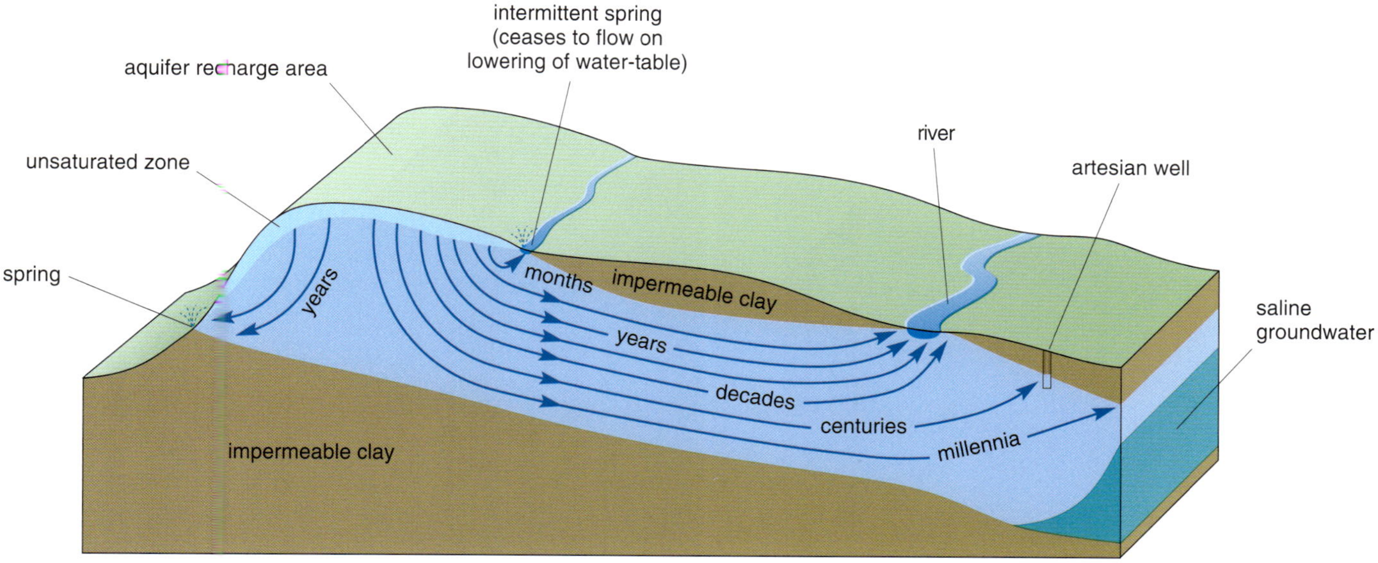

Figure 3.26 Age of groundwater. Groundwater in the upper part of the unconfined zone generally varies in age from months to years. As water penetrates deeper within the confined part of the aquifer, the age increases to decades, centuries or even millennia. The very saline water lying in aquifers below the zone of active freshwater circulation can be millions of years old.

Box 3.4 Atomic structure and isotopes

An **atom** is the smallest component (building block) of a chemical element that retains the chemical properties of that element. Atoms contain three fundamental particles: **protons** and **neutrons** (both contained within a dense nucleus), and **electrons**. Protons and neutrons have roughly the same mass, but the mass of an electron is negligible by comparison (about 2000 times smaller). The simplest atom of the element hydrogen has a single proton in its nucleus, which is orbited by a single electron. All atoms are electrically neutral because the positive charges on the protons balance the negative charges on the surrounding electrons. The number of protons within the nucleus defines the element and is called the **atomic number**, i.e. each element has a different atomic number. The atoms of all elements other than hydrogen also contain neutrons in their nuclei. These particles are electrically neutral. The total number of protons and neutrons in the nucleus is known as the **mass number** of the element.

Although all atoms of the same element have the same number of protons in their nuclei (i.e. they have the same atomic number), they don't necessarily have the same number of neutrons (i.e. they may have different mass numbers). These are said to be **isotopes** of the element. Isotopes of a particular element have identical chemical properties, because their properties are determined by the number of electrons around the nucleus (which is equal to the number of protons, i.e. the atomic number) and are independent of the number of neutrons in the nucleus.

If we were to look in detail at the hydrogen atoms present in water, we would find that a small proportion of them (about 0.02%) contain a neutron as well as a proton in the nucleus (Figure 3.27). This heavier isotope of hydrogen is called deuterium (and is given the symbol 'D' to distinguish it from the more common isotope, hydrogen 'H'). Hydrogen is, in fact, the only element that has named isotopes. Isotopes are usually denoted by writing the mass number as a superscript before the chemical symbol of the element or as a number after the element name, e.g. ^{2}H (for deuterium or hydrogen-2) or ^{12}C (for carbon-12).

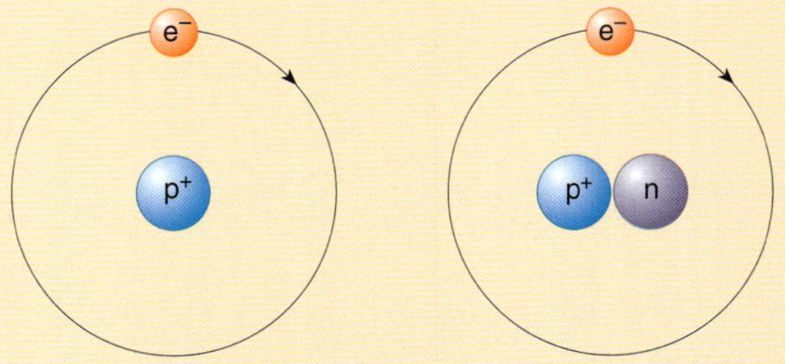

Figure 3.27 Schematic diagram of two isotopes of hydrogen. The hydrogen in naturally occurring substances on Earth is present as either hydrogen (^{1}H) or deuterium (^{2}H or D), with hydrogen predominating. p^+, n, and e^- represent a proton, neutron and electron, respectively. (Not drawn to scale.)

A third isotope of hydrogen is called tritium (^{3}H or hydrogen-3). It has one proton and two neutrons in its nucleus, which makes it unstable and therefore radioactive. It is produced in nuclear reactors and nuclear explosions. Unlike tritium, the two main isotopes of hydrogen (hydrogen and deuterium) are **stable isotopes**, and so are not radioactive and do not decay with time. Most elements have more than one isotope, some of which may be stable, others radioactive. For example, naturally occurring carbon contains three isotopes, ^{12}C, ^{13}C, ^{14}C (carbon-12, carbon-13 and carbon-14), the first two are stable, the third is radioactive.

- The atomic number of magnesium is 12. Magnesium has three isotopes, ^{24}Mg, ^{25}Mg, ^{26}Mg. How many neutrons does each isotope of magnesium contain?

- The number shown as a superscript before the symbol of an element is the number of neutrons plus the number of protons in the nucleus. The number of protons in the magnesium nucleus is 12. Therefore ^{24}Mg has 12 neutrons, ^{25}Mg has 13 neutrons, and ^{26}Mg has 14 neutrons.

Radioactive isotopes change by radioactive decay with a **half-life**, λ, which is constant for any particular isotope into stable isotopes of other elements. The half-life is simply the time taken for half of the number of atoms of the radioactive isotope (the 'parent' isotope) to decay into the stable isotope (its 'daughter' isotope). The half-lives of all commonly occurring radioactive isotopes are known from laboratory experiments; the range is very large. For example, the half-life of uranium-238 is 4.47×10^9 years, while the half-life of hydrogen-3 is only 12.4 years. The constancy of each half-life is the critical factor that allows us to use radioactivity as a natural clock. But to determine the ages of groundwaters we have to learn how to 'tell the time' using that clock.

The general equation for radioactivity can be expressed as:

$$P = \frac{P_0}{2^n} \qquad (3.8)$$

where P_0 is the number of parent isotope atoms initially present in the groundwater sample, and P is the number of parent isotope atoms remaining after n half-lives have elapsed. The value 2^n is simply the number two to the power n. For example, if $n = 3$, then 2^3 is $2 \times 2 \times 2$, which is 8.

- ❍ A groundwater sample is 10 years old. It contains 1×10^{24} atoms of a radioactive isotope that has a half-life of 2 years. How many atoms of the radioactive isotope did the sample initially contain?
- ● The radioactive isotope has a half-life of 2 years. If the sample is 10 years old, then 10/2 = 5 half-lives have elapsed (i.e. $n = 5$). Rearranging Equation 3.8, the number of atoms initially present in the sample (P_0) is:

$$P_0 = 2^n P = 2^5 \times 1 \times 10^{24} \text{atoms} = 3.2 \times 10^{25} \text{atoms}$$

As the number of parent atoms declines, so the number of daughter atoms increases, but the total number of atomic nuclei remain the same. For example, in the decay of ^{14}C to its daughter ^{14}N, each time a ^{14}C nucleus decays, a new ^{14}N nucleus is formed, so the total number of atomic nuclei in the system is constant. In the general case, this can be stated as:

$$D + P = P_0 \qquad (3.9)$$

where D is the number of daughter isotope atoms present. This can be rearranged to calculate D:

$$D = P_0 - P \qquad (3.10)$$

In a groundwater sample we usually measure the relative number of daughter and parent isotope atoms, which, in our terminology is D/P. From Equations 3.9 and 3.10 we can find an expression for the value of D/P in the following way.

Rearranging Equation 3.8 to make P_0 the subject we get:

$$P_0 = 2^n P$$

Substituting this expression for P_0 into Equation 3.10 leads to:

$$D = 2^n P - P = P(2^n - 1)$$

and so:

$$D/P = 2^n - 1 \qquad (3.11)$$

In other words, for every value of n, the number of half-lives elapsed since the parent isotope entered the water, there is a unique value of D/P (known as the daughter to parent ratio). The important observation from this is that the value of D/P depends on the number of half-lives, n, that have elapsed since the parent isotope entered the water, and this value can be used to calculate the age of groundwater. A plot of D/P versus n is given in Figure 3.28.

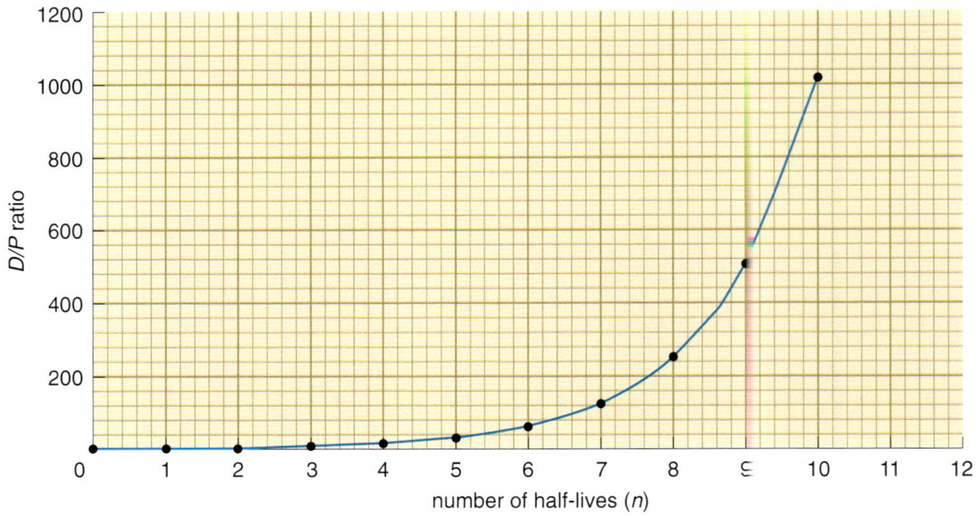

Figure 3.28 A plot of D/P value against number of half-lives, n, elapsed.

Activity 3.1 Calculating the value of *D/P* for a simple radioactive decay scheme

In this activity, you will calculate the number of atoms of parent and daughter isotopes present in a sample after a various number of half-lives, and will use these results to calculate the value of D/P at these times. You will then compare these D/P values with those calculated from Equation 3.11, derived above, and will plot a graph that shows how the value of D/P depends on the number of half-lives that have elapsed.

Task 1

Consider a sample that initially has 1024 radioactive parent atoms — a ridiculously small number, but one that lends itself to easy manipulation. Calculate the number, P, of parent atoms left after successive half-lives, and the number, D, of daughter atoms that have been produced, and enter these numbers in Table 3.3. Remember, after each successive half-life, the number of parent atoms remaining is half the number at the beginning of that half-life period. Then calculate the value of D/P at each time, and enter these values in the table.

The expression for D/P that was derived in Equation 3.11 is:

$$D/P = 2^n - 1$$

Use this expression to calculate the value of D/P for various values of n, and check that the values calculated from the equation are the same as those in the table.

Table 3.3 Numbers of parent and daughter atoms during radioactive decay.

No. of elapsed half-lives, *n*	No. of parent atoms, *P*	No. of daughter atoms, *D*	*D*/*P*
0	1024	0	0
1	512	512	1
2			
3			
4			
5			
6			
7			
8			
9			
10			

Task 2

(a) Use the grid in Figure 3.29 to plot a graph of the number of parent atoms *P* (on the vertical axis) against the number of half-lives *n* (on the horizontal axis).

(b) Then plot the value of *D*/*P* against the number of half-lives *n* on the grid in Figure 3.30.

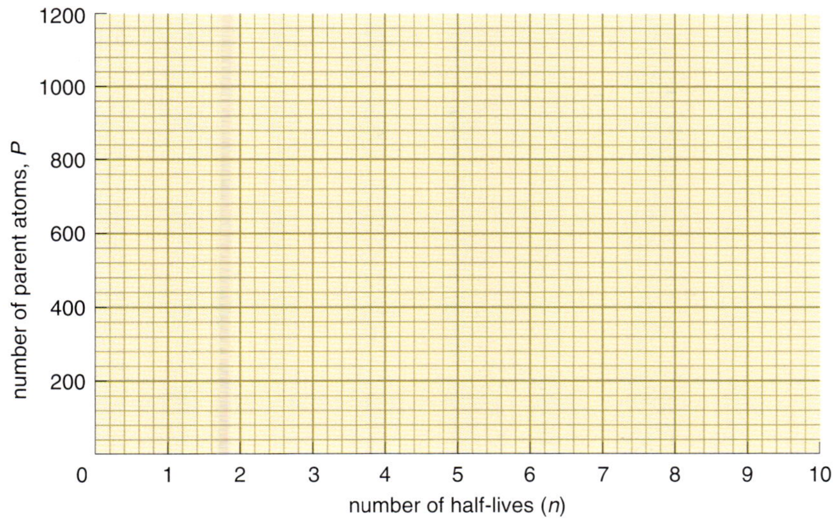

Figures 3.29 Grid for use with Task 2a.

Figures 3.30 Grid for use with Task 2b.

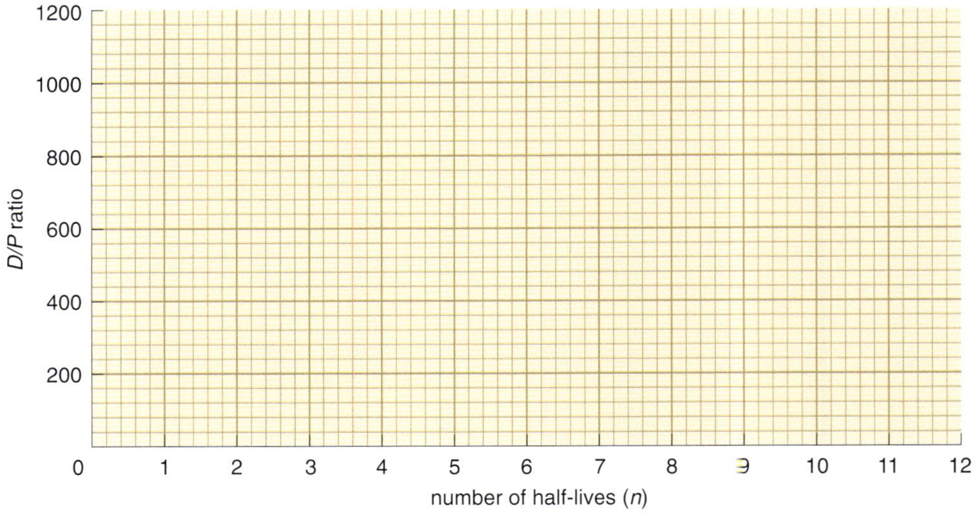

We have now shown in two ways, the first mathematical and the second graphical, that we can relate the values of *D*/*P* in groundwater to its age. However, certain conditions must be met for our simple version of a radioactive clock to work. First, the sample must act as a closed system — atoms of parent and daughter isotopes must neither enter nor leave the sample after its formation. Second, there must be no daughter isotope atoms present in the sample at the time of formation. For many radioactive systems this is not the case and corrections for initial concentrations of isotopes have to be applied.

Examples of radiometric dating of groundwater

The most common isotopes used for radiometric dating of groundwaters are carbon-14 and hydrogen-3 (tritium). Both isotopes are formed in the atmosphere by the tests of thermonuclear bombs in the 1950s and 1960s, but carbon-14 is also formed naturally by bombardment of nitrogen by cosmic rays. Both isotopes are incorporated into surface water, and their concentration in surface water is directly proportional to the atmospheric concentration. If this water infiltrates the ground, then the concentration of carbon-14 and hydrogen-3 starts to fall as both radioactive isotopes are isolated from their atmospheric source.

Carbon-14 has a half-life of 5700 years, and tritium has a half-life of 12.4 years. This means that carbon-14 is useful for dating waters that have been underground for some thousands of years, but tritium is only suitable for dating relatively young groundwaters. More specifically, the maximum reliable age estimate that can be calculated from any given isotope is ten times the half-life of that isotope, i.e. 10λ.

- ❍ What is the maximum age of groundwater appropriate for (a) carbon-14 dating, and (b) tritium dating?

- ● (a) The half-life of carbon-14 is 5700 years. Therefore carbon-14 can be used to date groundwater up to 10×5700 years = 57 000 years old. (b) The half-life of tritium is 12.4 years. Therefore tritium can be used to date groundwater up to 10×12.4 years old = 124 years old.

Analysis for tritium in the late 1960s revealed that water infiltrates through the Chalk in the unsaturated zone at a rate of about 1 metre per year. Thus where the unsaturated zone is 50 m thick, the water is some 50 years old when it reaches the water-table. However, the data also showed that where the Chalk is fractured, water travels much more rapidly through these fractures, at speeds of the order of 50 metres per day. About 10–15% of the infiltration through the Chalk flows through fractures in the unsaturated zone to the water-table.

- ❍ The *D*/*P* value of groundwater in the Chalk in the centre of the London Basin in the UK is 10.4. The parent isotope is ^{14}C. With reference to Figure 3.28, what is the age of this water?

- ● From Figure 3.28 we can see that a *D*/*P* value of 10.4 corresponds to 3.5 half-lives. As the half-life of ^{14}C is 5700 years, this is equivalent to an age of $3.5 \times 5\,700$ years = 20 000 years.

This means that groundwater in the London Basin fell as rain during the last ice age! An even more extreme example is the great Artesian Basin of Australia. Water in this basin has been in the ground for more than a million years. These groundwater age data are extremely important, because they tell us that water can remain underground for long periods of time; some parts of the water cycle are extremely slow moving.

3.4 The chemical evolution of groundwater

The chemical composition of groundwater is very variable but, on average, groundwaters have different proportions of dissolved substances relative to their surface water source (Figure 3.31). The purpose of this section is to investigate how and why the composition of water changes as it flows through the ground.

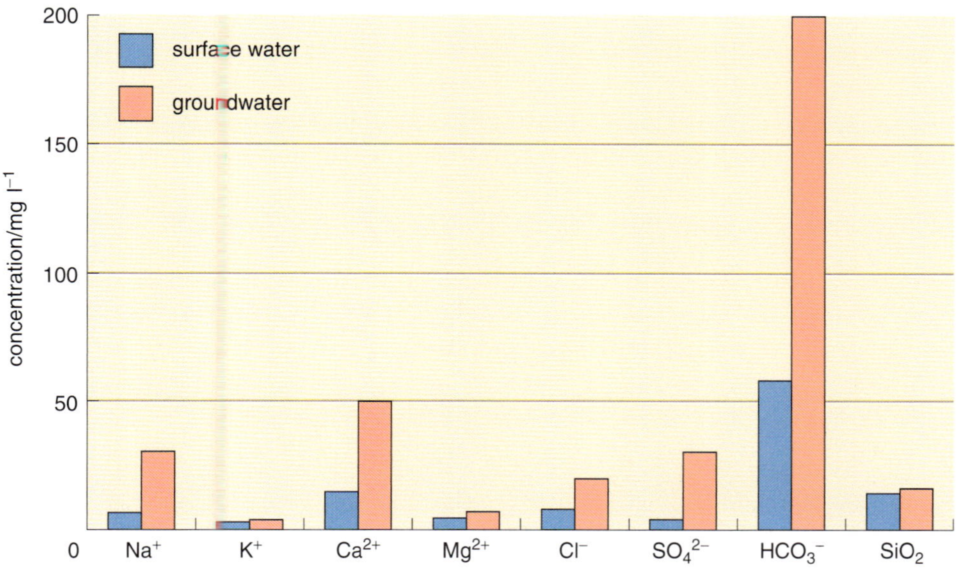

Figure 3.31 The average chemical composition of surface water and groundwater. Only the major dissolved constituents are shown, and dissolved gases are not included.

3.4.1 Reactions with surrounding rocks

Chemicals dissolved in groundwater are derived mainly from the dissolution of minerals in the soil and the rocks with which it is or has been in contact. However, the situation is a little more complicated than this as some rocks are more soluble than others. As you saw in the Block 2, Part 2, the ease with which a mineral dissolves can be expressed in terms of its ionic product, *K*. In general, the bigger the value of *K*, the more soluble the mineral. Table 3.4 gives values of *K* for a series of different minerals.

Table 3.4 Some common examples of mineral dissolution in pure water and their associated ionic product (*K*).

Mineral	Dissolution process	K
Chlorides and sulphates		
halite	$NaCl(s) = Na^+(aq) + Cl^-(aq)$	[illegible]$7.6\ mol^2\ kg^{-2}$
gypsum	$CaSO_4(s) = Ca^{2+}(aq) + SO_4^{2-}(aq)$	[illegible] $\times 10^{-5}\ mol^2\ kg^{-2}$
Carbonates		
magnesite	$MgCO_3(s) = Mg^{2+}(aq) + CO_3^{2-}(aq)$	[illegible] $\times 10^{-9}\ mol^2\ kg^{-2}$
calcite	$CaCO_3(s) = Ca^{2+}(aq) + CO_3^{2-}(aq)$	[illegible]$.5 \times 10^{-9}\ mol^2\ kg^{-2}$
Dolomite	$CaMg(CO_3)_2(s) = Ca^{2+}(aq) + Mg^{2+}(aq) + 2CO_3^{2-}(aq)$	[illegible] $\times 10^{-17}\ mol^4\ kg^{-4}$
Silicates		
quartz*	$SiO_2(s) = SiO_2(aq)$	[illegible] $\times 10^{-5}\ mol\ kg^{-1}$

*This is not an ionic product in the strictest sense because no dissolved ions are formed.

- ❍ Which of the minerals listed in Table 3.4 is likely to be the most soluble?
- ● Halite is likely to be the most soluble as it has the highest ionic product, *K*.

After halite, gypsum is likely to be the next most soluble, and after gypsum, quartz and then magnesite and calcite. However, as you saw in Block 2, Part 2, quartz crystals in a granite tend to survive chemical weathering to be transported away as solid grains, while calcite crystals in limestone and marble show visible evidence of solution. The explanation for this apparent paradox is as follows. It turns out that the *rate* of silicate dissolution is actually very slow (the value of *K* tells us nothing about the speed at which a mineral dissolves), while the rate of dissolution of calcite by subsurface waters is high because it is assisted by hydrogen ions. Recall:

$$CaCO_3(s) + H^+(aq) = Ca^{2+}(aq) + HCO_3^-(aq)$$

- ❍ Where do the hydrogen ions in subsurface water come from?
- ● The hydrogen ions can be generated when carbon dioxide dissolves in water, and by plant metabolism and decay. They are also found in acid rain.

The value of *K* varies with, for example, the chemical composition of the solution (e.g. calcite is more soluble in a weakly acidic solution than it is in pure water) and with temperature (minerals are usually more soluble at higher temperature). The values of K given in Table 3.4 are therefore only a *guide* to mineral solubility. However, in general terms, when a rock contains merely 1% of halite, gypsum or calcite, their dissolution will define water chemistry. Silicates and clay minerals less readily contribute dissolved ions to the water, when they are present.

3.4.2 Interaction with clay minerals

Groundwaters also interact with clays within which they are confined. Clay is made up of extremely fine minerals which have a high surface area per unit of mass. This area is the area of contact between water and solid, and cations located on the surface of mineral grains can exchange with cations in the water (Figure 3.32). This process of cation exchange has been described in detail for soils in Block 2, Part 2. In groundwaters, calcium and magnesium ions are typically adsorbed from groundwater onto the surface of clay minerals, and sodium (or potassium) is released in their place. Since Ca^{2+} and Mg^{2+} ions cause water 'hardness', ion exchange leads to 'softer' water.

- ❍ Why are double-charged cations, such as Ca^{2+} and Mg^{2+}, generally more attracted to clay mineral surfaces than single-charged cations, such as Na^+ and K^+?
- ● As you saw in the Block 2, Part 2, clay mineral surfaces carry a slight negative charge. The higher the charge on the cation, the more strongly attracted it is to this negatively-charged surface, so double-charged cations will tend to be found closer to the particle surface than single-charged cations.

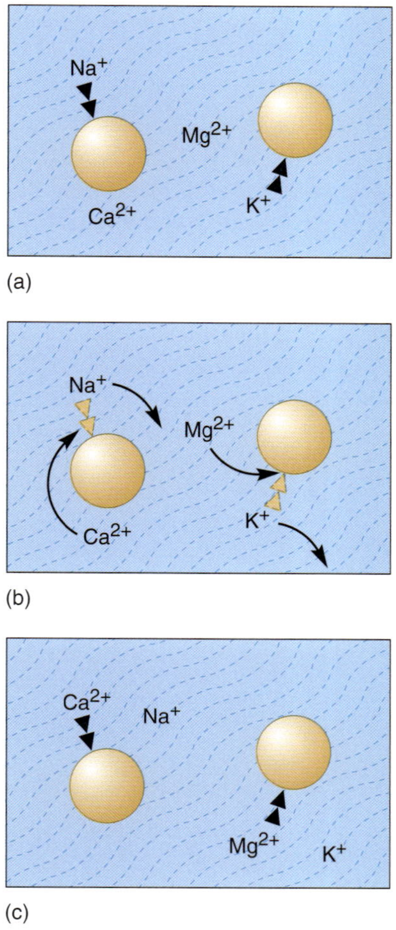

Figure 3.32 Exchange of ions between groundwater and clay minerals. (a) K^+ and Na^+ ions are adsorped on the surface of clay particles. Ca^{2+} and Mg^{2+} are in solution. (b) As double-charged cations are more attracted to clay surfaces than single-charged cations, Ca^{2+} and Mg^{2+} displace the Na^+ and K^+ cations. (c) The Ca^{2+} and Mg^{2+} cations are adsorbed on the clay surface, and Na^+ and K^+ are in solution.

3.4.3 Oxygen in groundwater

If fluid circulation is restricted as water flows further and further into a confined aquifer, then levels of oxygen in the water may gradually decrease. This is because (a) some bacteria, which are present in the aquifer, use oxygen as a source of energy, (b) oxygen is consumed in the decomposition of organic material, and (c) oxygen lost in these ways cannot be replenished by the atmosphere as the groundwater is no longer in contact with the atmosphere.

When oxygen levels are low, other chemicals will act to decompose organic material present in the aquifer. Most of these decomposition reactions are speeded up, or initiated, by bacteria present in the aquifer. Chemicals that help in the decomposition of organic matter include the oxygen-rich anions nitrate and sulfate, and iron in its oxidized form (Fe^{3+}). Levels of nitrate and sulfate are low in low-oxygen groundwaters; this is because nitrate is reduced to nitrite or nitrogen and sulfate is converted to hydrogen sulfide. Levels of dissolved iron increase in low-oxygen groundwaters because insoluble Fe^{3+} ions in iron minerals which form the aquifer walls are reduced to the more soluble Fe^{2+} form of iron. If the aquifer is very deep, then carbon dioxide may be reduced, producing methane gas in the process.

Box 3.5 The deep biosphere

For most of the 20th century, microbiologists believed that bacteria lived down to about 1.5 metres depth in soils or ocean mud, then quickly died out due to a lack of food sources. Any bacteria found at deeper depths were usually considered to be contaminants carried down from the surface with the sampling equipment.

However, drilling by the US Department of Energy (DOE) into sedimentary rocks beneath its nuclear site at Savannah River in South Carolina in 1987 provided indisputable evidence for a viable subsurface microbial community extending to at least 500 metres, where the deepest core ended (Figure 3.33). Microbiologists found over 10 000 strains of microbes in the cores; only about 400 have been characterized so far, but some seem to be new species which have adapted to the conditions of the rocks that they inhabit.

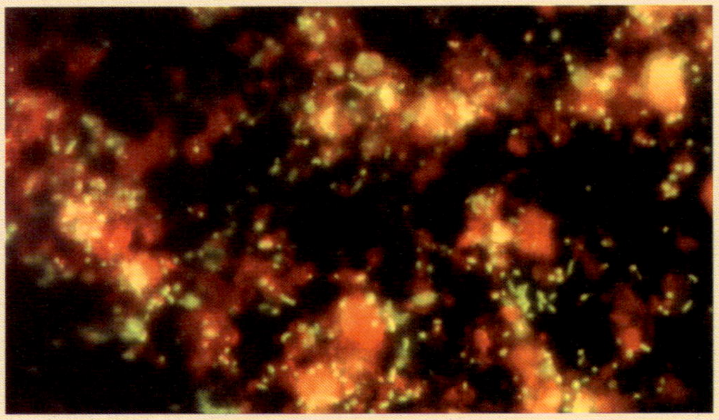

Figure 3.33 Bacteria in rocks. A chemical compound (acridine orange) has been added to the rock. Bacteria in the rock react with the compound and can be seen to fluoresce beneath a microscope. This enables microbiologists to estimate bacterial populations in the subterranean zone.

Light is not available in the deep subsurface, so photosynthesis is not possible. Therefore deep microbial life is dependent on energy sources that have been buried in sediments and laid down as rock, or enter as dissolved components in recharge water. These energy sources are organic matter and chemicals in their reduced form such as Fe^{2+} and hydrogen sulfide that can be oxidized with the release of energy. As long as appropriate energy sources are available, the primary factors limiting the distribution of microbial life in the deep subsurface are likely to be temperature and moisture content (Figure 3.34). The generally accepted upper limit for microbial life is approximately 110 °C, but some microbiologists speculate that the actual upper limit may be closer to 150 °C.

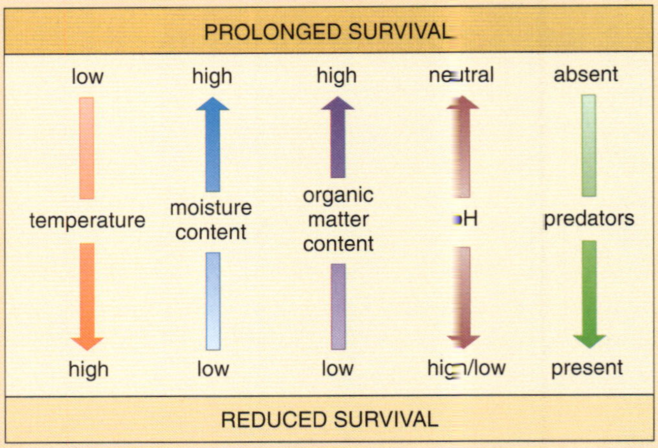

Figure 3.34 Factors affecting microbial survival in subsurface systems, including groundwater and aquifers.

The nature of the aquifer substrate also affects levels of microbial activity. Organisms can only pass through aquifer spaces that are physically larger than the organism itself (Figure 3.35). Studies have shown that although bacteria are found in rocks with pore sizes of <0.2 μm in diameter, they are dormant. Relatively rapid rates of microbial activity are found in rocks with interconnected pore spaces >0.2 μm in diameter; they are sustained by nutrients supplied by groundwater flowing through the rock.

Rates of microbial activity in deep aquifers are at least 10^3–10^5 times slower than those in modern surface sedimentary environments; bacteria do not exactly thrive under these harsh conditions where energy supplies are limited.

Microbes both depend upon and modify the chemistry of their environment. For example, fermentative bacteria (which convert large organic compounds into smaller organic compounds) are usually present in all groundwaters and are responsible for most of the primary degradation of organic carbon. Iron-utilizing bacteria are abundant where reactive iron oxide clays and organic carbon are present in favourable concentrations, but sulfate-utilizing bacteria are dominant only deep within an aquifer where sulfate levels are high. This 'microbial succession' is due to the isolation of the groundwaters from the surface, so that

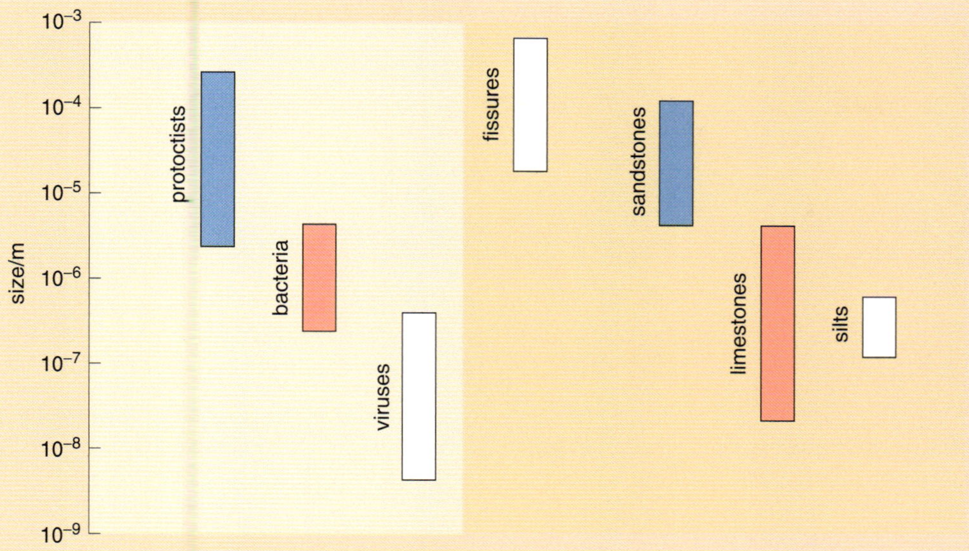

Figure 3.35 Comparison of organism size, and size of pore spaces in rocks present in the subsurface. Protoctists are unicellular organisms (and also multicellular algae); viruses are disease-producing particles which can only grow within a living host cell.

levels are high. This 'microbial succession' is due to the isolation of the groundwaters from the surface, so that nutrients are not readily replenished. Microbial diversity along groundwater flowpaths often decreases because needed nutrients are consumed and not replaced.

There is great interest in the study of the microbiology of groundwater not only because of its influence on groundwater chemistry. Microbes are hungry consumers of natural organic carbon, but they are also happy to consume organic contaminants. For example, bacteria capable of converting organic contaminants to harmless minerals have been isolated from boreholes heavily contaminated with fuel oils. The introduction of bacteria to contaminated sites can be an effective and cheap way of cleaning them up.

3.4.4 Journey through a groundwater system

Some of the chemical processes described above are most important in the early stages of the groundwater journey, while other processes become more important deep within the system. Figure 3.36 shows a typical sequence of the chemical evolution of groundwater.

In the unconfined part of the aquifer, where water is continually recharging and discharging, dissolution processes dominate and the soluble minerals are flushed out. Oxygen levels are high, but levels of dissolved ions (collectively known as 'total dissolved solids') are relatively low. With increasing age, concentrations of hydrogen carbonate ions increase, but later sulfate increases and becomes the major anion. Sulfate is derived from the mineral pyrite (an iron sulfide) which is sometimes found in sedimentary rocks. Ion exchange between the fluid and clay minerals softens the water, replacing calcium and magnesium ions with sodium and potassium. Deeper groundwaters are slow moving and often have high levels of total dissolved solids. Sodium and chloride are usually major constituents of the fluid. They may be derived from dissolution of salt deposits, or from seawater trapped in the rocks as they formed. Levels of oxygen can be low, which leads to the subsequent disappearance of sulfate ions as they are reduced to hydrogen sulfide and high levels of methane which is generated by the reduction of carbon dioxide.

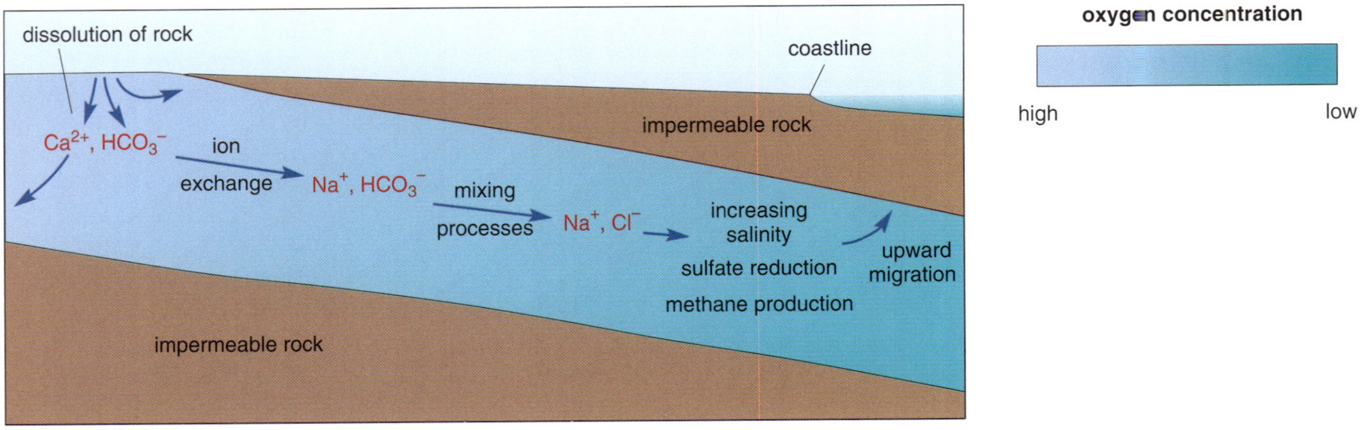

Figure 3.36 Schematic diagram of downgradient chemical changes in groundwater. The sequence shown here is referred to as the Chebotarev sequence: dominant ions at each stage are shown in red.

Box 3.6 Mineral waters: Bath Spa

Legend has it, that in the last millenium BC, Prince Bladud returned from his travels a leper. Because of this illness he was confined, but escaped in disguise from his father's court and came to a place called Swainswick where he was employed as a swineherd. In cold weather he saw his pigs wallowing in a mire (Figure 3.37). He found that the mud was warm and the pigs enjoyed the heat. Noticing that the pigs which bathed in the mire were free of scurfs and scabs, he reasoned that he might benefit from the waters. Cured of his leprosy, he returned to his father's court where he was restored to his inheritance. He succeeded to the throne on his father's death, whereupon he founded the City of Bath around the hot springs and built the baths so that others might benefit as he had done.

The mineral waters at Bath have total dissolved solids of more than 1000 milligrams per litre — they taste salty and can be rather pungent due to the presence of hydrogen sulfide. The waters are probably over 10 000 years old, and they have a temperature of 46.5 °C. The waters have descended from their source in the Mendip Hills to a depth of about 2 km before rising through fractures to discharge as springs at Bath. In the UK, the temperature below the ground increases with depth at an average rate of 25 °C per kilometre. Thus as the water flows through the rocks it not only becomes rich in minerals, but it also becomes hotter.

Figure 3.37 The Prince and his pigs.

It must be emphasized that this 'typical' sequence is subject to many variations. However, the general idea is a useful one. It is sometimes called the **Chebotarev sequence,** after the scientist who first proposed it.

3.4.5 Describing chemical data

The results of an analysis of groundwater chemistry are most commonly presented as a table of data. However, for many purposes it is much more informative to display data in a graphical form. A **Piper diagram** is particularly useful for classifying groundwater and defining a pathway of chemical evolution. The general form of the Piper diagram, which actually consists of three separate diagrams, is shown in Figure 3.38.

As shown in Figure 3.31, the major chemical species in groundwaters are Na^+, K^+, Ca^{2+}, Mg^{2+}, Cl^-, HCO_3^-, and SO_4^{2-}. These data are plotted on the Piper diagram in the following way. First, the relative abundances of the cations (Na^+, K^+, Ca^{2+} and Mg^{2+}) are plotted on the bottom left triangle. You have already come across a triangular diagram in Block 2, Part 2 — data are plotted on the Piper diagram in exactly the same way. Relative abundance is calculated assuming that the total number of moles of charge of Na^+, K^+, Ca^{2+} and Mg^{2+} is equal to 100% (see Box 3.7). Na^+ and K^+ are grouped together as both elements tend to behave similarly in groundwater systems. Second, the relative abundances of the anions (Cl^-, HCO_3^-, CO_3^{2-}, and SO_4^{2-}) are plotted on the bottom right triangle. The carbon species (HCO_3^- and CO_3^{2-}) are grouped together. Finally, data are plotted in the diamond-shaped field between the two triangles. This represents the composition of water with respect to both cations and anions.

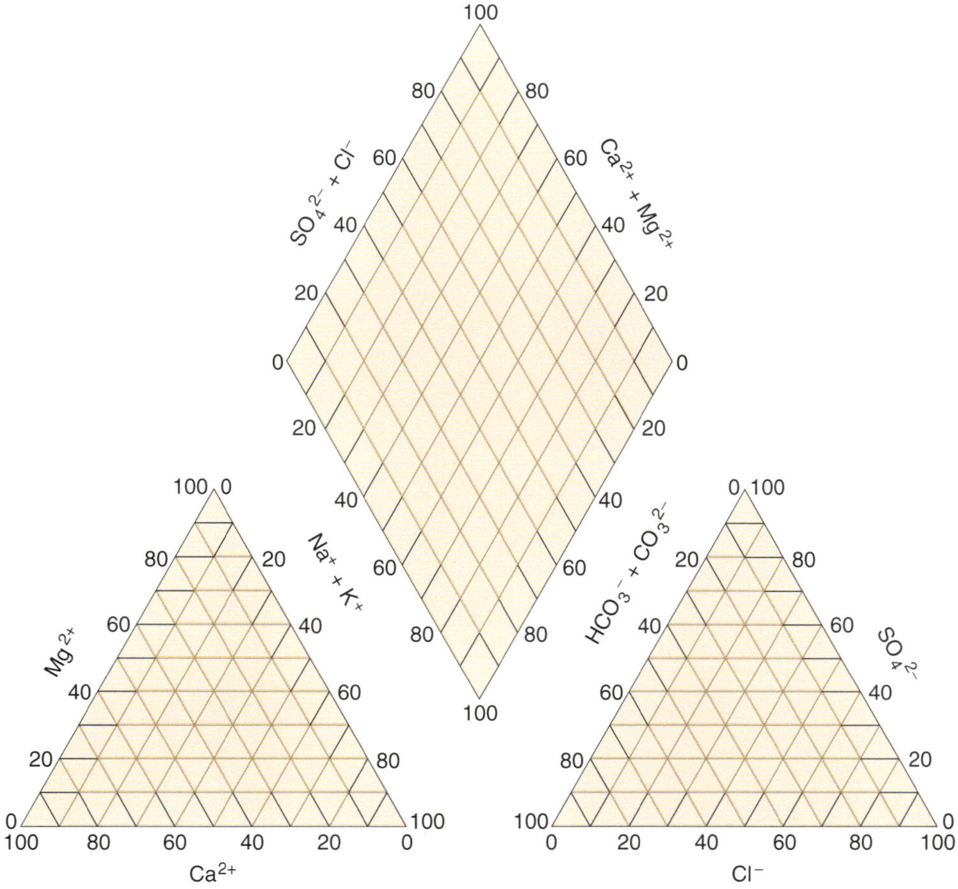

Figure 3.38 Form of the Piper diagram that is used to display the results of water chemistry studies. The values on the axes are expressed as a percentage of the total cation or total anion charge (see Box 3.7). The black lines within the diamond and triangles indicate the direction followed by the axis they meet.

Box 3.7 Calculating relative abundance of charge

As you saw in the Block 2, Part 2, it is often useful for chemists to express concentration in terms of moles of charge because natural waters are **charge-balanced** — the number of moles of positive charge is equal to the number of moles of negative charge. To recap, a solution containing 420 mmol l^{-1} Mg^{2+} has 2 × 420 = 840 mmol of positive charge (840 $mmol_c$). If magnesium only had a single charge, then the solution would contain 420 mmol of positive charge.

In order to plot a Piper diagram, concentrations need to be expressed in terms of relative abundance of either positive or negative charge. By way of example, Table 3.5 lists results of a chemical analysis of groundwater, and Table 3.6 calculates the percent of each cation and anion group as a percentage of the total (respectively, positive or negative) charge. These data are shown on a Piper diagram in Figure 3.39.

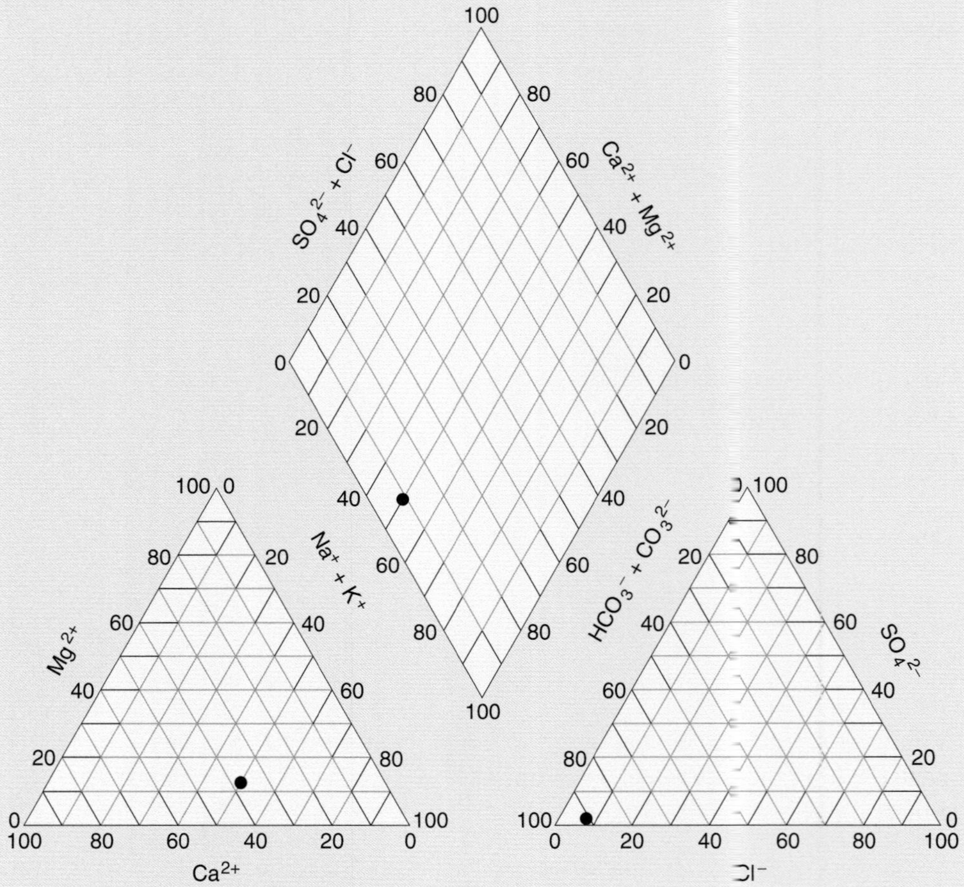

Figure 3.39 Data from Table 3.6 plotted on a Piper diagram.

You should satisfy yourself that you can convert from moles per litre to moles of charge, and that you can calculate the percentage of cation and anion charge. You can do this by completing Activity 3.2. Note that the total moles of positive and negative charge exactly match; groundwater carries an overall neutral charge.

Table 3.5 Results of chemical analysis of a groundwater sample expressed in moles, and moles of charge (mol_c) per litre. A $mmol_c$ is 10^{-3} moles of charge.

	Ca^{2+}	Mg^{2+}	Na^+	K^+	HCO_3^-	CO_3^{2-}	SO_4^{2-}	Cl^-
$mmol\ l^{-1}$	0.58	0.19	1.43	0.12	2.80	0	0.01	0.27
$mmol_c\ l^{-1}$	1.16	0.38	1.43	0.12	2.80	0	0.02	0.27

Table 3.6 Calculation of the relative percentage of cation and anion charge in groundwater.

Cations	$mmol_c\ l^{-1}$	% of total	Anions	$mmol_c\ l^{-1}$	% of total
Ca^{2+}	1.16	37.5	Cl^-	0.27	8.7
Mg^{2+}	0.38	12.3	SO_4^{2-}	0.02	0.65
$Na^+ + K^+$	1.55	50.2	$CO_3^{2-} + HCO_3^-$	2.80	90.6
total	3.09	100	total	3.09	100

Activity 3.2 Plotting a point onto a Piper diagram

Below are a series of measurements of anion and cation concentrations in a groundwater sample (Table 3.7).

(a) Convert these data into units of $mmol_c\ l^{-1}$.

(b) Calculate the relative percentage of each cation (Ca^{2+}, Mg^{2+}, $Na^+ + K^+$) and each anion (Cl^-, SO_4^{2-}, $HCO_3^- + CO_3^{2-}$) group.

(c) Plot these points onto Figure 3.38.

Table 3.7 Anion and cation concentrations in a groundwater sample.

	Ca^{2+}	Mg^{2+}	Na^+	K^+	Cl^-	SO_4^{2-}	CO_3^{2-}	HCO_3^-
concentration/ $mmol\ l^{-1}$	0.40	0.10	1.90	0.02	1.50	0.09	0.12	1.00

A Piper plot of chemical analyses from a series of groundwater supply sites located within the catchment areas of the Rivers Loddon and Wey in the UK (Figure 3.40) is given in Figure 3.41.

Groundwater is held within the Chalk at these sites. The Chalk is exposed at the surface to the south of the area boundary in an east–west belt from just beyond Guildford to Basingstoke (Figure 3.40). This Chalk is unconfined. To the north of this area, the Chalk is overlain by impermeable London Clay and this part of the aquifer is therefore confined. The Chalk is also exposed in the north of the area boundary close to Wargrave.

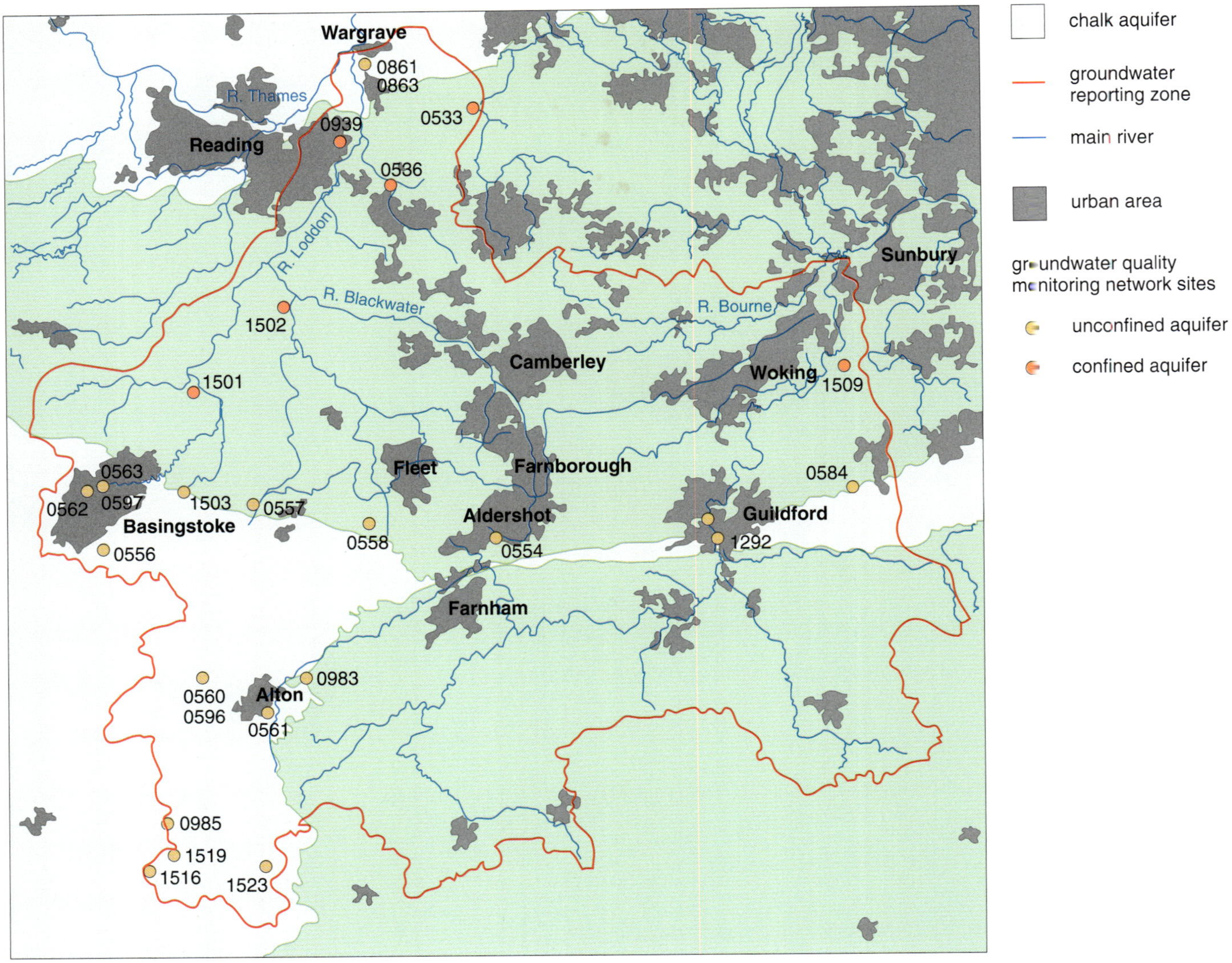

Figure 3.40 Map showing groundwater quality monitoring points in the Chalk aquifer in the Loddon and Wey catchment areas. The Chalk aquifer is shown only where it is exposed at the surface. At sites 0533, 0536, 0939, 1501, 1502, and 1509 the Chalk aquifer is overlain by impermeable London Clay; thus this part of the aquifer is confined.

The diagram shows at a glance that the groundwater chemistry of the confined and unconfined sites is distinctly different. Groundwater from the unconfined sites is rich in calcium and the carbonate species with low concentrations of other ions. Groundwater from sites confined below the London Clay have waters with elevated concentrations of sodium and potassium, magnesium, sulfate and chloride.

❍ Why do groundwaters from confined and unconfined sites show differences in their chemistry?

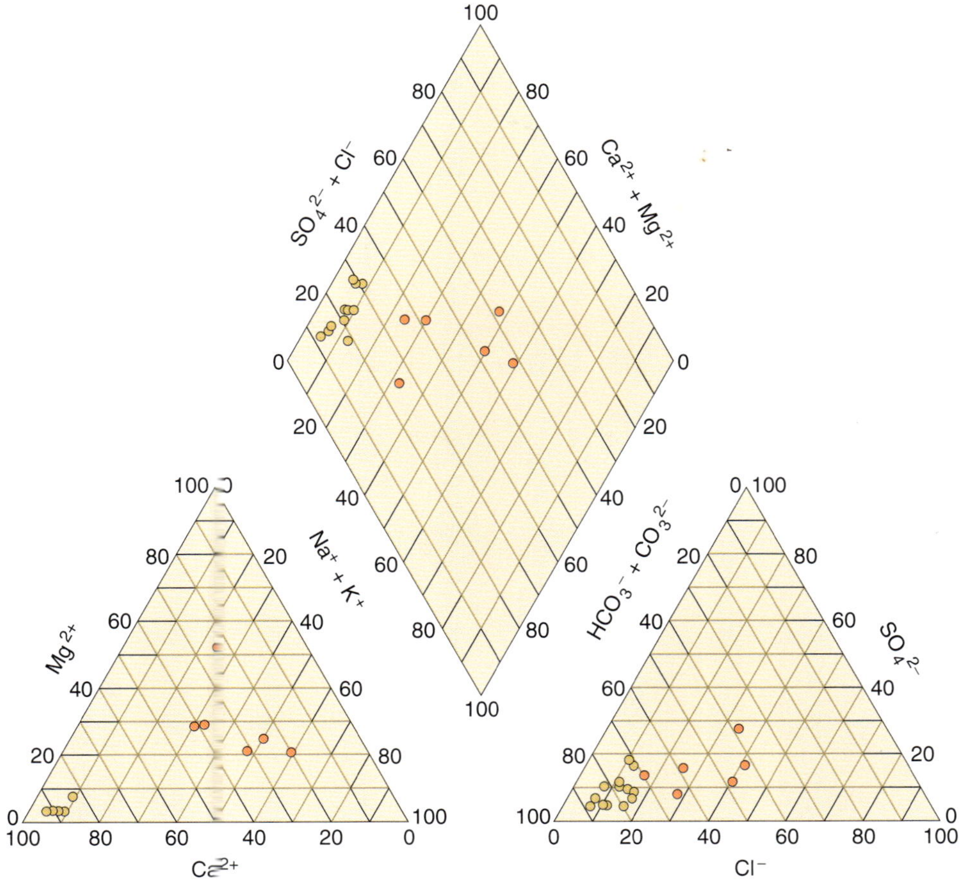

Figure 3.41 A Piper plot of chemical analyses of groundwater monitoring sites in the Loddon and Wey catchment areas. A representative point is plotted for each site. (Sites from the unconfined part of the aquifer are shown by yellow circles; sites from the confined part of the aquifer are shown by red circles.)

- Groundwaters are continually recharging and discharging in the unconfined part of the aquifer, dissolving soluble calcium carbonate which forms the Chalk. Therefore levels of calcium and carbonate species are high. In the confined zone water movement is sluggish and the waters become enriched in sodium and potassium due to cation exchange. Levels of sulfate also increase due to oxidation of pyrite, and levels of chloride increase due to dissolution of salt deposits, or addition of seawater trapped in the rocks as they formed (see Section 3.4.2).

In fact, although groundwater ages are not given here, it turns out that the major ions become progressively more concentrated in the confined part of the aquifer with increasing age.

3.4.6 Groundwater contamination

Because aquifers are both porous and permeable, they are vulnerable to contamination from human activities associated with agricultural practices, urbanization, industrial processes, disposal of wastes, and spillages of chemical such as solvents and fuel oils (Figure 3.42). The threat can either be from a single source, or from a widespread use of an area of land (for example, agriculture). Once in an aquifer, contaminants are difficult to remove because flow rates are low and residence times are long, so the effects of this pollution may continue for a long time.

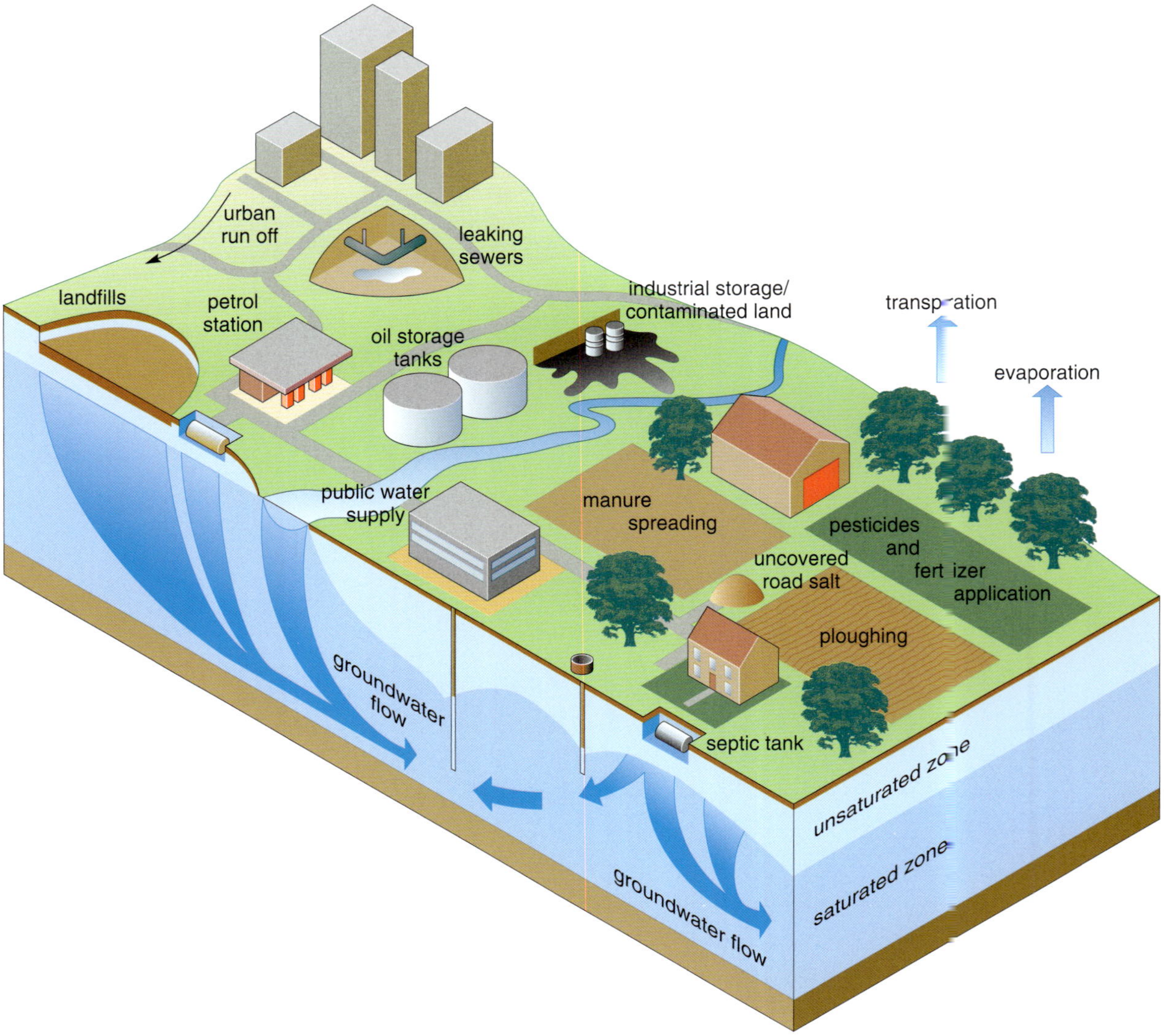

Figure 3.42 The hazards posing a threat to the quality of groundwater. The main sources of pollution are from industry, agriculture, domestic waste and urban areas.

The vulnerability of an aquifer to contamination depends on the composition of any overlying deposits, the nature and thickness of the unsaturated zone, and the speed with which water flows through it. Aquifers are best protected by a thick cover of clay and a thick unsaturated zone. Water infiltrates only slowly, allowing the soil to absorb pollutants. Fractured limestones with a thin soil cover and a shallow water-table are very vulnerable (Figure 3.43).

In the UK, the Environment Agency publishes maps of groundwater vulnerability. Aquifers are first classified into major, minor or non-aquifers on the basis of geological information (specifically, in terms of their lithology and permeability characteristics). Each of these so-called 'geological' classes is then subdivided in terms of the properties of the overlying soil. Soils can have high, intermediate or low leaching potential; that is, soils with high leaching potential

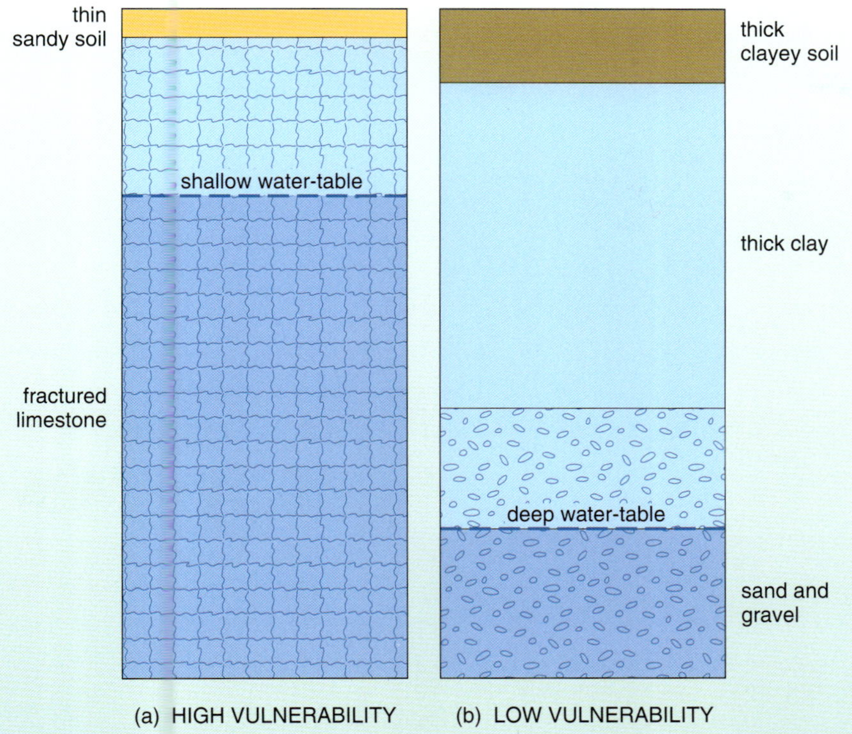

Figure 3.43 Vulnerability of aquifers to pollution. Aquifer (a) is overlain by relatively thin soil, and water percolates rapidly through the highly fractured limestone rock. This aquifer is very vulnerable to contamination. Aquifer (b) is overlain by thick clay so water percolates only slowly through to the aquifer. This allows pollutants to be absorbed, so the aquifer is less vulnerable to contamination.

have little ability to absorb pollutants so the pollutant can pass freely into the saturated zone, while soils with a low leaching potential strongly absorb pollutants so they are unlikely to reach the saturated zone. A groundwater vulnerability map of the Teign catchment, Devon, UK is given in Figure 3.44. Major aquifers are highly permeable formations with significant fracturing. They are considered to be significant for water supplies. Minor aquifers can be fractured rocks which do not have a high primary permeability, or other formations of variable permeability including unconsolidated deposits. These aquifers seldom produce large quantities of water for extraction, but they can be important for local water supplies and in supplying baseflow to rivers. Non-aquifers (there are none in the Teign catchment) have negligible permeability. Soils of high leaching potential (H) have little ability to absorb pollutants which can move rapidly to underlying strata or shallow groundwater. Soils are usually shallow, and/or coarse textured. Soils of intermediate leaching potential (I) have moderate ability to attenuate pollutants. Soils of low leaching potential (L) are unlikely to transmit pollutants to the saturated zone. They generally have a high clay or organic matter content.

- ❍ According to Figure 3.44, where are the major aquifers covered by soil with low leaching potential concentrated?
- ● These aquifers are concentrated in the eastern and southeastern sections of the catchment.

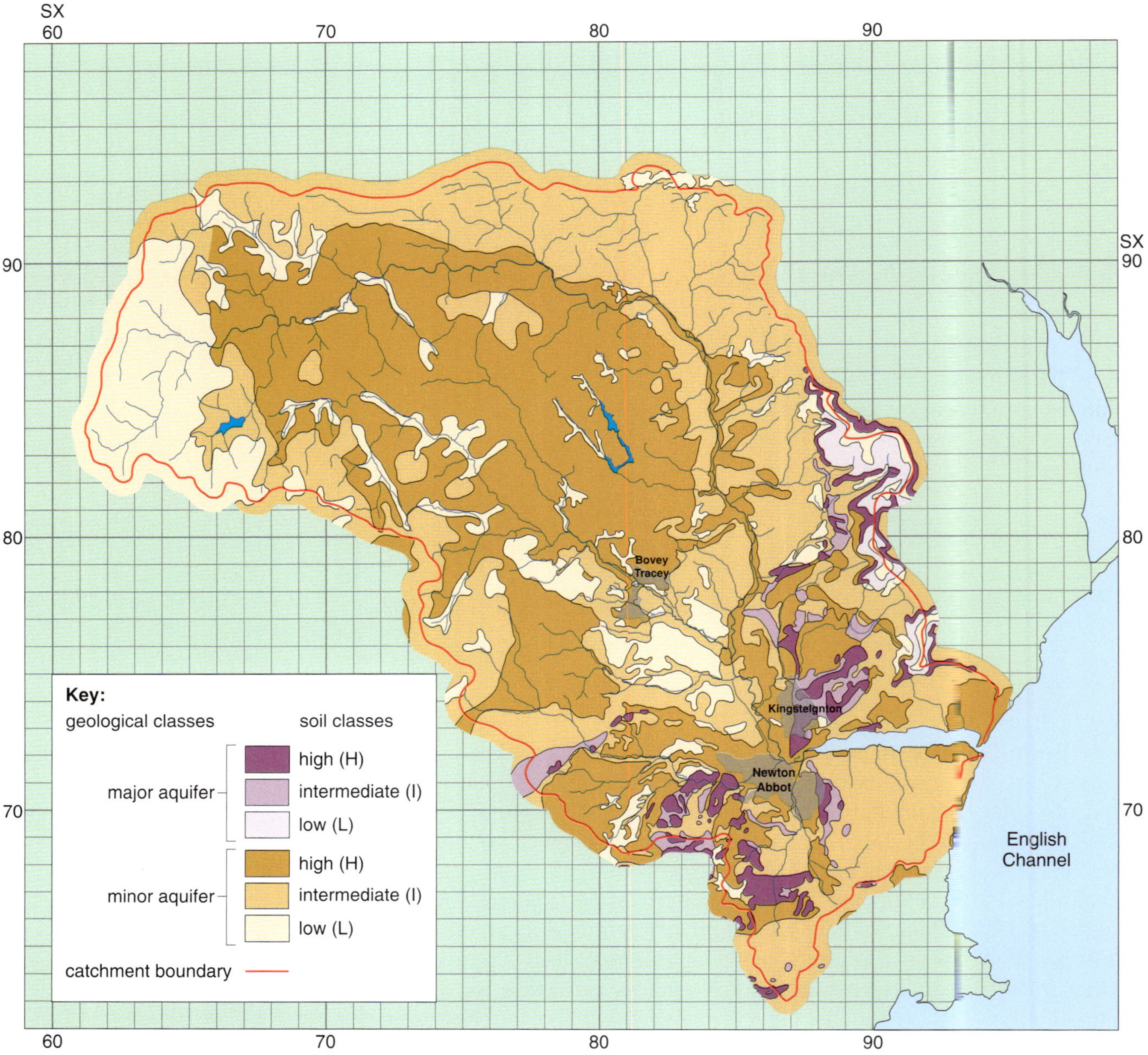

Figure 3.44 Groundwater vulnerability map of the Teign catchment, Devon, UK.

Groundwater contamination is not necessarily irreversible. There are natural conditions that act to remove contaminants, such as dilution and absorption on to soil particle surfaces, and organic contaminants may be broken down by bacteria (Section 3.4.3). In recent years a number of techniques have been developed for restoring the quality of groundwater that has been contaminated. These include:

- Containment of the pollutant, that is, prevent it spreading by using physical barriers such as sheet piles, cement grouts and impermeable membranes.
- Removal of the contaminated water by pumping and then treating it to an appropriate quality. The water can then be either re-injected into the aquifer or used directly for water supply.

- Removal of volatile contaminants by vacuum pumping from a borehole.
- Bioremediation. Nutrients are added to the contaminated zone to encourage bacteria to grow. In turn, the bacteria decompose the complex organic contaminants.

Despite the availability of these techniques, restoring an aquifer to its original pristine state after a pollution event is difficult. When a contaminated aquifer is pumped, the groundwater flows preferentially through the more permeable horizons and consequently these are cleaned relatively easily and quickly. However, contaminants that have penetrated into less permeable horizons and the smaller pore spaces may be relatively unaffected by pumping. When pumping stops, contaminants from these zones diffuse into the permeable zones that have been cleared. The more permeable fractures in the Chalk commonly represent only 1–2% of the total rock volume while the relatively immobile water held within the pore spaces represents 25 to 40%; in this situation the cleaning process can take a very long time. Similarly, if nutrients are injected to encourage biodegradation they may penetrate only slowly into the less permeable horizons.

It is important to note that many pollution incidents from landfills, or contaminated land, or leaking pipes are only local problems. Contamination from diffuse sources, such as agricultural land and the use of pesticides, creates more widespread difficulties, but the fact remains that groundwater is basically water of very good quality. As such, it is a key natural resource.

3.5 Summary of Section 3

1. Groundwater is held in rocks in the saturated zone which lies below the water-table. It is the Earth's largest accessible store of freshwater.
2. The amount of water that can be held in a rock depends on its porosity. The ease with which water can flow through a rock is determined by its permeability. It is possible for rocks to have high porosity, but low permeability.
3. Layers of rock that are porous and permeable enough to store water and let it flow through them easily are called aquifers. Unconfined aquifers are underlain by impermeable rock; above this layer rain soaks down to recharge the aquifer. Confined aquifers are sandwiched between two layers of impermeable rock. Water in a confined aquifer is under pressure; if a borehole is drilled into the aquifer, water will rise up until the weight of the column of water balances the water pressure in the aquifer. The level to which the water rises is called the potentiometric surface.
4. If the potentiometric surface lies above ground level then water will naturally flow from a borehole.
5. In an unconfined aquifer, water storage is mainly provided by the pore space. The amount of water that can be released from storage is determined by the specific yield. In a confined aquifer, water can also be released from or taken into storage as a result of decompression or compression of both artesian water and the aquifer skeleton. However, the amount of water stored in this way is insignificant relative to the amount held in the pore spaces.

6 Groundwater moves in response to pressure gradients. The speed of groundwater movement is determined by Darcy's law:

$$v = K\frac{h}{l}$$

where h/l is the hydraulic gradient and K is the hydraulic conductivity. K is dependent on the permeability of the rock, and also on the density and viscosity of the fluid flowing through it.

7 The effectiveness of an aquifer as a water supply depends on K and also on the thickness of the aquifer. The product $K \times b$, where b is the aquifer thickness, is called the transmissivity of the aquifer.

8 Pumping of water from boreholes lowers the water level (or potentiometric surface) in the surrounding area. Water flows into the borehole, and this creates a cone of depression around it.

9 Aquifers can be characterized in terms of their geological, hydrological and hydraulic properties. Measurements of these properties can be conducted in the field and in the laboratory.

10 Groundwaters move very slowly, particularly where they are confined. The age of a groundwater sample can be determined from levels of the radioactive isotopes, and also other tracers (such as CFCs) present in groundwaters.

11 The chemical composition of groundwater is altered by reactions with surrounding rocks, ion exchange with clay minerals, and microbiological processes. Older waters generally have higher levels of dissolved constituents. The chemical composition of a groundwater sample can be represented by a Piper diagram.

12 Groundwaters are vulnerable to contamination from human activities. They can be an important water resource, so prevention and remediation of contamination is critical.

Question 3.4

Groundwater quality in the London Basin aquifer is monitored using a network of public supply and private abstraction boreholes. The aquifer is located in the Chalk, and it is unconfined towards the outer basin but it is confined within London clay in the centre of the basin, where groundwater ages in excess of 25 000 years have been recorded.

Table 3.8 reports concentrations of a number of chemical species in groundwater sampled at four different locations within the aquifer.

(a) Convert the data given in Table 3.8 into units of mmol l^{-1} of charge (mmol$_c$ l^{-1}) and fill in Table 3.9. Data for Site 0693 are given by way of example. Note that the HCO_3^- and CO_3^{2-} data have already been converted into mmol$_c$ l^{-1}. Details on unit conversion are given in Section 3.6.

(b) Calculate the relative percentage abundance of each of the cations, and each of the anions, and complete Table 3.9. This has been done for you for Site 0693.

(c) Plot the data in Table 3.9 on the Piper diagram in Figure 3.45. Data for Site 0693 are plotted by way of example.

(d) By referring to your diagram, suggest which of the groundwater samples is likely to be the oldest. Give reasons for your allocation.

Table 3.8 Results of chemical analysis of groundwater sampled from four boreholes in the London Basin. (Ion concentrations are given in mmol l^{-1}, except $HCO_3^- + CO_3^{2-}$ which is given in $mmol_c\ l^{-1}$.)

Site	Ca^{2+}	Mg^{2+}	Na^+	K^+	$HCO_3^- + CO_3^{2-}$	SO_4^{2-}	Cl^-
0693	0.14	0.11	6.60	0.10	2.60	1.05	2.50
1171	2.80	0.21	1.21	0.17	6.20	0.50	0.20
1406	2.40	0.20	0.35	0.04	5.20	0	0.39
1512	0.50	0.30	2.80	0.26	1.90	1.10	0.56

Table 3.9 Relative percentage of mmol l^{-1} of cation and anion charge in groundwater.

Site 0693

Cations	$mmol_c\ l^{-1}$	% of total	Anions	$mmol_c\ l^{-1}$	% of total
Ca^{2+}	0.28	3.9	Cl^-	2.50	34.7
Mg^{2+}	0.22	3.1	SO_4^{2-}	2.10	29.2
$Na^+ + K^+$	6.70	93.0	$CO_3^{2-} + HCO_3^-$	2.60	36.1
total	7.20	100	total	7.20	100

Site 1171

Cations	$mmol_c\ l^{-1}$	% of total	Anions	$mmol_c\ l^{-1}$	% of total
Ca^{2+}			Cl^-		
Mg^{2+}			SO_4^{2-}		
$Na^+ + K^+$			$CO_3^{2-} + HCO_3^-$	6.20	
total		100	total		100

Site 1406

Cations	$mmol_c\ l^{-1}$	% of total	Anions	$mmol_c\ l^{-1}$	% of total
Ca^{2+}			Cl^-		
Mg^{2+}			SO_4^{2-}		
$Na^+ + K^+$			$CO_3^{2-} + HCO_3^-$	5.20	
total		100	total		100

Table 3.9 (continued)

Site 1512

Cations	$mmol_c\,l^{-1}$	% of total	Anions	$mmol_c\,l^{-1}$	% of total
Ca^{2+}			Cl^-		
Mg^{2+}			SO_4^{2-}		
$Na^+ + K^+$			$CO_3^{2-} + HCO_3^-$	1.90	
total		100	total		100

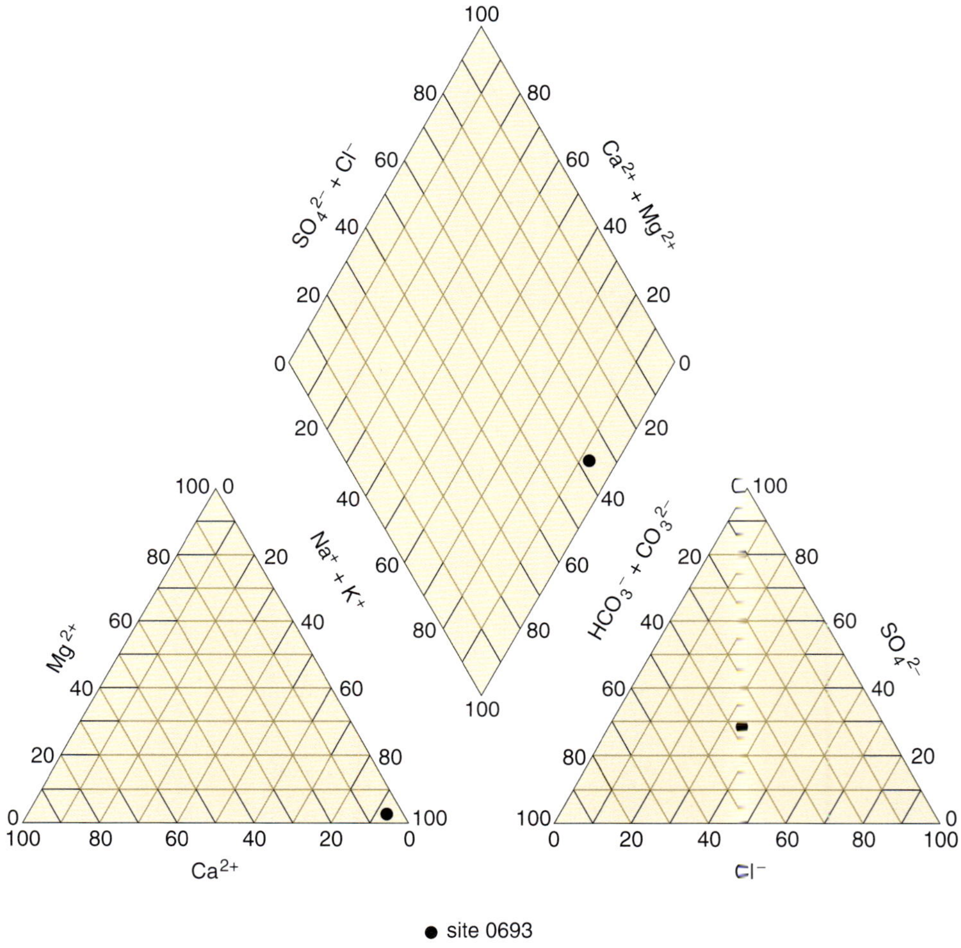

Figure 3.45 Piper diagram for Question 1.

Question 3.5

Figure 3.46 shows hydrographs of groundwater levels in a wel for the period 1988–1991. The well is located in a chalk aquifer in Yorkshire and water levels have been recorded continually on a weekly basis since 1889. Water levels are not influenced by human activity.

(a) Describe the average annual variation in groundwater level and give reasons for this variation.

(b) Describe the annual variation in groundwater levels. Would you regard groundwater levels for the years 1988–1991 as typical for the entire period for which records are available?

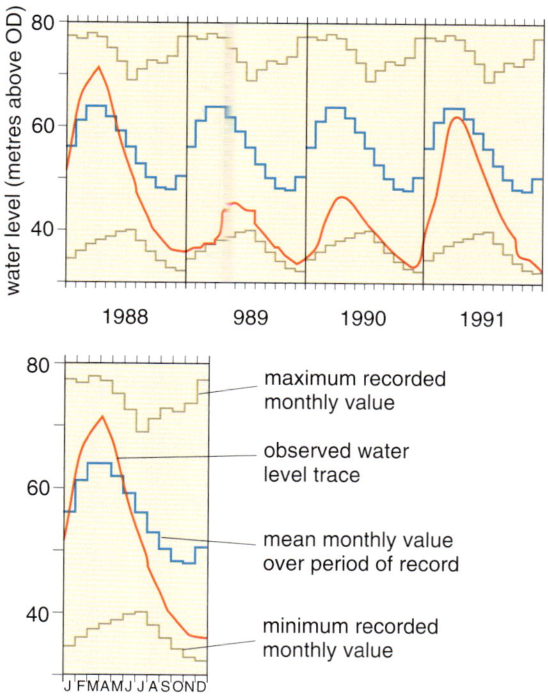

Figure 3.46 Hydrographs of groundwater levels in a well located in the Yorkshire Chalk for the period 1988–1991.

Question 3.6

Figure 3.47 shows photographs of (a) a sandstone, and (b) shale. A 100 cm^3 sample of each rock is weighed first dried, and then saturated with water (Table 3.11).

(a) (b)

Figure 3.47 Photographs of (a) a sandstone and (b) a shale, for use in Question 3.6

Table 3.11 Masses of 100 cm^3 of shale and sandstone: (a) dry; (b) saturated with water.

Sample	Dry weight/g	Wet weight/g
shale	200	205
sandstone	250	270

(a) Calculate the porosity of the shale and sandstone. Assume that the density of water is 1 g cm^{-3}.

(b) By reference to Figure 3.47, are these rocks heterogeneous or homogenous, and isotropic or anisotropic?

(c) Which of the two rocks is likely to show the highest permeability? Justify your answer.

(d) What are the potential problems with applying Darcy's law to study the flow of water through shale?

Question 3.7

A potentially harmful substance has entered groundwater and is travelling towards a borehole used for the supply of drinking water (Figure 3.48). The aquifer is a homogenous sandstone with a hydraulic conductivity of 30 m d^{-1}.

(a) Calculate the average hydraulic gradient between the pollutant source and the borehole.

(b) Calculate the speed at which water is travelling towards the borehole. How many years will it take a water molecule, and therefore presumably for the pollutant, to reach the borehole?

(c) Suppose that the aquifer in Figure 3.48 is a limestone in which all the permeability is contributed by horizontal fissures such that the *effective* hydraulic conductivity of the rock is 5000 m d^{-1}. How long would it take for the pollutant to reach the borehole?

(d) The pollutant takes 5 years to decompose into a harmless by-product. Does the pollutant pose a threat to the water supply in (i) the sandstone aquifer, and (ii) the limestone aquifer?

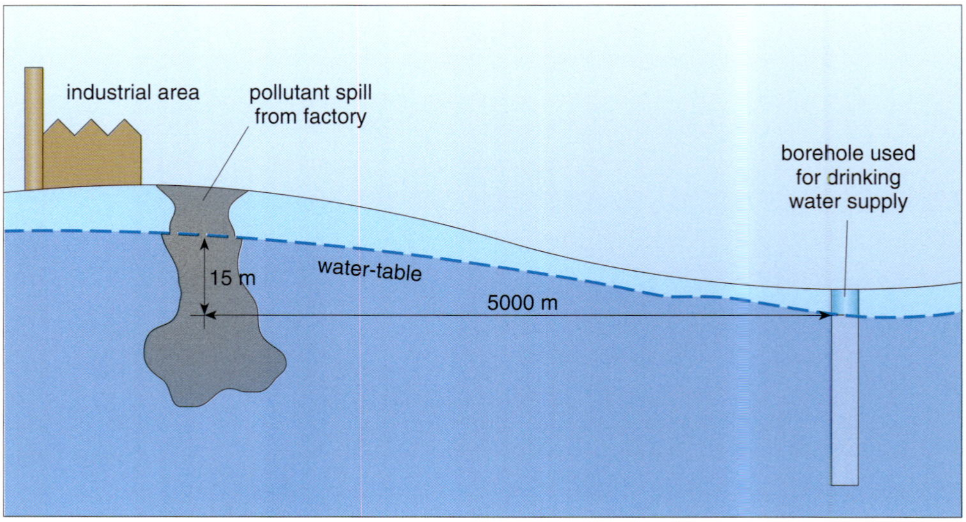

Figure 3.48 Flow of a pollutant towards a borehole.

Question 3.8

^{85}Kr (krypton-85) is a radioactive, chemically unreactive gas, with a half-life of 10.76 years. It is produced in nuclear reactors, and concentrations in the atmosphere have increased since 1950 due to bomb testing. ^{85}Kr is useful as a groundwater dating tool as there are no significant sources or sinks of ^{85}Kr in the subsurface.

Table 3.12 lists measurements of ^{85}Kr at different depths in the Bordon aquifer, Ontario. Measurements are given in units of becquerels per m^3 (Bq m^{-3}); you do not need to understand the precise definition of this unit, but it is equivalent to the number of atoms of ^{85}Kr per m^3.

Table 3.12 ^{85}Kr measured in groundwater samples from different depths in the Bordon Aquifer, Ontario.

Depth/m	^{85}Kr/Bq m^{-3}	D/P	No. of elapsed half-lives, n	Age/yr
0 (soil gas/atmosphere)	0.88	0	0	0
1.1	0.75			
4.2	0.60			
6.4	0.45			

(a) Assuming that the number of ^{85}Kr atoms present when the water entered the ground (i.e. P_0) is equivalent to 0.88 Bq m^{-3}, use the data in Table 3.12 to calculate the D/P ratio at each depth. Remember that $D + P = P_0$. Write your answer in the appropriate column in Table 3.12.

(b) With reference to Equation 3.11, find the number of elapsed half-lives corresponding to your calculated D/P ratios. Calculate the age of groundwater at each depth (fill in Table 3.12), and plot this as a function of depth.

(c) With reference to your graph, can you see any evidence for a change in the hydraulic properties of the aquifer over the depth range for which data are available?

Learning outcomes for Section 3

When you have completed this section, you should be able to:

3.1 Describe and account for the annual variation in baseflow on a hydrograph.

3.2 Differentiate between porosity and permeability.

3.3 Outline the different types of aquifer.

3.4 Describe how water is stored in an aquifer.

3.5 Calculate groundwater speed and transmissivity of groundwater flow using Darcy's law.

3.6 Point out the limitations of Darcy's law.

3.7 Give examples of methods used to determine the age of groundwater, and to calculate groundwater age from measurements of *D*/*P* ratios given the half-life of the radioactive parent isotope.

3.8 Outline and account for variations in the chemical composition of groundwater with (a) rock type, and (b) age.

3.9 Plot the results of chemical analysis of a groundwater sample onto a Piper diagram.

Journey down a river

4

We have seen that the bulk of water that falls on the land returns directly to the atmosphere. Some water reaches the ocean by underground routes as throughflow or groundwater flow but the most visible route that water can take from the land to the ocean (and thence back to the atmosphere) is along stream and river channels (Figure 4.1).

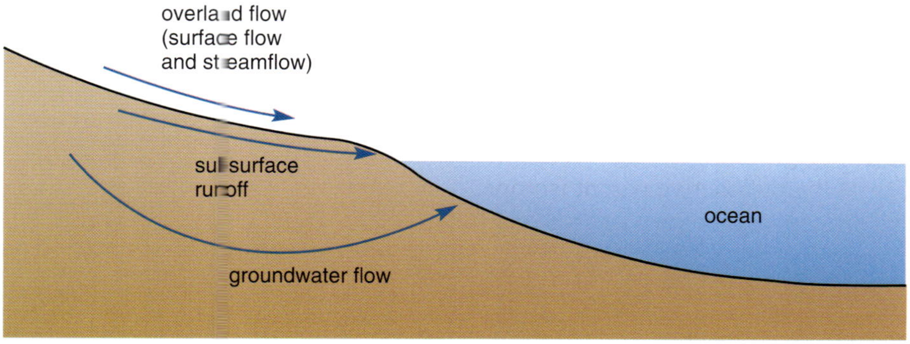

Figure 4.1 Routes for water from land to ocean.

In this section, we shall follow a river from source to ocean taking a look at the physical form of the river, how the water flows and how a river reacts to inputs of material and the influence of dissolved materials on life in a river.

To get a feel for some of the features of a river, we shall focus on a real river. The source of the River Sprint is at an elevation of 640 m above the head of Longsleddale in Cumbria in the UK and the river flows in a generally southerly direction, joining the River Kent en route to the Irish Sea at the north-eastern part of Morecambe Bay. It is not a long river at 33.7 km but it does exhibit a range of features, some of which are illustrated in Figure 4.2. The progression of the river from a steep, turbulent mountain stream, the engineered river through the town of Kendal to the wide tidal estuary can be seen.

(a)

(b)

(c)

(d)

Figure 4.2 The River Sprint from its source to its mouth: (a) source at British National Grid Reference (GR) NY463085; (b) at GR NY477083; (c) at GR NY479073; (d) at GR NY483057;

(e)

(f)

(g)

(h)

(i)

(j)

(e) at GR NY489500; (f) at GR NY515006; (g) at GR NY524993; (h) confluence with River Kent at GR NY508953; (i) through Kendal at GR NY519929 and (j) the Kent Estuary looking northwards from the eastern end of Arnside railway viaduct at GR NY457792.

4.1 River channels

The flow of water in rivers is motivated by gravity and resisted both by friction with the river bed and banks and also by the structure of water itself. All rivers have the source at higher altitude than the mouth but the gradient, the ratio of the drop in elevation divided by the distance over which the drop occurs, changes along the route.

To see how the gradient changes from source to mouth for the River Sprint (and River Kent), you will need the map in Figure 4.3. Identify a number of points along the river where you are able to estimate the elevation of the river. It makes sense to choose points where contour lines cross the river although further downstream, where the gradient is less steep (and few contours cross the river), your estimate will be more difficult to make. For each point (you should choose at least ten points), you will also need to estimate the distance from the river source. (Remember that the side of each grid square is 1.00 km in length and a diagonal across the square is 1.41 km. Should you find that the contour lines are difficult to see, this Ordnance Survey map is available in the Landranger Series Sheets 90 and 97.) Construct a table with headings: *grid reference*, *elevation*, and *distance from source of river*. Then plot *elevation* against *distance from source* to give a representation of the **longitudinal profile** of the river.

Compare your data with the data that we obtained shown in Table 4.1 and the derived plot in Figure 4.4.

Table 4.1 Elevation data for the River Sprint.

Grid reference/NY	Elevation/m	Distance from source/km	Comments
463085	640	0.0	source between The Knowe and Brown How
471086	530	0.9	
477083	400	1.9	
479073	230	3.0	
483057	200	4.5	Sadgill
489047	170	6.1	
510010	150	10.6	
523984	100	14.9	Thornybank
508953	50	18.0	south of Burneside
516919	44	22.1	Kendal
512891	34	25.9	
480844	0	33.7	Halforth, Kent estuary

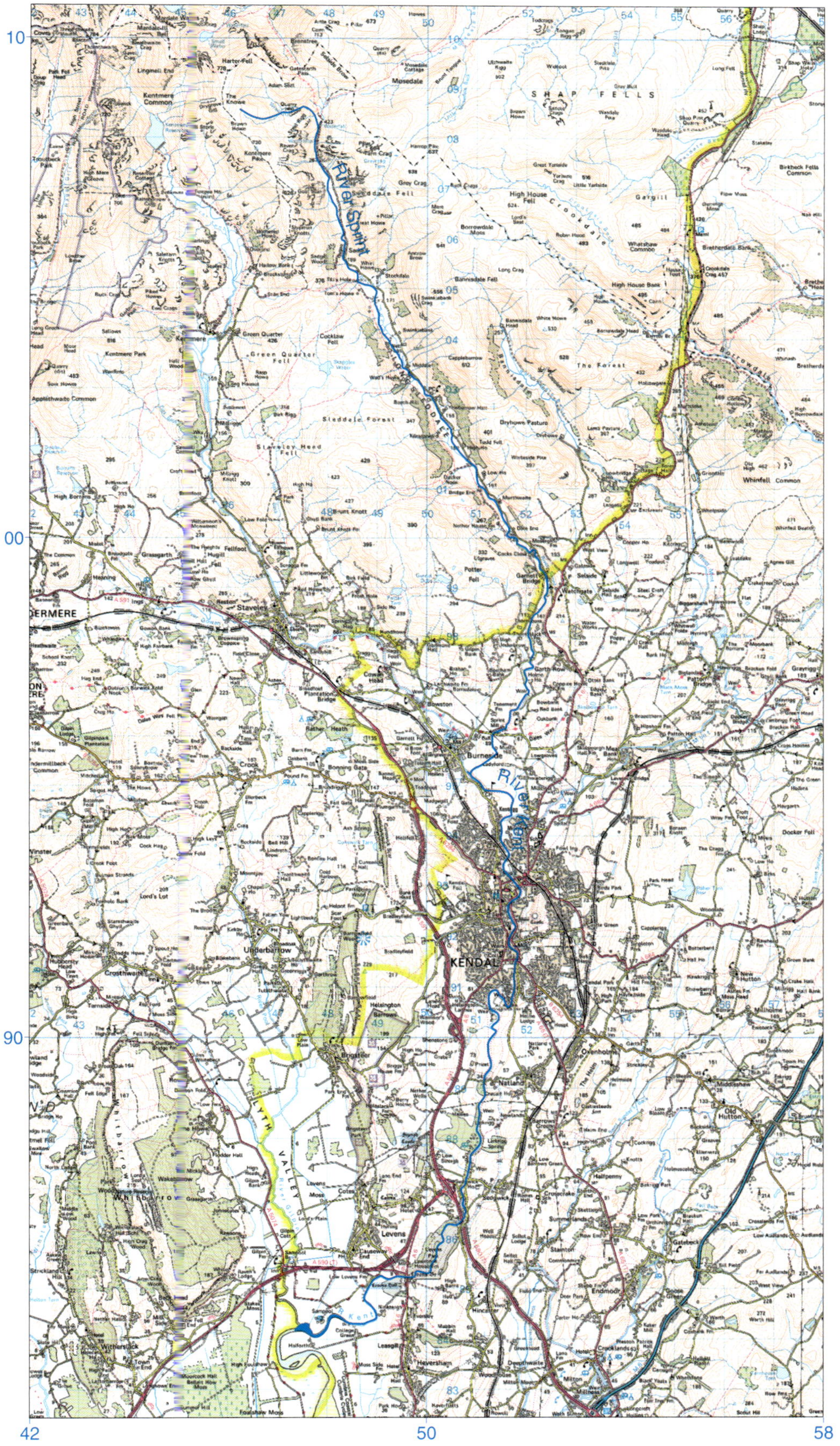

Figure 4.3 Strip map of the River Sprint from its source to its tidal estuary (as the River Kent).

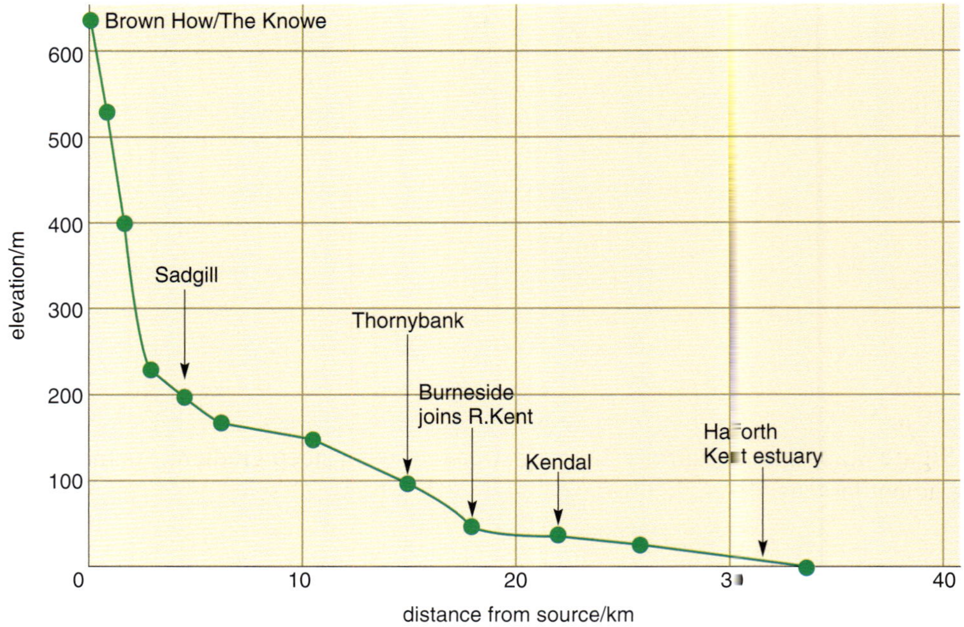

Figure 4.4 Long profile of River Sprint (and River Kent) constructed from data in Table 4.1.

- ❍ Use Figure 4.4 to estimate the gradient of the river over 5 km intervals from its source.

- ● You can do this by finding the difference in elevation between the start and the end of the interval, and dividing by the interval distance (in this case, 5 km). For example, the elevation at the source is 640 m, and at 5 km from the source it is ~190 m. The average gradient over this interval is therefore:

$$\frac{640\text{ m} - 190\text{ m}}{5\text{ km}} = \frac{640\text{ m} - 190\text{ m}}{5 \times 10^3\text{ m}} = 9.0 \times 10^{-2}$$

The results of this exercise are shown in Table 4.2.

Table 4.2 Average gradient over intervals along the River Sprint.

Interval (distance from source)/km	Gradient/10^{-3}
0.0–5.0	90
5.0–10.0	7
10.0–15.0	11
15.0–20.0	12
20.0–25.0	1

Although the gradient does not decrease smoothly from source to mouth, there is a clear overall fall, the river becoming less steep towards the river mouth. Irregularities in gradient often arise when the river crosses a boundary such as that between a harder and a more easily erodable rock. On a more detailed scale, there are clear, sudden changes in gradient such as those high up the river (Figure 4.2b) or the constructed weirs in Kendal (Figure 4.2i). All streams, from small rills to large rivers, exhibit this general shape of an upwardly concave curve, known as a **graded stream**. The idealized profile of graded stream is shown in Figure 4.5. The lower level of the river is fixed where it flows into the ocean (or lake) and the river cannot go lower than this **base level**.

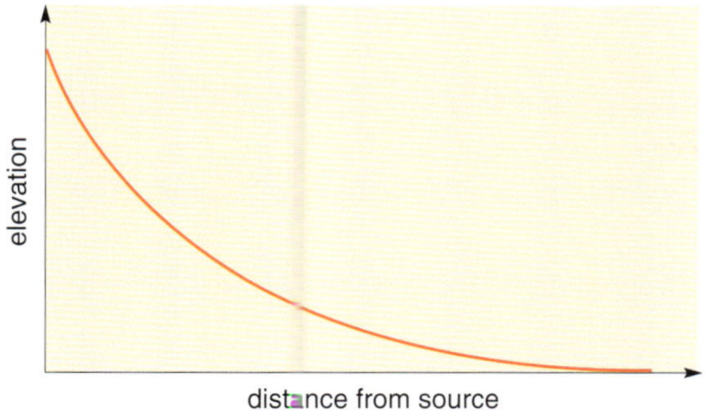

Figure 4.5 The idealized long profile of a graded stream.

The River Sprint changes its form along its length (Figure 4.2). Near its source (Figure 4.2a) it is a small fast flowing stream down a steep gradient. As the gradient lessens, the valley widens and now the stream channel occupies only a part of the valley bottom (Figure 4.2f). The channel carries all of the water in non flood times. Increased water flow causes the flat area either side of the stream to flood. As the flow rate falls, sediment in the water is deposited and a flood plain is built up. Finally the river develops a sinuous course near its mouth. Change in gradient is an important influence on these different forms of the river but it is not the only one. To see why these changes occur, we need to explore how water flows.

4.2 How water flows

There are two basic ways in which water can flow. Simply looking at water coming out of a tap can illustrate this. It is possible with relatively low flow rates for the water to emerge as a steady stream. The water stream is clear and this is known as laminar flow. The molecules of water are all moving parallel to one another and at the same speed. If the tap is put full on, the flow of water increases, becomes less steady and is no longer clear. With turbulent flow, there are eddies in the water, the water molecules are not all moving on parallel paths and have different speeds, (Figure 4.6).

(a) (b)

Figure 4.6 Water flowing from a tap showing (a) laminar flow and (b) turbulent flow.

Water can adopt laminar or turbulent flow characteristics in rivers. One of the factors that affects the flow type is the speed of the water, v. This is apparent from Figure 4.7 where laminar flow occurs at a low flow speed and the higher speed gives turbulent flow. For a river, another factor is the depth of the water, d and the third variable is a function of water itself, the **kinematic viscosity**, ν. This is a measure of how easily a fluid (in this case water) flows. Motor oil or syrup has a higher value of ν than does water.

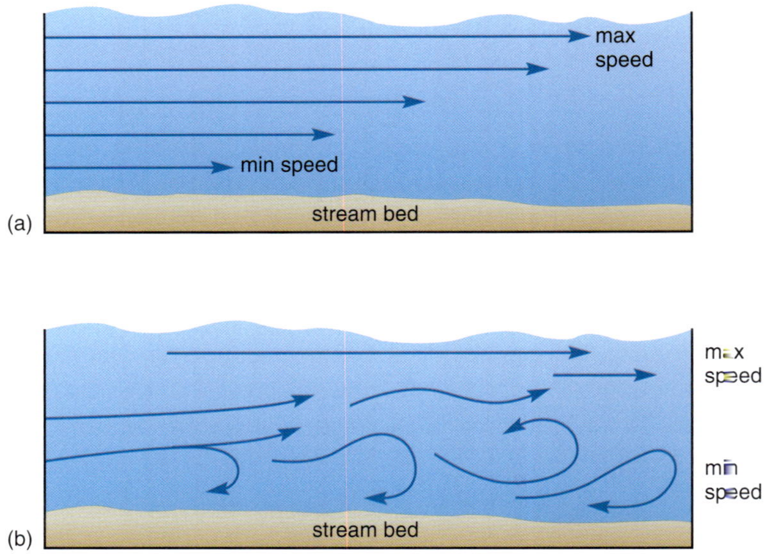

Figure 4.7 Flow along a stream showing (a) the variation in flow speed with depth and (b) the region of turbulent flow near the stream bed.

These three variables are related to the Reynolds number, Re, by the relationship,

$$Re = \frac{vd}{\nu} \qquad (4.1)$$

where v is measured in cm s^{-1}, d in cm and ν has units of $cm^2\,s^{-1}$. For Reynolds numbers below about 500, flow is laminar but much above this value, turbulent flow results. Let's see what this means for real rivers.

At 20 °C, the kinematic viscosity of water has a value of $1.00 \times 10^{-2}\,cm^2\,s^{-1}$. We shall assume a very modest flow rate of $5.00\,cm\,s^{-1}$. The depth of water corresponding to the laminar/turbulent threshold of 500 is given by:

$$500 = \frac{5.00\ \text{cm s}^{-1} \times d}{1.00 \times 10^{-2}\ \text{cm}^2\ \text{s}^{-1}}$$

$$D = \frac{5.00 \times 10^2 \times 1.00 \times 10^{-2}}{5.00}\ \text{cm} = 1.00\ \text{cm}$$

So even for this relatively low flow rate, laminar flow only occurs at very shallow depth. True laminar flow is seldom significant in streams. Turbulent flow is the dominant type of flow.

There is some variation of kinematic viscosity, ν, with temperature, as is evident from Table 4.3.

Question 4.1

Estimate the depth of the laminar/turbulent flow boundary for water at 30 °C with a speed of 3.2 cm s^{-1}.

Table 4.3 Variation in the kinematic viscosity of water with temperature.

Temperature/°C	Kinematic viscosity, ν /10^{-2} cm^2 s^{-1}
0	1.79
5	1.52
10	1.31
15	1.14
20	1.00
25	0.890
30	0.798

4.2.1 Speed of water flow

The speed of water flow can be measured using a current meter; this is a device equipped with a vane which turns in the current and an electrical device which counts the number of revolutions of the vane in a fixed time period (usually 30 or 60 seconds). A calibration curve is used to relate the number of counts per unit time to speed of flow. The speed of water in a river changes with distance from the bank and with depth. The flowing water is subject to friction as it comes into contact with the river bottom and its banks. In general terms, an estimate of the average speed for a vertical section is the speed measured at 0.6 times the total depth of the section.

❍ A series of speed measurements have been made in a section of a stream with a total water depth of 50 cm. These are reported below. What is the average speed of the flow in this section?

Water depth/cm	Speed/m s^{-1}
0	0.21
10	0.20
20	0.15
30	0.10
40	0.03

● The total water depth of the section is 50 cm. Therefore the average speed is measured at 0.6 × 50 cm = 30 cm depth. This is 0.10 m s^{-1}.

Look at the data in Table 4.4 which represents water speed over a cross-section of a stream which has a maximum depth of 0.60 m and a width of 18 m. Use the data in Table 4.4 to sketch a cross-section of the river (use graph paper) and then draw lines that connect points of the same water speed, **isovel lines**. You will have to estimate speed values between the given data points.

Table 4.4 Water speed data (m s^{-1}) across a straight section of a stream.

Depth/m	Horizontal distance across river/m									
	0	2	4	6	8	10	12	14	16	18
0.0	0.00	0.55	1.55	2.60	3.10	3.55	3.05	2.20	1.10	0.00
0.10		0.00	1.30	2.30	3.00	3.05	3.00	2.10	1.00	0.00
0.20			1.00	2.10	2.70	2.80	2.60	2.00	0.00	
0.30			0.00	1.90	2.30	2.25	2.20	1.70		
0.40				1.20	1.95	2.00	1.90	0.90		
0.50				0.00	1.30	1.40	1.10	0.00		
0.60					0.00	0.00	0.00			

Your sketch should be similar to that in Figure 4.8. To produce an accurate drawing of the speed profile, you would need more data than appear in Table 4.4. However, it is clear that the region of maximum speed is close to the middle of the surface of the stream. Nearer the stream bed (indicated by zero values for the water speed in Table 4.4), the water speed is much slower.

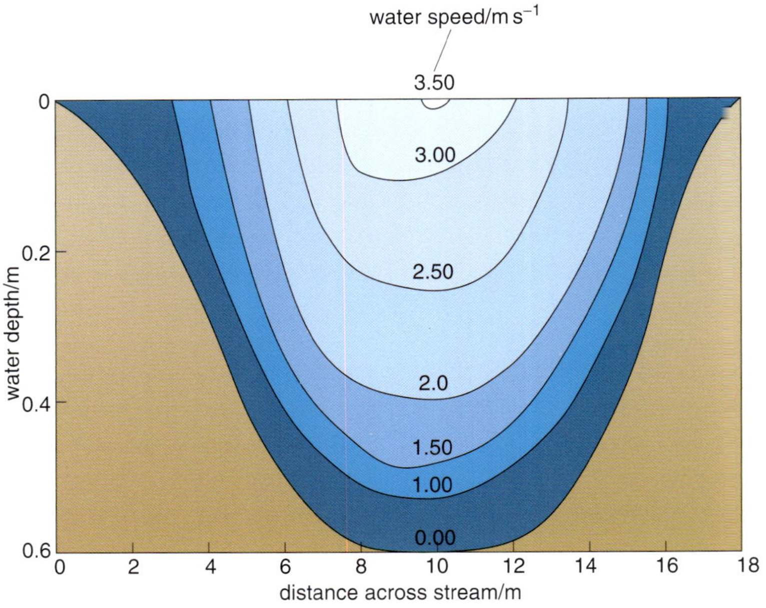

Figure 4.8 Sketch of the water speed profile from the data in Table 4.4.

You may recall from Section 2 that runoff (or stream discharge), Q (m^3 s^{-1}) is found from the Equation $Q = vA$, where A is the cross-sectional area (m^2) and v is the average speed of river flow (m s^{-1}). You can now see that establishing the average speed of river flow is not straightforward. It is most easily done by assigning a speed to each of the grid squares in Figure 4.8 (this requires some judgement!), summing the values and dividing by the total number of grid squares. In this case, the average speed is about 1.95 m s^{-1}.

The observation from the data in Table 4.4 that water speed is low near the stream bed and highest in the centre of the channel suggests that channels with relatively small bed areas present a smaller frictional area to the moving water. The important factor is the **hydraulic radius**, R, of the channel which is defined by

$$\text{hydraulic radius} = \frac{\text{cross-sectional area of the channel}}{\text{wetted perimeter}} \qquad (4.2)$$

Calculate the hydraulic radius of the channel shown in Figure 4.9. You should assume that the channel has a semicircular cross-section.

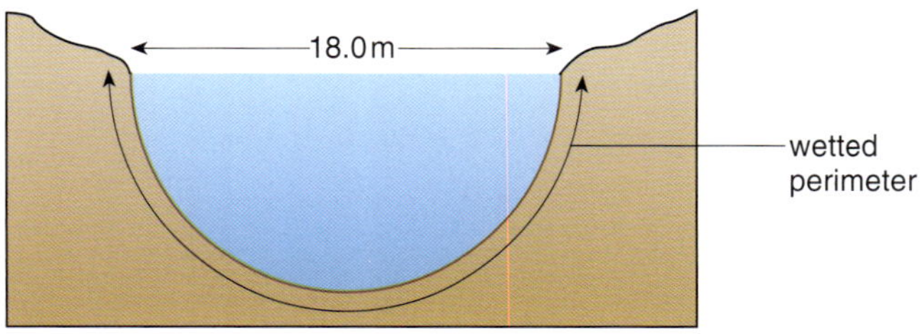

Figure 4.9 Semicircular cross-sectional stream channel.

Area of cross-section $= \frac{1}{2}\pi r^2$

(where r is the channel radius)

$= \frac{1}{2} \times \pi \times (9.00\,\text{m})^2 = 127\,\text{m}^2$

Wetted perimeter $= \pi r = \pi \times 9.00\,\text{m} = 28.3\,\text{m}$

Hydraulic radius, $R = \dfrac{\text{cross-sectional area of the channel}}{\text{wetted perimeter}} = \dfrac{127\,\text{m}^2}{28.3\,\text{m}}$

$= 4.49\,\text{m}$

Now calculate the hydraulic radius for the channel of rectangular cross-section shown in Figure 4.10.

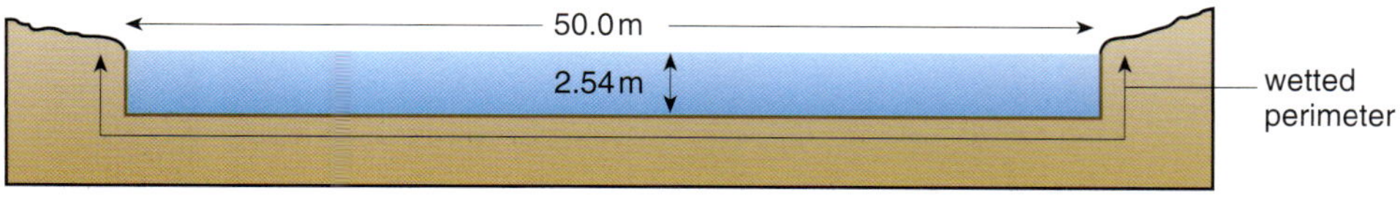

Figure 4.10 Shallow, wide stream channel.

Area of cross-section $= 2.54 \times 50.0 = 127\,\text{m}^2$

Wetted perimeter $= 2.54 + 50.0 + 2.54 = 55.1\,\text{m}$

Hydraulic radius, $R = \dfrac{127\,\text{m}^2}{55.1\,\text{m}} = 2.30\,\text{m}$

The semicircular channel has a much higher value for the hydraulic radius than does the wide shallow channel even though our two illustrations have the same cross-sectional area. Friction with the river bed is much greater with the shallow, wide channel. The region of high water speed is certainly smaller with this kind of channel (Figure 4.11) and the channel is much less efficient in transporting water.

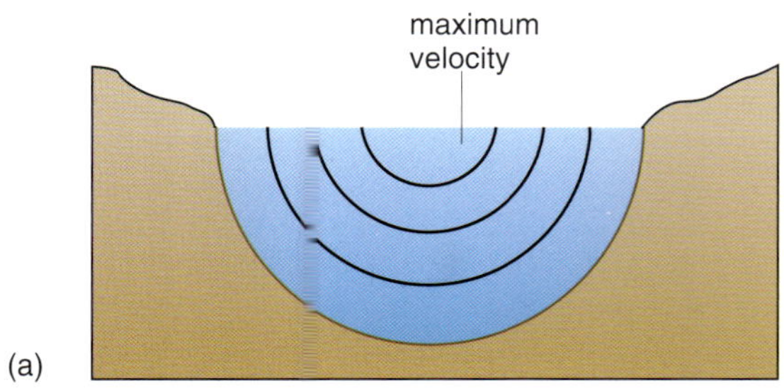

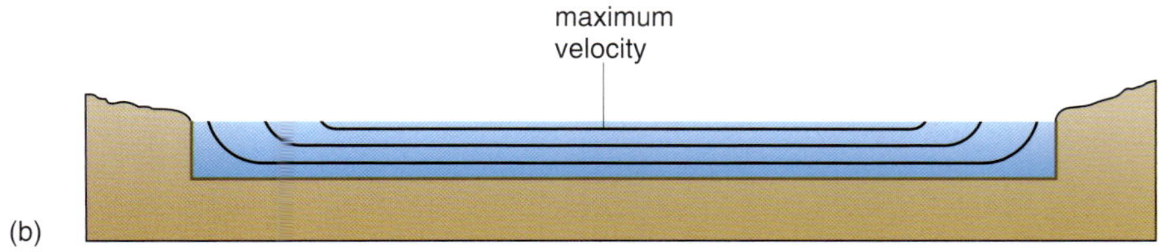

Figure 4.11 Stream channels (a) of Figure 4.10 and (b) of Figure 4.11 showing isovel lines.

Question 4.2

Calculate the hydraulic radii of two rivers each of rectangular cross-section: the first being 12.0 m wide and 3.20 m deep and the second 20.0 m wide and 1.92 m deep.

The empirical **Manning equation** has been used since its introduction in 1889 and still provides a convenient way to relate the three factors that affect the average speed of water flow in a stream channel.

$$v = \frac{R^{\frac{2}{3}}s^{\frac{1}{2}}}{n} \qquad (4.3)$$

where v has units of m s^{-1} and R is in metres.

The speed increases as the gradient of the water surface, s, increases but not in a linear manner. A doubling of s results in an increase in speed of about 1.41 ($2^{\frac{1}{2}}$). Increasing the hydraulic radius, R, of the channel, increases the speed but again not in a linear manner. Doubling R results in the speed increasing by about 1.59 ($2^{\frac{2}{3}}$). The third factor is the **Manning roughness coefficient, n**, which depends on the size and shape of the bed material and vegetation in and at the edges of the channel. An indication of the range of n is shown in Table 4.5.

Table 4.5 Relation between the Manning roughness coefficient, n and nature of the channel bed.

Channel bed	Range of n / m$^{-\frac{1}{3}}$ s
large rivers	0.025–0.060
flood plain river in woodland	0.100–0.150
flood plain river in pastureland	0.025–0.035
small, weed-clogged stream	0.075–0.150
small, open stream	0.025–0.033
small mountain stream	0.040–0.070

Note the rather curious units for n. For empirical relationships, it is not uncommon for the proportionality constant to have unusual units, in this case, m$^{-\frac{1}{3}}$ s.

Question 4.3

Use the Manning equation to estimate the water flow speed of a stream at a point where the cross-section is semicircular of width 2.42 m, the gradient of the water surface is 0.00122 and there is extensive weed growth in the channel.

Such is the difficulty in quantifying relationships in the field that several relationships for finding the average speed of water flow in channels have been developed. Some are better applicable in some areas rather than others but for general applicability, the Manning equation has stood the test of time.

4.3 What is in stream water?

Although water in stream channels comes from precipitation, there are often major differences in the chemical composition of precipitation and of stream water. We have already seen that the interception of precipitation by vegetation can change significantly the dissolved material in water, even on a seasonal basis. Precipitation reaches stream channels by falling directly into the channels, by surface runoff, or by shallow or deep throughflow. During the latter three routes, the chemical composition of the water is changed by interaction with the rock and soils through and over which it flows. The actual composition of the dissolved material, the **dissolved load**, of any stream is particular to the region through which the stream flows; its vegetation cover, the type of rocks and soils, the extent of weathering and, of course, direct human intervention.

So, what exactly is in stream water? The answer is that *all* of the chemical elements occur in stream water to some extent with the exception of those rare radioactive elements that have only ever been made within the laboratory and not released in some way into the environment. In stream water, the elements can appear as ions or molecules, and as charged or neutral species. The elements can have both natural (rock weathering and gas emissions) and anthropogenic (pollutant) sources.

With regard to the natural sources, gases derived from the early degassing of the Earth's interior are an important source of carbon, nitrogen, sulfur and chloride, but most other elements are derived from rock weathering. Some rocks and minerals dissolve far more readily than others. The extent of solubility depends on a number of factors. These include the chemical properties of the elements that make up the rock, the concentration of different elements in the rock and the length of time the water is in contact with the rock. Because of this, waters can vary in composition enormously. For example, highly acidic ($pH < 3$) and metal-rich waters occur in areas where sulfide-rich soils and rocks are oxidized (generating sulfuric acid), while calcium hydroxide-rich spring waters of pH 12 can occur in areas associated with the weathering of ultramafic rocks (Figure 4.12). Stream waters can also be either very dilute, for example in remote inland areas where rainfall has a low concentration of dissolved ions and mineral weathering and pollution are very low, or highly concentrated where, for example, surface and groundwaters leach salt-bearing deposits or in arid areas where evaporation is very high.

Figure 4.12 A spring discharging into a wadi channel in Northern Oman. The spring waters are alkaline (pH ~12) and rich in calcium hydroxide. The white deposits are calcium carbonate formed by reaction of the calcium hydroxide in the spring water with carbon dioxide from the atmosphere.

Life as we know it would not occur if exotic waters like those described above predominated on the Earth's surface. What then, is the composition of 'regular' river water and how, and more importantly why, does it vary in time and by location? We will address these questions in the following sections.

4.3.1 The composition of 'average' river water

The relative amounts of the different chemical elements varies considerably in river water. Table 4.6 lists the elements in order of decreasing concentration. Elements at the top of the list are the **major elements**: sodium, potassium, calcium, magnesium, carbon, chlorine, sulfur, nitrogen and silicon. These major elements have concentrations typically over 0.5 mg l^{-1} and often one or two orders of magnitude higher. Four of these major elements occur as positively-charge ions (cations) in river water and these are known as the **major cations** (Na^+, K^+, Ca^{2+} and Mg^{2+}). There are three major elements that have a negative charge in solution, these are known as the **major anions**: carbon (mainly in the form of hydrogen carbonate, HCO_3^-), sulfur (which occurs mainly as sulfate, SO_4^{2-}), and nitrogen (mainly in the form of nitrate, NO_3^-). The remaining major element is silicon. The rest of the elements listed in Table 4.6 have concentrations below about 100 µg l^{-1} and these are known as the **trace elements**. Some elements are used in biological processes, and these form a sub-category called the **nutrients***. The nutrients include nitrate, phosphate, bromine and iodine.

If the composition of river water were largely determined by weathering of rocks and soils, then we would expect to see a relationship between the concentration of an element dissolved in river water and its concentration in crustal rock. This is explored in Table 4.6.

○ How closely does the composition of river water match that of the crust?

● Not very closely at all. Table 4.6 shows that some elements are enriched in river water relative to the crust, while other elements are depleted in river water relative to the crust.

The element with the highest river water:crust ratio is chlorine, which is a major element in river water, but is considered a trace element in the crust. In contrast, iron and aluminium are major components of the Earth's crust but are trace components of river water. The elements with the highest river water : crust ratio are all anions and they have relatively high solubilities in surface waters. The elements with the lowest river water:crust ratios all form dissolved ions with a high positive charge (iron, Fe^{3+}, and aluminium, Al^{3+}).

*The term nutrient can be applied to both the elements necessary for life and to the compounds of those elements that are taken in by the organism.

Table 4.6 The average composition of river waters (μg l^{-1}) and crustal rocks (μg g^{-1}). The ratio of the concentration of an element in river water relative to its concentration in the crust is also given (g l^{-1}). A value of zero denotes a concentration of less than one μg l^{-1} in river water.

Element	Concentration in river water, [*R*]	Concentration in the crust, [*C*]	[*R*]/[*C*]
Ca	19 185	29 540	0.651
Cl	13 030	640	20.4
Na	7125	25 670	0.278
Mg	5510	13 510	0.408
C	5154	3240	1.59
S	2684	953	2.82
Si	1078	303 480	0.004
N	887	83	10.7
K	775	28 650	0.027
F	100	611	0.164
Sr	60	316	0.190
Al	50	77 440	0.001
Ba	50	668	0.075
Fe	40	30 890	0.001
Zn	30	52	0.577
P	24	665	0.036
B	20	17	1.18
Br	20	1.6	12.5
Li	10	22	0.455
Mn	8	527	0.015
I	7	1.4	5.00
Cu	7	14.3	0.490
As	2	2	1.00
Ni	2	18.6	0.108
Cr	1	35	0.029
Pb	1	17	0.059
Rb	1	110	0.009
Mo	1	1.4	0.357
Co	0	11.6	0.017
Se	0	0.083	2.41
U	0	2.5	0.040
Y	0	20.7	0.003

4.3.2 The Hubbard Brook catchment

So, to what extent does the composition of river water reflect that of rainwater then? Let's look at the Hubbard Brook Experimental Forest once again. Evapotranspiration for the HBEF area has been estimated as representing an overall loss of about 38% so the volume of water in the streams is necessarily going to be less than the volume of precipitation. We might then expect a concentration increase of the ions in stream water by a factor of 1.6. However, Table 4.7 shows that the concentrations of some ions (H^+, NH_4^+, PO_4^{3-}) are actually lower in streamwater and one ion (Al^{3+}) that was not detected in precipitation appears in streamwater. Clearly, just as there was interaction between precipitation and the forest canopy, there is an interaction between water and the ground as it makes its way to and along the streams, as you saw in the Block 2, Part 2.

Table 4.7 Mean concentrations of some dissolved ions in precipitation and streamwater for forested catchments at the HBEF.

Ion	Concentration in bulk precipitation/mg l^{-1}	Concentration in streamwater /mg l^{-1}
Ca^{2+}	0.17	1.65
Mg^{2+}	0.05	0.38
K^+	0.07	0.23
Na^+	0.12	0.88
Al^{3+}	0	0.23
NH_4^+	0.22	0.04
H^+	0.074	0.013
SO_4^{2-}	2.87	6.23
NO_3^-	1.43	1.93
Cl^-	0.51	0.54
PO_4^{3-}	0.008	0.002
HCO_3^-	0.006	1.62

Streamwater can be thought of as having two main components. Firstly, there is the water that has flowed through the ground. This water contains relatively high concentrations of ions arising in the main through the weathering of the soils and rocks (particularly Ca^{2+}, Mg^{2+} and Al^{3+}). The second water source is that which has come directly from precipitation or snowmelt and this has a much lower concentration of ions. In periods of high rainfall and snowmelt, the major component of stream water is this latter dilute source. In summer, when rainfall is low, the major source of water to streams is water that has flowed through the ground.

Superimposed on these annual changes are patterns associated with the demand for nutrients by vegetation in seasonal growth cycles and any seasonal changes in atmospheric inputs. A convenient way to see what is going on is to examine data that reflect catchment inputs and outputs for a range of ions. This may seem to be an easy task but meaningful data are difficult to obtain and have to be gathered over a number of years.

Monthly input and output data for the magnesium cation in the HBEF are shown in Figure 4.13. Note that these data are presented not in concentration units but as quantities per unit area of land. For every month, the catchment output of Mg^{2+} is greater than the input. This is not altogether surprising in that Mg^{2+} is a weathering component of rocks and what we see is a net increase in the quantity of magnesium as water passes through the catchment. However, what is not immediately clear is why there is a monthly variation with two maxima; one around April and the other in November/December. These periods correspond to increased stream discharge. In the November/December period, it is the higher rainfall that generates the increased discharge. However, by April there is also snowmelt, the snow having accumulated over the winter. The more water, the greater the leaching of weathered material (including magnesium).

The relationship between precipitation, streamflow and ion concentrations is complicated. If more water is added to a stream the concentration of dissolved ions would be expected to fall. However, a rainstorm can displace water that has been in contact with soils and rocks for some time (and has a significant concentration of dissolved ions) and produce an initial *increase* in the ion concentration of a stream.

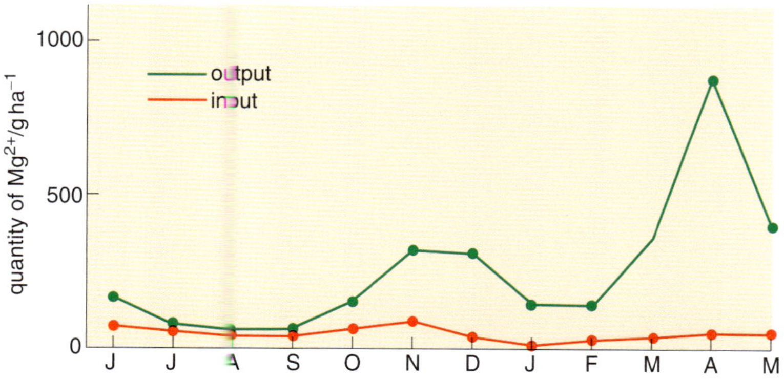

Figure 4.13 Average monthly input and output data for magnesium cation in HBEF catchments.

Given that we are dealing with a forested (both coniferous and deciduous) ecosystem, we might have expected that there would be a demand for magnesium by trees during the growing season. Magnesium is a vital component of chlorophyll which all green plants use in the photosynthetic conversion of carbon dioxide to carbohydrate. In Figure 4.13, the vertical distance between the input and output plots represents the net output of magnesium. As trees have increasing demand for magnesium during the April to July growing season, one might expect the net catchment output of magnesium to decrease. However, it turns out that the quantity of magnesium required for chlorophyll synthesis is very small relative to the changes in magnesium availability brought about by the spring and early summer rains and snow melt. Support for this also comes from the input/output plot for calcium. The calcium ion is not a significant factor in tree growth and yet the plot (Figure 4.14a) shows a similar seasonal pattern to that for magnesium (and also sodium, Figure 4.14b). This situation of empirical observations being a composite of a number of (difficult to separate) factors is very typical of studies in biogeochemistry.

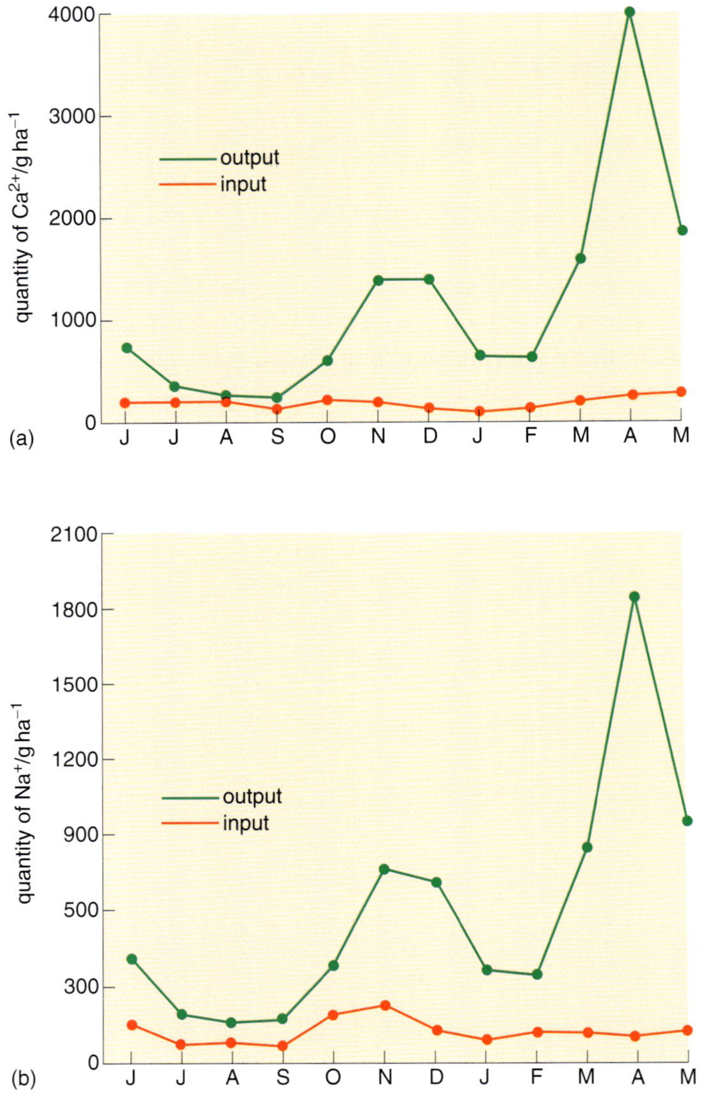

Figure 4.14 Average monthly input and output data for (a) calcium and (b) sodium cations in HBEF catchments.

A very different situation arises with the ammonium ion (NH_4^-) where the input is higher than the output at all times of the year. Ammonium ions come from the atmosphere and from biological activity in the soil. Ammonia is one of the products from the breakdown of protein from plants and animals. The fall in input of ammonia during the winter may reflect the reduced input from the atmosphere when the ground is 'insulated' by snow cover and biological activity in the soil is low (Figure 4.15). In the growing season, ammonia is consumed and this is reflected by large differences between input and output at these times. The data in the summer in particular are difficult to fully explain as ammonia is rapidly converted to nitrate and other nitrogen compounds. Another nutrient that is often in limited supply is phosphate (PO_4^{3-}) and although there is essentially no atmospheric input, it shows a similar pattern to ammonia.

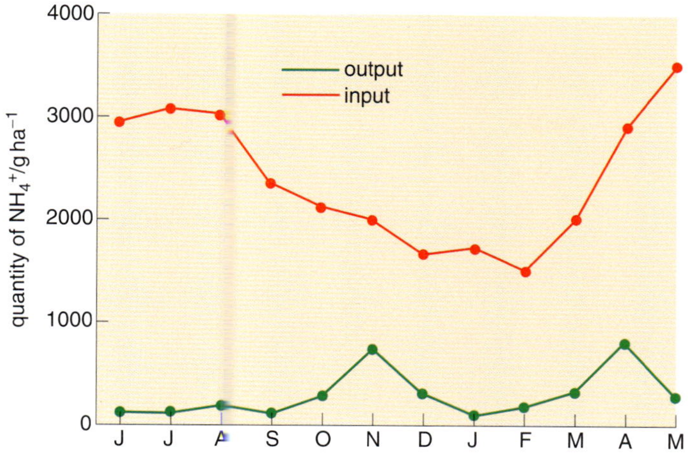

Figure 4.15 Average monthly input and output data for the ammonium cation in HBEF catchments.

Two more ions that are essential for growth are potassium (K^+) and nitrate (NO_3^-) and the seasonal patterns for these ions are shown in Figure 4.16. These ions show a different pattern. With magnesium, calcium and sodium, output exceeded input at all times of the year, for ammonia and phosphate input exceeds output at all times and for potassium and nitrate, input exceeds output during the summer and becomes less than output during the late winter. How do we explain this crossover?

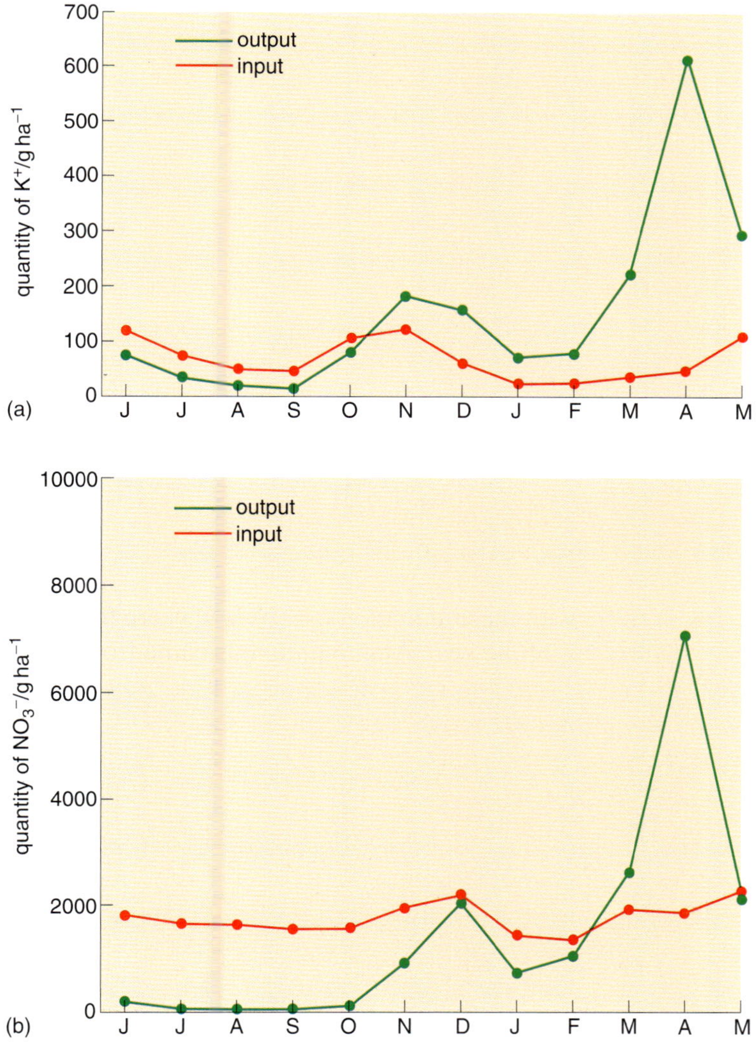

Figure 4.16 Average monthly input and output data for the (a) potassium cation and (b) nitrate anion in HBEF catchments.

A major factor is the demand for both of these ions during the growing season when plant growth results in their extraction from water on and near the ground surface. During the winter, the demand falls and in the late winter, output exceeds input.

The HBEF data that we have been considering have been averaged over a period of several years. There is variability between years particularly if there is unusually heavy rainfall or periods of drought when the quantities of ions leached from the soil and put into water on the ground can change.

4.3.3 Aluminium solubility in natural waters

Aluminium is an environmentally harmful element in natural waters (see Box 4.1) and it is one of the main elements mobilized with the acidification of surface waters due to acid rain. Aluminium is present in soil and rocks as aluminium oxides and hydroxides as well as aluminosilicate minerals and secondary clays. Aluminium occurs dissolved in water as the trivalent ion Al^{3+}, but it also occurs as a **complex**. A complex is made when an Al^{3+} ion forms a chemical bond with anions or neutral molecules also present in the solution. Hydroxide, sulfate and fluoride ions all form complexes with aluminium ions, as do silica molecules and organic material. The concentrations of the 'free' Al^{3+} ion and the different aluminium complexes vary with pH (Table 4.8). As pH declines, the total aluminium concentration in river water increases considerably. Below pH 4.5, the main species present in solution is Al^{3+}, and this is probably the most environmentally harmful form of aluminium. At higher pH levels, Al^{3+} and the total concentration of aluminium decline and hydroxide complexes become more significant. The reason for the increase in total aluminium concentration with decreasing pH is partly because the solubility of aluminium oxides and hydroxides increases as pH decreases, and partly because more acidic waters occur during stormflow conditions when the waters are mainly derived from highly acidic soils that contain aluminium oxide and hydroxide minerals. At times of low flow, waters are mainly derived from groundwater where weathering reactions release calcium and magnesium to solution, resulting in neutral waters that contain hydrogen carbonate ions and are low in aluminium.

This type of behaviour is typical of most of the transition metal elements (such as copper, iron and zinc) but the proportion of the various complexes will of course differ. For both aluminium and many of the transition metals, complexation by organic material can be very important, and if river waters are organic rich, then organic complexes can dominate. Aluminium is however unusual in that under alkaline conditions ($pH > 8$) it can form the aluminate complex $Al(OH)_4^-$ which can be the dominant complex in solution. Because of the nature of aluminium complexes, the total concentration of aluminium in solution is a minimum at around pH 7.

In the case of the HBEF, an increase in acid deposition (due to acid rain) would lead to increased soil and stream acidity with elevated aluminium concentrations. Deposition of acidic oxides from industrial emissions can constitute a major environmental threat in acidic and acid-sensitive areas. In a study of acidification in UK surface waters, pH values as low as 3.7 have been observed with aluminium concentrations of up to almost $2\,mg\,l^{-1}$; damage to fish stocks can occur at aluminium concentrations as low as $50\,\mu g\,l^{-1}$.

Table 4.8 Average concentrations and percentages of aluminium in all its forms dissolved in river water at various pH. HBA is hydroxy-bound aluminium, SBA is sulfate-bound aluminium, FBA is fluoride-bound aluminium, OrgBA is organic-bound aluminium, and SiBA is silica-bound aluminium. These data are from river waters in upland Wales, UK. All concentrations are reported in units of nmol l^{-1}.

	pH range					
	4.0–4.5	4.5–5.0	5.0–5.5	5.5–6.0	6.0–6.5	6.5–7.5
Concentration in solution						
Al^{3+}	10 254	3631	987	72	8	2
HBA	635	619	480	244	287	357
SBA	600	205	52	4	0	0
FBA	2866	2849	2030	1290	662	340
OrgBA	2491	2367	1677	899	418	191
SiBA	832	1713	1608	574	407	434
Total	17 678	11 384	6834	3083	1782	1324
Percentage of total concentration in solution						
Al^{3+}	58	31	14	2	0	0
HBA	4	6	7	8	17	27
SBA	3	2	1	0	0	0
FBA	16	26	32	43	38	26
OBA	14	20	24	28	23	16
SiBA	5	15	23	18	22	32

Box 4.1 The Camelford incident

Aluminium does not simply cause harm in water due to acidification processes, direct contamination of drinking water and river waters by human error can also be important. For example, in July 1988 aluminium-contaminated water affected 20 000 homes in Camelford, Cornwall, producing Britain's worst mass water poisoning incident. This occured when aluminium sulfate, used as part of a clean-up procedure to clarify water, was accidentally added directly to the mains water supply at the local water treatment works. This produced a toxic mix of aluminium sulfate and a cocktail of other chemicals including lead and copper from piping in people's homes. Residents in the Camelford community suffered a range of effects including dyed green hair from copper residues. In 1994, 148 victims accepted damages totalling almost £400 000 in a settlement approved by a High Court judge, but many people are still complaining today of chronic side effects including short term memory loss, joint pains, lethargy and multiple allergies.

4.3.4 Human impacts on river waters

Humans can impact on surface waters in many different ways. For example, the construction of dams and reservoirs increases surface water storage and this can reduce the fluctuations seen in the chemistry of river waters. In the case of acidic environments, this can be a good thing as the acidic events associated with high flows can be damped out. However, humans can also have a detrimental effect on river water chemistry. For example, inputs of organic material due to the discharge of sewage and inputs of nutrients from fertilizers can lead to exotic algal blooms and anoxic conditions. Changes in water chemistry (or 'water quality') due to human activity results from pollution from either diffuse or point sources.

Diffuse sources of pollution are those that come from contamination of broad areas of land. One example is acid rain, which affects much of the UK uplands. Diffuse pollution is also associated with agricultural, urban and industrial activity. In agriculture, the main sources of pollution come from the widespread application of fertilizers and from organic compounds used as herbicides and pesticides. The main fertilizer of concern is nitrate; phosphates are also used as fertilizers, but they tend to be more strongly bound to the soil. Nitrate pollution occurs on two timescales. Firstly, there is leaching from the land during periods of high flow in the autumn and winter (Figure 4.17a). Secondly, there is long-term build up of nitrate within catchments that contain major aquifers; in this case (as you saw in Section 3) the pollution may not occur in a stream for years to decades, and may take many more decades to recover.

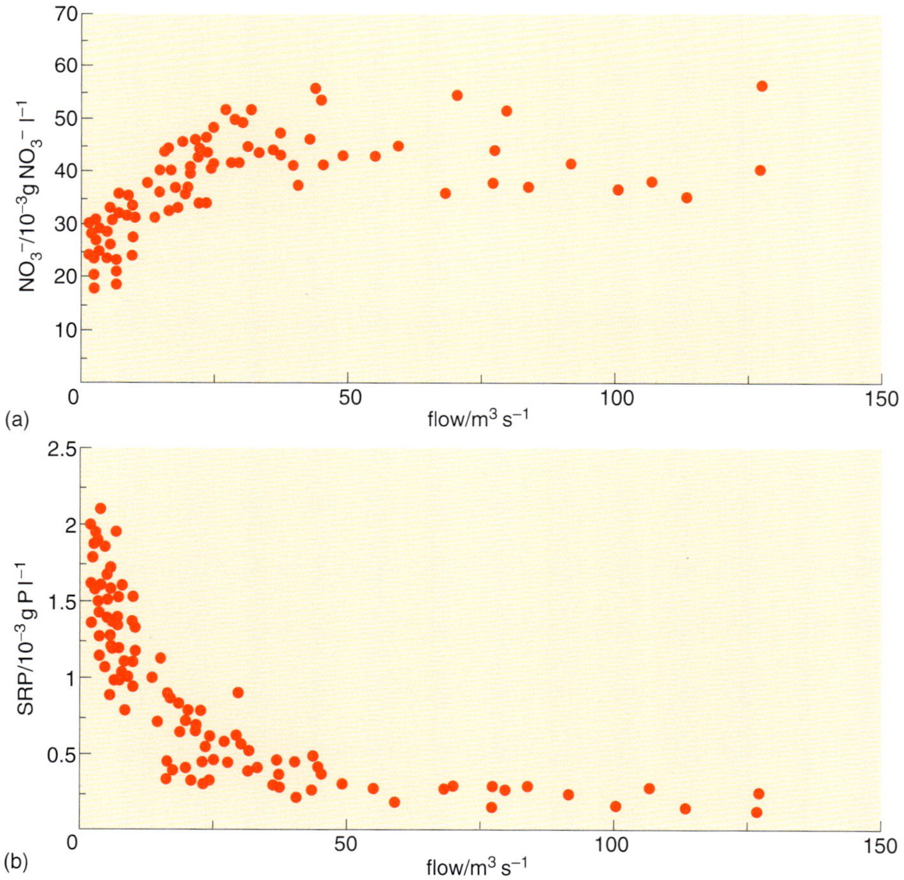

As well as nitrate, carbon-based compounds from animal slurry can also lead to the contamination of rivers. High levels of ammonium are produced and there is loss of dissolved oxygen. For herbicides and pesticides, pollution occurs soon after application as organic compounds are either adsorbed by the soils or biologically degraded. There are also diffuse sources of pollution from industry and these come from landfill sites, contaminated land and contaminated floodplain sediments. This type of contamination can lead to a variety of problems linked to anoxic conditions and organic and trace metal pollution.

Figure 4.17 Variation in (a) nitrate and (b) phosphate (expressed here as soluble reactive phosphate, SRP) with flow in the River Thames, UK.

In contrast to diffuse sources, point sources of pollution are usually associated with the direct discharge of chemicals to a river. These discharges can be intermittent in terms of spillage of chemicals for urban, industrial and agricultural use, but often they are associated with regular discharges of industrial and sewage effluent. Point source inputs cover an extremely wide range of substances that include:

- Many of the major elements (e.g. chlorine, nitrogen, sulfur, sodium and potassium) at extremely high levels (tens to hundreds of mg l^{-1})
- Trace elements such as boron (which is used in washing powder) and the transition metals (from industry) at levels of hundreds to several μg l^{-1}
- A wide range of organic compounds. These include detergents, petrochemicals, herbicides, pesticides and chlorinated hydrocarbons. Within this group are the steroid hormones (see Box 4.2).

For point source inputs, concentrations often dilute with increasing flow. This is shown in Figure 4.17b where phosphate (which is derived from sewage) decreases with increasing flow. This is typical for systems where the input of a pollutant is relatively constant.

Box 4.2 Effeminate fish

In the 1980s male roach (a type of coarse fish) on the River Lea in North London were found to show signs of part-feminization. Further work confirmed that around a third of the male roach in the river were affected, and studies on caged fish in other UK rivers have shown that the degree of feminization is linked to effluent discharge from sewage treatment works (Figure 4.18). These effluents contain both natural (oestrone and oestradiol) and artificial (ethinyl oestradiol) steroid hormones, as well as alkylphenol-based detergents which are known to have endocrine (hormone) disrupting properties.

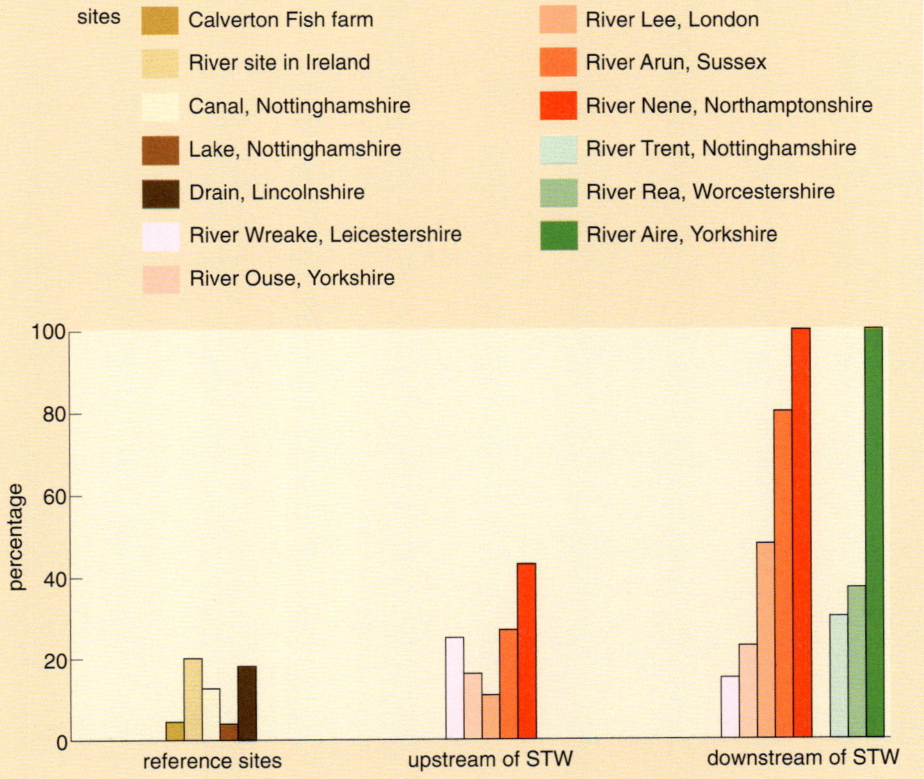

Figure 4.18 Percentage of male roach showing signs of feminization (possession of female reproductive cells and/or female reproductive tracts) in a series of UK rivers. Data are provided for river sites both upstream and downstream of sewage treatment works (STWs), and for rivers that do not receive sewage effluents (so-called reference sites).

Fish are not the only organisms affected by contamination of river waters by steroid hormones; levels of < 1 ng l^{-1} can also cause part-feminization of crocodiles.

Activity 4.1 Modelling the results of discharges of materials into rivers

This DVD activity requires you to use the HERMES system which models the progress and dilution of a pollution input to a river. You will be able to predict from the model the concentration of a pollutant at a particular time and point down river from its introduction. The model has been developed to represent real rivers and has provided valuable estimates of the outcomes of pollution incidents.

The program guides you through an initial exercise which shows you how to use the model. Then you will be asked to feed data into the system to model incidents. You will need to be able to print out data and displays.

4.3.5 Dissolved carbon dioxide and oxygen

Stream waters contain not only dissolved ions, they also contain dissolved gases. Gases from the atmosphere, such as nitrogen (N_2), oxygen (O_2) and carbon dioxide (CO_2), all dissolve in water, and their solubility increases as temperature decreases (Table 4.9). Gases such as nitrogen and oxygen dissolve in water to give a solution that contains the original diatomic molecules. The dissolution of carbon dioxide in water, however, results in the formation of both dissolved carbon dioxide, CO_2 (aq) (often represented as carbonic acid, H_2CO_3), and hydrogen carbonate ions (HCO_3^-). At higher pH carbonate ions (CO_3^{2-}) are also formed:

Table 4.9 The effect of temperature on the solubility of oxygen in water.

Temperature/°C	Solubility of oxygen in water/ mg l^{-1}
0.0	14.6
5.0	12.8
10	11.3
15	10.1
20	9.07
25	8.24
30	7.54
35	6.93
40	6.41
45	5.93

$$CO_2(g) + H_2O(l) = H_2CO_3(aq) = H^+(aq) + HCO_3^-(aq)$$

and at higher pH,

$$HCO_3^-\ (aq) = CO_3^{2-}\ (aq) + H^+(aq)$$

The change in the proportion of the components of these equilibria with pH can be seen in Figure 4.19. The pH of rainwater is usually less than 5.6, which suggests that dissolved carbon is in the form of carbon dioxide and the hydrogen carbonate anion HCO_3^-. At pH > 10.5 the carbonate anion CO_3^{2-} becomes the dominant form of dissolved carbon.

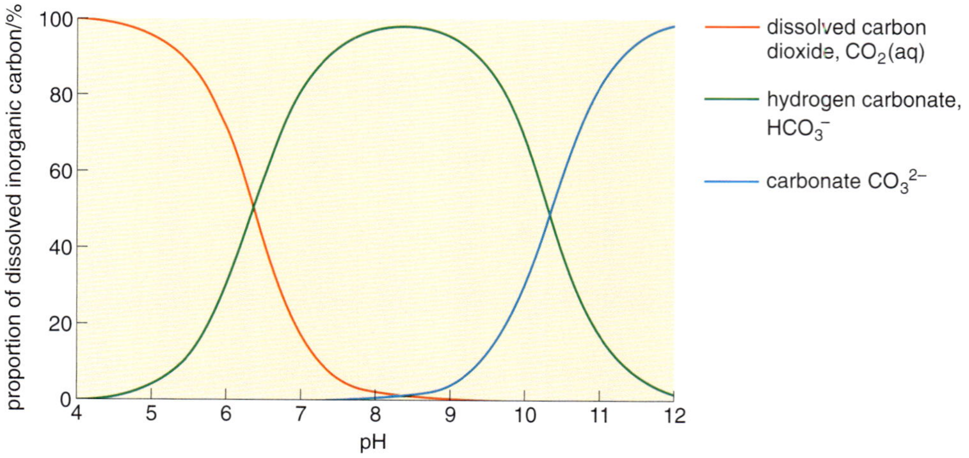

Figure 4.19 Dissolved carbon dioxide, hydrogen carbonate and carbonate anions as a mass proportion of dissolved inorganic carbon in water at 25 °C. (Proportions vary slightly with temperature.)

Question 4.4

From the data in Figure 4.19, estimate the molar concentration of the hydrogen carbonate anion in water at pH 10 containing a total of 2.0×10^{-4} g l^{-1} of dissolved inorganic carbon.

Dissolved carbon dioxide in river water also varies in response to biological activity. Carbon dioxide is consumed by photosynthesizing plants in the day and converted into organic material, and carbon dioxide is released during respiration by microbes and by plants at night. The processes of photosynthesis and respiration can be represented by the following equations:

Photosynthesis $\quad$ light energy + $6CO_2(g) + 6H_2O(l) \rightarrow C_6H_{12}O_6(s) + 6O_2(g)$

Respiration $\quad$ $C_6H_{12}O_6(s) + 6O_2(g) \rightarrow$ energy + $6CO_2(g) + 6H_2O(l)$

In these equations, organic material is represented by the molecule glucose, $C_6H_{12}O_6(s)$.

- ❍ What effect would photosynthesis by aquatic plants have on levels of dissolved carbon dioxide in a river?
- ● Photosynthesis consumes CO_2, so levels of dissolved carbon dioxide would fall. This would more than offset respiration.

Depending on the relative importance of photosynthesis versus respiration, river water (and groundwater) can contain either more or less CO_2 than waters that are in equilibrium with the atmosphere. Such waters are, respectively, said to be supersaturated or undersaturated with respect to CO_2. The ratio of the measured value of dissolved CO_2 relative to the equilibrium value is called the **saturation state**, and it is given the symbol Ω. A value of $\Omega = 1$ indicates that CO_2 in the water is in equilibrium with CO_2 in the atmosphere, while values <1 and >1 represent undersaturation and oversaturation, respectively. Table 4.10 shows the average value and the range of values of the saturation state of CO_2 for a number of eastern UK rivers which are shown in Figure 4.20. Data for dissolved phosphorus and dissolved organic carbon (DOC) are also shown to indicate levels of pollution in the river, and the amount of carbon available for respiration. The data show a broad increase in dissolved phosphorus and Ω towards the south which corresponds to the progressive increase in pollution from the rural northern rivers to the rivers draining urban and industrial areas to the south. The industrial rivers of the Aire and Calder contain highest levels of DOC, and they have highest average values of Ω.

- ❍ Why do high DOC values correspond to high values of Ω?
- ● High DOC means that there is lots of organic material available for respiration. Respiration results in the production of CO_2 so waters with high DOC are likely to be highly saturated with CO_2.

In the less polluted waters to the north, dissolved CO_2 concentrations are largely controlled by photosynthesis and respiration. For example, waters in the River Tweed show a large range in Ω because they are often undersaturated in CO_2 during the day when photosynthetic activity is high (and CO_2 is converted into organic material), but oversaturated at night when photosynthesis ceases and respiration (which produces CO_2) dominates.

Figure 4.20 Location map of various rivers in the eastern UK.

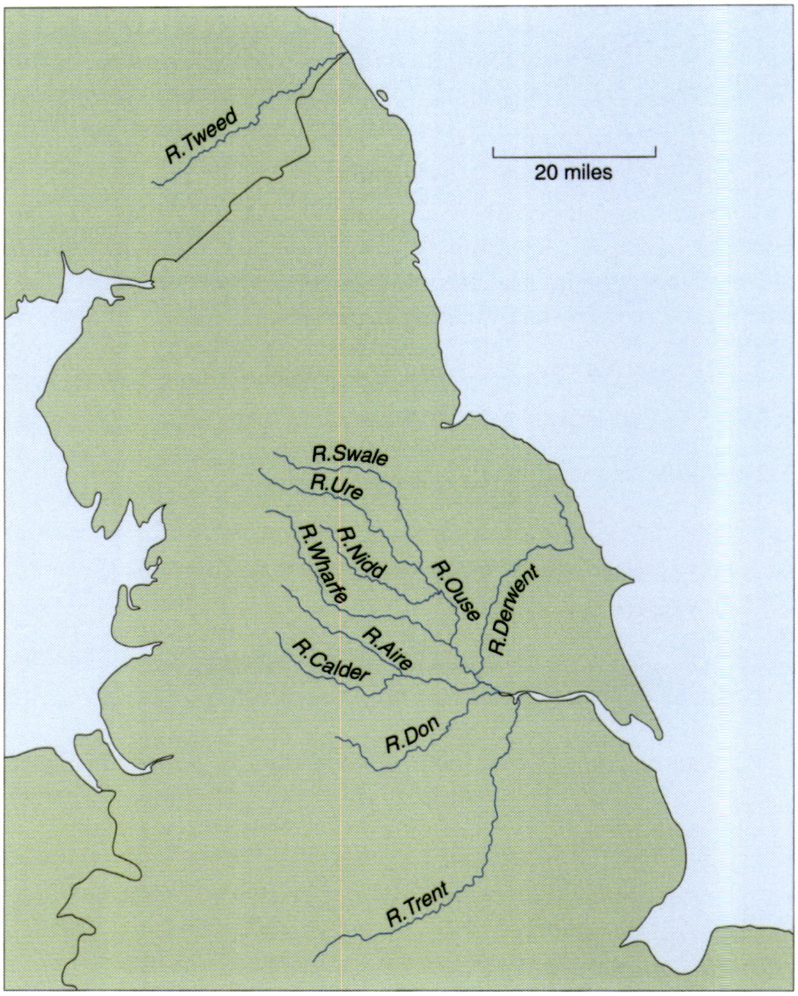

Question 4.5

The concentration of dissolved carbon dioxide has been measured in two different river water samples in order to determine their saturation state. Sample A has an Ω value of 0.5, and sample B has a value of 0.8. Which sample contains the higher level of dissolved carbon dioxide, given that the temperature of both samples is 10 °C?

Dissolved oxygen levels in rivers are also strongly influenced by the processes of photosynthesis and respiration. A few microorganisms (those that do not photosynthesize) are able to live in the absence of oxygen but if dissolved oxygen levels fall below threshold values, many organisms die. Many fish become distressed at oxygen concentrations of 5 mg l^{-1} and lower values are often fatal. Even amongst relatively similar creatures, tolerance to low oxygen levels varies significantly (Table 4.11).

A tumbling mountain stream is likely to have oxygen concentrations close to saturation but the level of dissolved oxygen can fall markedly in slow moving rivers. There is often a change in oxygen concentration from the surface to the lower levels in lakes and rivers. Photosynthesizing plants increase oxygen levels (and decrease carbon dioxide levels, as you saw above) but at depth or in murky

Table 4.10 The variation in the saturation state of dissolved carbon dioxide (Ω) and the concentrations of dissolved phosphorus and dissolved organic carbon (DOC) in eastern UK rivers. Average values are given first, followed by the range of values. Dissolved phosphorus is expressed as soluble reactive phosphorus (SRP). The location of the rivers is shown in Figure 4.20.

River	Ω	SRP/μg l^{-1}	DOC/mg l^{-1}
Tweed	2.7 < 0.1–23.5	34 1–366	3.8 1.0–13.8
Swale	7.9 0.5–57.6	176 1–578	4.3 1.0–19.1
Ure	7.0 0.4–51.5	53 1–170	4.3 1.0–18.9
Nidd	8.9 0.3–46.0	505 1–1687	5.2 1.0–11.8
Derwent	11.3 1.5–159	121 1–314	3.8 1.0–63.8
Wharfe	6.8 0.4–51.9	170 1–1084	4.9 1.0–73.6
Ouse	11.3 0.4–82.8	393 1–1150	8.9 1.0–14.1
Aire	18.6 0.6–92.6	1013 1–2570	8.9 1.0–18.1
Calder	17.3 0.4–66.4	1076 1–2765	10.3 1.0–26.3
Don	12.5 1.0–51.7	1396 1–3505	5.4 1.0–13.2
Trent	9.7 0.7–113	1520 1–3104	5.4 1.0–11.0

water where light penetration is minimal, photosynthesis is limited and oxygen is reduced by respiration. Dissolved oxygen concentrations can change on a diurnal basis (in the opposite sense to carbon dioxide) as photosynthetic processes dominate during the day and respiratory processes dominate during the night. The diurnal pattern is most pronounced during the summer months when biological activity is particularly high, but during the winter low biological activity ensures that the diurnal pattern is small. For example, small streams within the catchment of the River Thames show a variation in dissolved oxygen of between 0.8 and 1.2 Ω in the summer, and of between only 0.95 and 1.05 Ω in the winter (in this case Ω represents the measured value of oxygen relative to the equilibrium value).

Falls in dissolved oxygen are sometimes caused by the input of sewage into a waterway, as you saw in your study of the Teign catchment. In an attempt to break down the organic matter, large populations of microorganisms consume oxygen and its concentration in the water can fall sharply (Figure 4.21). As the sewage passes downstream and is oxidized, the demand for dissolved oxygen decreases and the river slowly recovers. However, in the region indicated in Figure 4.21, a considerable dent would have been made in the fish stock.

Table 4.11 Minimum concentrations of dissolved oxygen at which individuals survived for seven days at 16 °C.

Freshwater fish	Dissolved oxygen concentration/ mg l^{-1}
rainbow trout	3.4
perch	1.2
roach	0.8
tench	0.7

Figure 4.21 Change in dissolved oxygen concentration with time at a point in a river downstream of a sewage input. The shaded region represents conditions that would put aquatic life under stress.

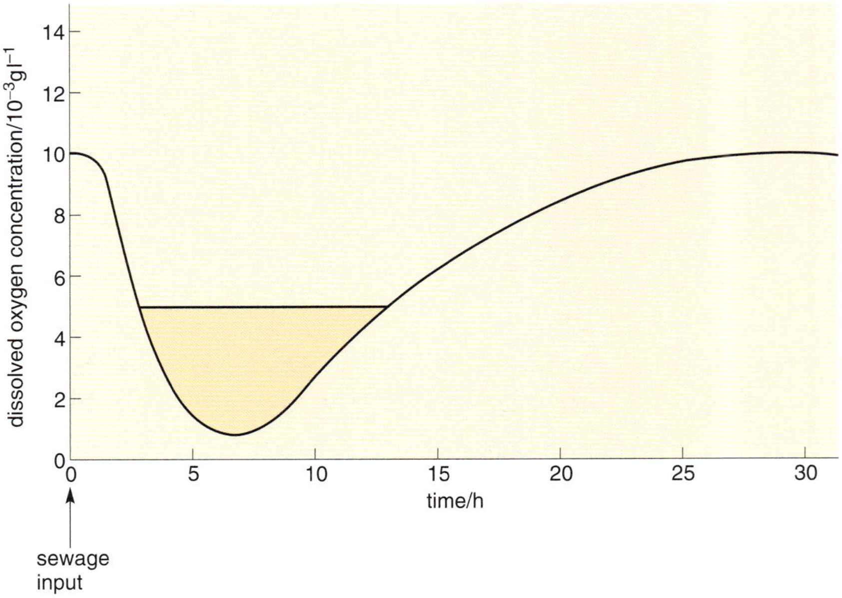

4.3.6 Diseases

When considering the composition of water, it is easy to overlook what is actually one of its most important constituents in relation to human health — water-borne and water-related diseases. The most serious hazards to health are associated with bacteria, parasites, viruses and rotaviruses. These risks relate to inadequate sanitation or contact with contaminated water and with hosts that either live in water or require water for part of their life cycle (for example, diseases spread by insects that breed or feed near contaminated water, Figure 4.22). They account for around six million deaths globally each year; the biggest killers are associated with diarrhoea and malaria. While many of these diseases have been overcome in the developed world, they are of major concern for the economically poorest nations. Furthermore, the disabling effects of these water-related diseases will affect a far greater proportion of the population, but the extent and degree of harm is unknown.

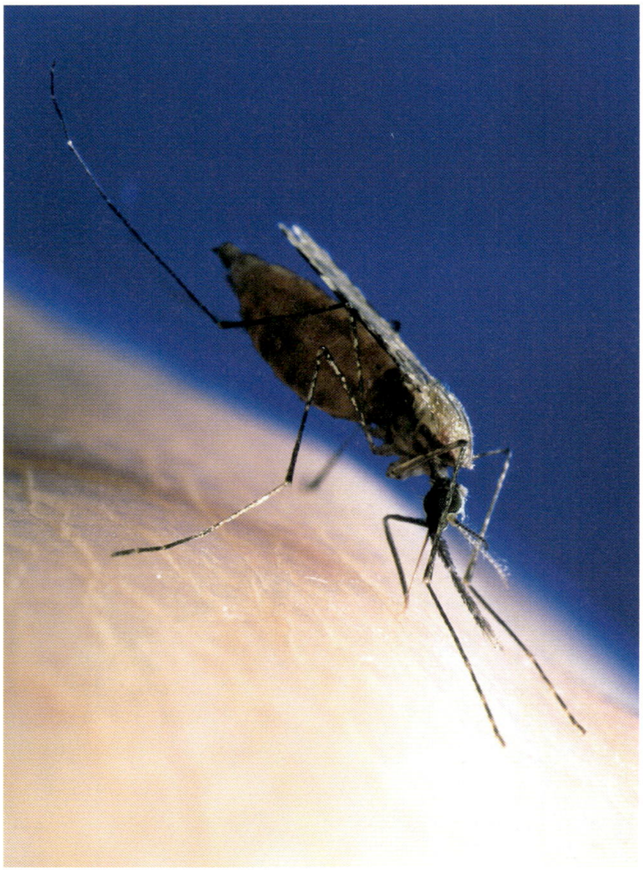

Figure 4.22 Female *Anopheles* mosquito which transmits malaria.

4.4 Summary of Section 4

1. During its journey from source to ocean, a river changes its characteristics. Generally the slope decreases and the upwardly concave profile of a graded stream is approached.
2. The speed of water in a river channel is not uniform over a cross-section being slower where the water is in contact with the river bed and faster nearer the centre of the channel. The nature of the river bed also affects the water speed as does the gradient and these variables are quantitatively linked through the Manning equation.
3. Stream water contains a different cocktail of dissolved material from that of precipitation, partly reflecting the terrain over and through which the water has flowed. The concentration of dissolved oxygen in water is critical for most aquatic life and the levels can be reduced by temperature rise or sewage input. Other inputs can also affect river life and the progress of these inputs can be modelled and the effect downstream can be estimated.

Learning outcomes for Section 4

When you have completed this section, you should be able to:

4.1 Use a topological map to plot a longitudinal profile of a river.

4.2 Identify conditions when laminar or turbulent flow is likely to occur.

4.3 Construct isovel lines over the cross-section of a river using positional water speed data.

4.4 Calculate values for the hydraulic radius of a river channel.

4.5 Apply the Manning equation to estimate speed of flow in a river.

4.6 Comment on and suggest reasons for the differences in dissolved ion concentrations in rainwater and streamwater.

4.7 Account for the seasonal variation in dissolved ion concentrations in streamwater in vegetated areas.

4.8 Estimate the relative proportions of dissolved carbon-containing species in water of different pH values.

4.9 Suggest reasons for changes in dissolved oxygen in a river and the potential effect on aquatic life.

4.10 Use the HERMES model to estimate the downstream profiles of point chemical discharges into a river. (*Activity 4.1*)

5 The hydrological cycle

So far we have looked in detail at water in rivers, and water in the ground. But as you saw in Section 1, there are many other reservoirs of water and water does not remain in any one particular reservoir. Rather, it moves over, on and through the Earth in a continuous cycle, driven by the Sun's energy and gravity. This is known as the **water cycle** or the **hydrological cycle** (Figure 5 1).

- ❍ Can you name some of the processes by which water is exchanged between reservoirs?
- ● Exchange occurs through a number of processes, including evaporation, infiltration, precipitation, and surface runoff to the oceans

Each of these processes also involves the transfer of energy. For example, as you saw in Section 2, water requires energy (latent heat) in order to evaporate, and energy is released as water vapour condenses. Thus the energy cycle of the Earth and the water cycle are intimately interconnected.

Water can take many paths through the hydrological cycle; for example, water may evaporate from the ocean and simply fall as precipitation back into it, or it may evaporate from the ocean, fall as precipitation over land, and either return to the ocean via a river or stream or evaporate once again.

You will see from Figure 5.1 that evaporation exceeds precipitation over the oceans; on average, 336×10^{15} kg of water evaporates annually from the oceans,

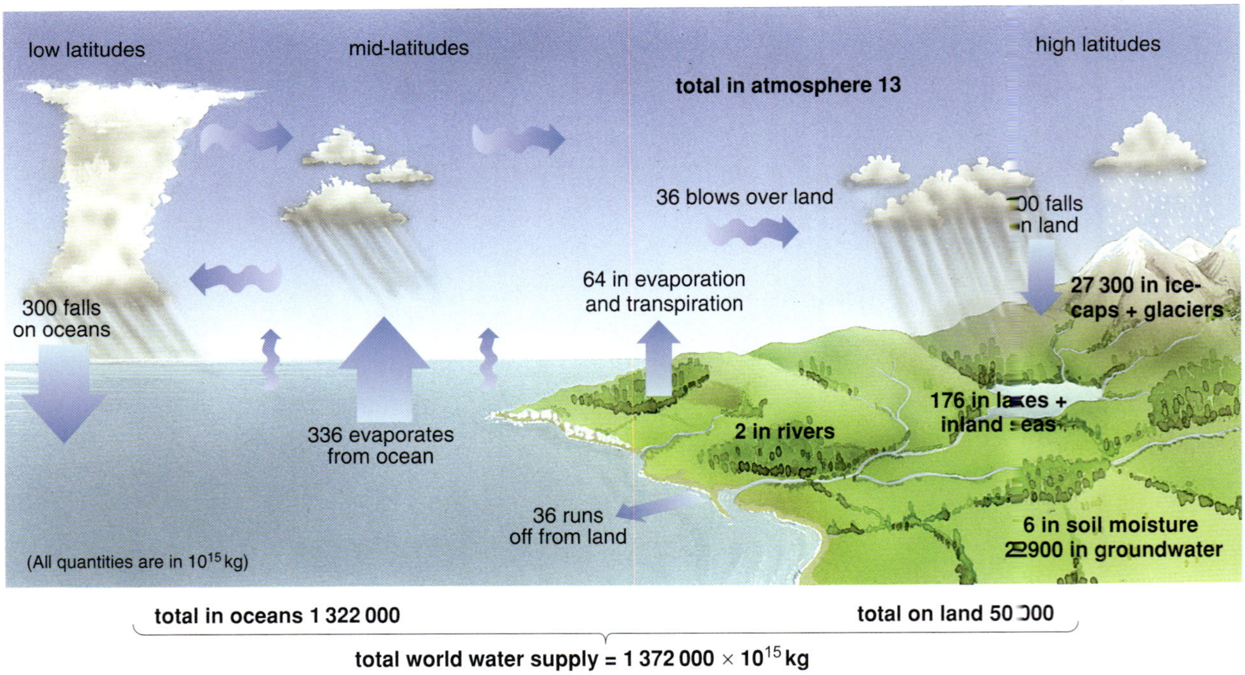

Figure 5.1 The hydrological cycle. Water cycles (arrows) between different reservoirs (bold type) of the hydrosphere. The rate of movement of water (usually referred to as the flux of water) is given in 10^{15} kg yr^{-1}; reservoir masses are in 10^{15} kg.

but only 300×10^{15} kg of water falls over the oceans each year which means that evaporation exceeds precipitation by 36×10^{15} kg per year.

- ❍ Can you explain, with the help of Figure 5.1, why is it then that the volume of water stored in the oceans is not decreasing with time?
- ● Over land, precipitation (100×10^{15} kg yr^{-1}) exceeds evaporation (64×10^{15} kg yr^{-1}). The 'excess' water (36×10^{15} kg yr^{-1}) runs off the continents into the ocean and balances the 'excess' water lost from the ocean by evaporation.

Box 5.1 The origin of water

The Earth is thought to be around 4.6 billion years old, based on the ages of meteorites that presumably formed at the same time. The oldest known minerals (zircons) on Earth are 4.4 billion years old, and their chemical composition indicates that they may have formed in the presence of water. Of the period between 4.6 and 4.4 billion years ago, there is no evidence for water. Thus, we can only really make an educated guess about its formation, based on our ideas about the origin of the Earth.

The Earth aggregated from a cloud of gas and dust around the Sun, and its growth was dominated by asteroid and comet impacts and planetary collisions. Because rocks of the Earth's outer layers do not contain significant amounts of water, scientists were originally at a loss to explain where the surface water came from. However, it has been discovered that both comets and asteroid material which bombarded the early Earth contain appreciable water.

Computer models predict that comets brought down to Earth at least four times as much water as the present mass of the oceans; much of this water was volatilized during bombardment. Water was also brought to Earth by one group of meteorites, the so-called carbonaceous chondrites (Figure 5.2). These contain up to 20% water by mass; this water is locked up in clay minerals. Heating of this material during bombardment released the water from the clay minerals which could accumulate at the Earth's surface.

Figure 5.2 Photograph of the Orgueil carbonaceous chondrite.

5.1 How long does water remain in a reservoir?

Table 5.1 Residence time of water in different reservoirs of the hydrosphere. Note that residence times can vary even within reservoirs; this is because water fluxes, and water storage can also vary. Only *average* values are shown in Figure 5.1.

Reservoir	Residence time
ocean	about 4000 years
polar icecaps	about 800 years
groundwater	a few weeks to 10 000 years
lakes	a few years
soil moisture	a few weeks to 1 year
atmosphere	about 11 days
rivers	a few weeks

The average length of time that water stays in a reservoir before moving to another is called the **residence time** for that reservoir, as shown in Table 5.1. If the reservoir empties and replenishes at the same rate (so that the volume of water within it remains constant), the residence time can be calculated from the flux of water transferred into (or from) the reservoir in relation to the mass of that reservoir. Note that there may be several ways in which water is transferred to and from a reservoir (Figure 5.1).

$$\text{residence time} = \frac{\text{mass in reservoir}}{\text{flux into or from reservoir}} \tag{5.1}$$

For example, from Figure 5.1, the residence time for water in a lake is:

$$= \frac{176 \times 10^{15}\ \text{kg}}{36 \times 10^{15}\ \text{kg yr}^{-1}} = 4.9\ \text{years}$$

The residence time gives an indication of how quickly a water molecule in a reservoir can be renewed. Water vapour in the atmosphere has the shortest residence time; it is rapidly renewed by evaporation from the oceans and land and lost by precipitation. This is a rapid *subcycle* of the water cycle. Subcycles involving the oceans, the icecaps, and groundwater are much slower.

5.2 The water balance

If we consider any hydrologic system — for instance, a lake — it contains a certain volume of water at a given time. Water can be added to the lake from several sources: precipitation onto its surface, streams that flow into it, groundwater that seeps in, and overland flow from nearby land surfaces. Water also leaves the lake through: evaporation, transpiration by emergent aquatic vegetation, outlet streams, and groundwater seepage from the lake bottom. If, over a given period of time, the total inflow to the lake is greater than the total outflow, then the lake level will rise as more water accumulates. If the outflow exceeds the inflow over a given period, then the lake level will fall. Hence any differences between the outflow and inflow to a hydrologic system will result in a change in the volume of water stored within it. Formally, this is expressed by the **water balance** equation:

$$\text{inflow} = \text{outflow} + \text{change in storage} \tag{5.2}$$

We can attempt to solve the water balance equation for any hydrologic unit with quantifiable inputs and outputs of water; it is most commonly used in studies of drainage basins. Figure 5.3 is a **box model** of a drainage basin system. Each reservoir within the system is symbolized by a box, and flow of water between each reservoir is indicated by arrows that point from the inputs of water into the reservoir to the outputs of water from the reservoir.

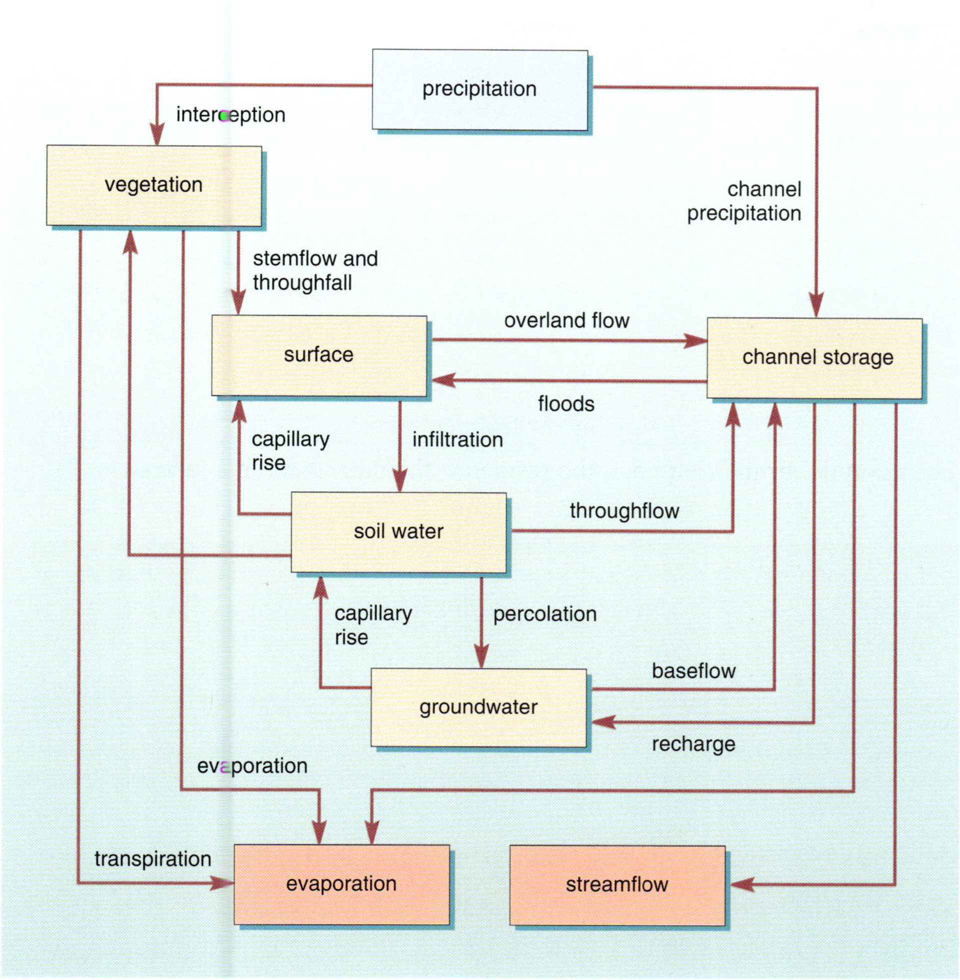

Figure 5.3 The drainage basin hydrological system. Light brown boxes show water storage.

- ❍ By referring to Figure 5.3, list the inputs and outputs of water to the drainage basin.
- ● Precipitation adds water to the drainage basin. Water is removed via evaporation and as streamflow.

Question 5.1

List (a) the inputs and (b) the outputs of water to the groundwater reservoir.

If the water balance equation can be solved, then it means that we can *quantitatively* assess the movement of water through the drainage basin. In turn, this enables us to predict future changes in hydrology (for example, streamflow) as a result of changes in climate or land use, for example. However, as we shall see, solving the water balance equation can be a difficult and complex task. In the next two sections we take a look at some of the methods employed by hydrologists to solve the water balance of a drainage basin.

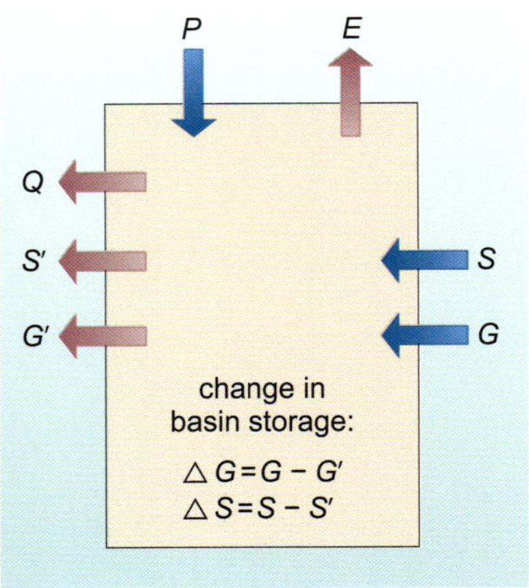

Figure 5.4 Components of the water balance for Catchwater Drain drainage basin, Yorkshire, UK. See text for explanation of terms.

5.2.1 Field studies

One approach to solving the water balance equation is to make field measurements of all the input and outputs (e.g., precipitation, evapotranspiration, infiltration) and water storages (e.g. groundwater, channel storage) within the drainage basin. Clearly, this is an intensive and time-consuming task, but there are a number of catchments for which this has been attempted. One example is given below, another is described in Box 5.2.

A pioneering example is the Catchwater Drain experimental drainage basin in East Yorkshire, UK. Within the 16 km^2 drainage basin, precipitation and streamflow are measured continuously, hourly evaporation is calculated from meteorological data, groundwater storage is estimated from a network of observation wells, and soil water storage is determined from tensiometer measurements. As this is a small drainage basin, the water balance may be adequately defined in terms of inputs of precipitation and groundwater flow, outputs of evaporation, streamflow and groundwater flow, and changes in water storage (principally groundwater and soil water) within the drainage basin (Figure 5.4). The water balance calculation is therefore given by the following equation:

$$P = Q + E - \Delta G - \Delta S \qquad (5.3)$$

where P is precipitation, Q is streamflow, E is evapotranspiration, and ΔG and ΔS are, respectively, changes in groundwater and soil water storage. It is usual to quote each variable in terms of water volume divided by the surface area of the catchment over the selected time period; thus data are given in mm.

Table 5.2 gives values for these variables measured over the period 1–12 June 1969.

- ❍ Use the data in Table 5.2 to show that the inputs and outputs of water to the Catchwater Drain drainage basin are in balance for this time period.

- ● If the inputs and outputs of water are in balance, then $P - Q - E + \Delta G + \Delta S = 0$ (Equation 5.3). Substituting the values in Table 5.2 into this equation:

  $29 - 12 - 25 + 6 + 2 = 0$

  Thus the inputs and outputs of water are in balance for this time period.

Table 5.2 Measurements of total precipitation, runoff, evapotranspiration, and changes in groundwater and soil water storage for the Catchwater Drain drainage basin for the period 1–12 June 1969. Negative values indicate loss of water from a reservoir.

Variable	Water transfer or storage/mm
precipitation	29
runoff	12
evaporation	25
groundwater storage	6
soil water storage	2

Although such results are encouraging, it is not always possible to establish a satisfactory water balance. There are many reasons for this. These include a lack of understanding of hydrological processes, a lack of reliable data, or modification of the hydrologic system by human activity (irrigation, for example).

Box 5.2 The Plynlimon experimental catchment

Plynlimon is an experimental catchment which was set up in mid-Wales in the late 1960s to assess the impact of land use change in upland areas, from moorland or pasture to coniferous forestry, on streamflow. The experiment was designed to test the hypothesis that planting of coniferous forests in moorland catchments increases evapotranspiration rates and reduces streamflow, and that the reduction in useable water resources results in considerable economic cost to the water industry as well as having water quality and ecological impacts.

A network of instruments was installed in the grassland upper River Wye and the predominantly forested upper River Severn catchments (10.55 and 8.7 km^2, respectively; Figure 5.5).

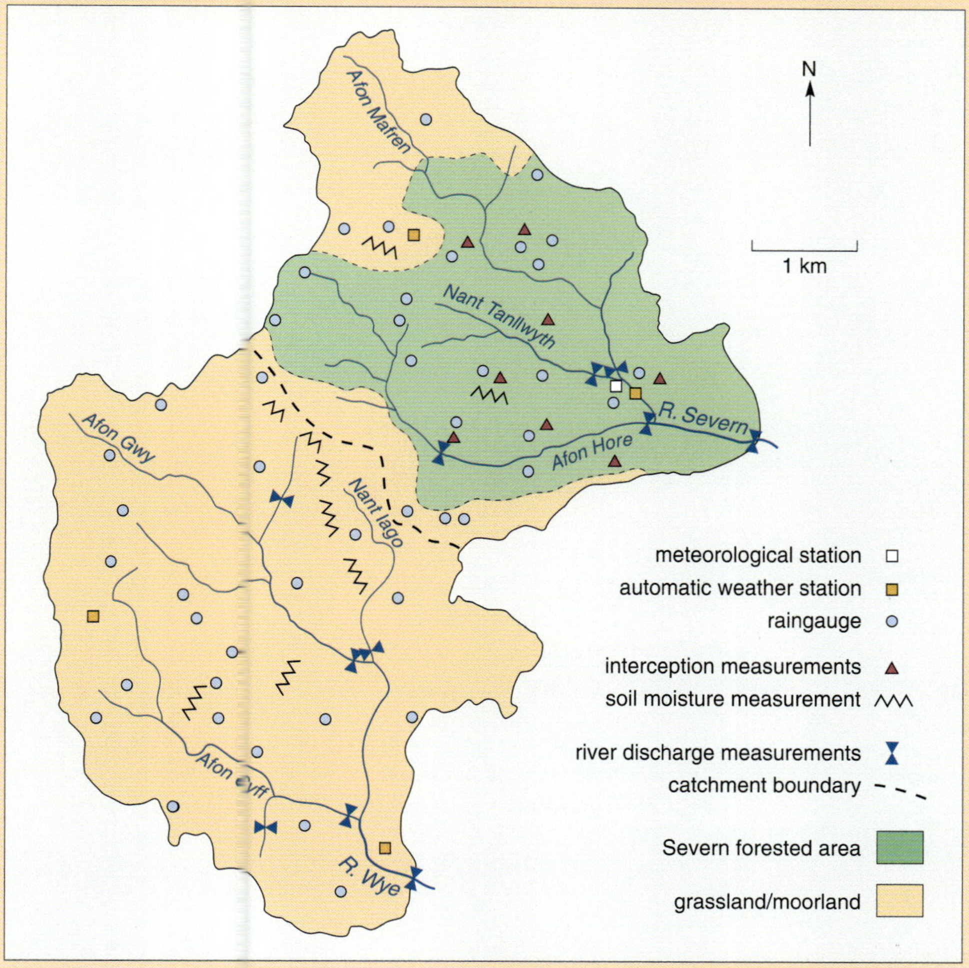

Figure 5.5 Instrumentation of the upland section of the Severn and Wye catchments, Plynlimon, mid-Wales.

❍ Given that the overall change in groundwater and soil storage is insignificantly small relative to precipitation (P) and streamflow (Q) over a long time period (usually more than 3 years), write an equation for the water balance in terms of evapotranspiration.

● The water balance equation is: $P = Q + E - \Delta G - \Delta S$ (Equation 5.3). If ΔG and ΔS are insignificant, then the equation becomes: $P = Q + E$. In terms of evapotranspiration, $E = P - Q$.

Figure 5.6 shows measured values of $P - Q$ for both the Wye and Severn catchments for the period 1972–1995. In 1985, clear felling of economically-mature trees was started in the Severn catchment and the area was replanted a short time later. The diagrams indicate that prior to 1985, $P - Q$ was larger for the Severn (average 632 mm yr^{-1}) than for the Wye (average 491 mm yr^{-1}). However, values of $P - Q$ appear to have fallen in the Severn post-1985 (average 518 mm yr^{-1}). Hence prior to 1985 there was greater evapotranspiration loss from the forested catchment, which concurs with the initial hypothesis. However, closer inspection of the data indicates that evapotranspiration losses from the Severn declined relative to the Wye even before clear felling started. The onset of clear felling reduced evapotranspiration from the Severn to less than the Wye for a few years, but evapotranspiration increased again in the Severn as newly-planted trees began to mature.

Results from Plynlimon indicate that afforestation and water supply may be incompatible where demand is high relative to available sources. However, close inspection of the data suggests that the forest does not maintain its highest rates of evapotranspiration throughout the growth cycle, taking some years to reach maximum rates and then declining as the trees approach their economic zenith.

Studies such as this help to improve understanding of hydrologic processes, and make it possible to develop and calibrate hydrological models which can be applied to other catchments with only minimum data input.

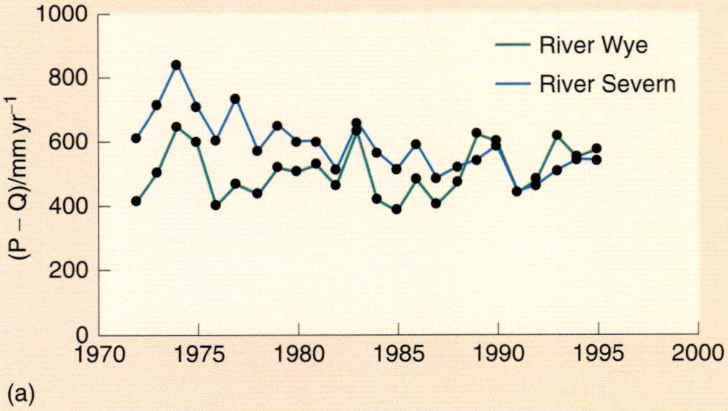

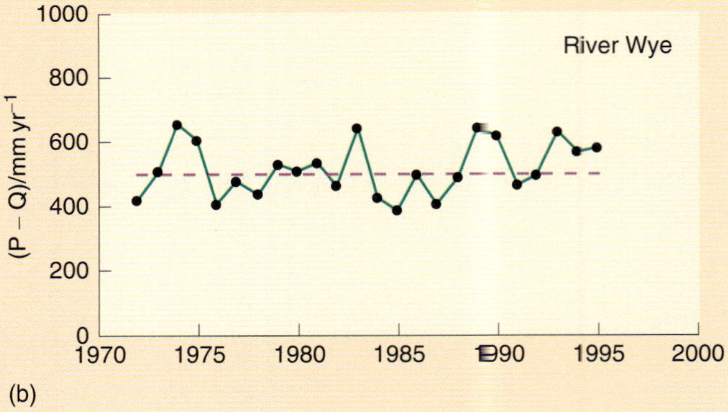

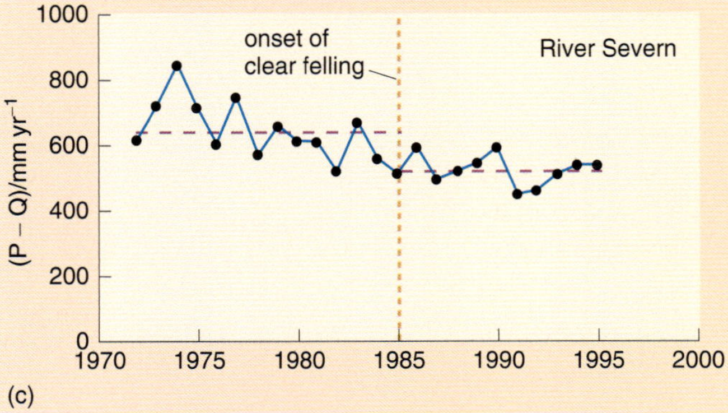

Figure 5.6 Time-series of $P–Q$ measurements for the upper River Wye and upper River Severn catchments. Measurements are shown for both catchments (a), and each catchment separately ((b) and (c)). Horizontal dashed lines denote average values over the indicated time period. The vertical dashed line denotes the onset of clear felling in the Severn catchment.

5.2.2 Modelling hydrological systems

Increasingly, computer models are used to simulate hydrological processes within the drainage basin. Some of these models are complex and require a rigorous understanding of all the relevant physical processes acting within the drainage basin, whereas other models are simpler and consider only those processes that act to produce outputs of water from the drainage basin. These latter, so-called conceptual models are the subject of this section.

Most conceptual models take input (precipitation) data, and transform these data via water movement and storage to streamflow. Water movement is simulated using many of the terms you have already met in this block and in Block 2, Part 2. For example, subsurface flow is usually modelled by Darcy's law, and evapotranspiration is commonly modelled using the Penman–Monteith equation. Parameters within these equations, such as temperature or hydraulic conductivity, can be obtained from meteorological data, soils maps, vegetation indexes and topographical maps. These can be backed up by additional field measurements. The success of the model is usually judged by assessing the empirical fit between the modelled and observed streamflow data. If the fit is poor, the parameters in the individual equations can be modified to produce an 'optimum' fit. Once an acceptable fit is obtained between modelled and observed data over a reasonable time period, the model can then be used to forecast changes in streamflow as a result of future deforestation, for example. You should now test one such model by completing Activity 5.1.

Activity 5.1 A rainfall–runoff model

In this activity you will use a simple model to simulate runoff in different river catchments from a suite of rainfall data. You will be able to compare your modelled data with the actual measured runoff. You will then use the model to predict changes in runoff as a result of greenhouse warming.

The program guides you through an initial exercise which shows you how to use the model. Then you will be asked to feed rainfall data into the model to obtain runoff. You will need to be able to print out data and displays.

5.3 The global picture

In this section, we have mainly focused on hydrological systems at the drainage-basin scale. This is extremely useful, but it is important to recognize that there are many purposes for which the drainage basin is too small a unit of study. For example, some of the emerging problems of environmental change associated with large-scale forest clearance or climate change can only be resolved by better understanding of hydrological processes on the global scale. Such studies require international, as well as interdisciplinary collaboration, and frequently large data networks derived from satellite-based remote-sensing techniques; these studies are driven by the recognition that water is a priceless resource for all inhabitants of Earth.

5.4 Summary of Section 5

1 The hydrological cycle describes the movement of water between the Earth's various reservoirs. The hydrological cycle is driven by the sun's energy, so the hydrological cycle and the energy cycle are intimately linked.

2 The residence time of a water molecule within a reservoir is given by:

$$\text{residence time} = \frac{\text{mass in reservoir}}{\text{flux into or from reservoir}}$$

3 The balance of inputs and outputs of water to a hydrological system, and any resulting change in the mass of water stored within that system, is described by the water balance equation:

inflow = outflow + change in storage

4 Inputs, outputs, and changes in storage of water can be assessed by (a) making field measurements, and (b) hydrological modelling.

Question 5.2

Figure 5.7 shows a box model of a river system. Arrows into the box (which depicts the river) represent addition of water to the river whereas arrows leading from the box represent removal of water from the river. Numbers alongside the arrows represent the mass of water added to, or removed from, the river *over a year-long period*. The total mass of water in the river is 125×10^6 kg.

(a) Assuming that there is no change in the amount of water stored in the river over the year-long period, set up a water balance equation for this river system and use your equation to show whether the inputs and outputs of water to the river are in balance.

(b) Calculate the residence time of water in the river.

(c) In part (a) you assumed that there was no change in the amount of water stored in the river over the year-long period. Is this assumption more likely, or less likely, to be valid for shorter time periods? Give your reasons.

(d) In addition to those illustrated in Figure 5.7, what other outputs and inputs of water are there likely to be to a river system?

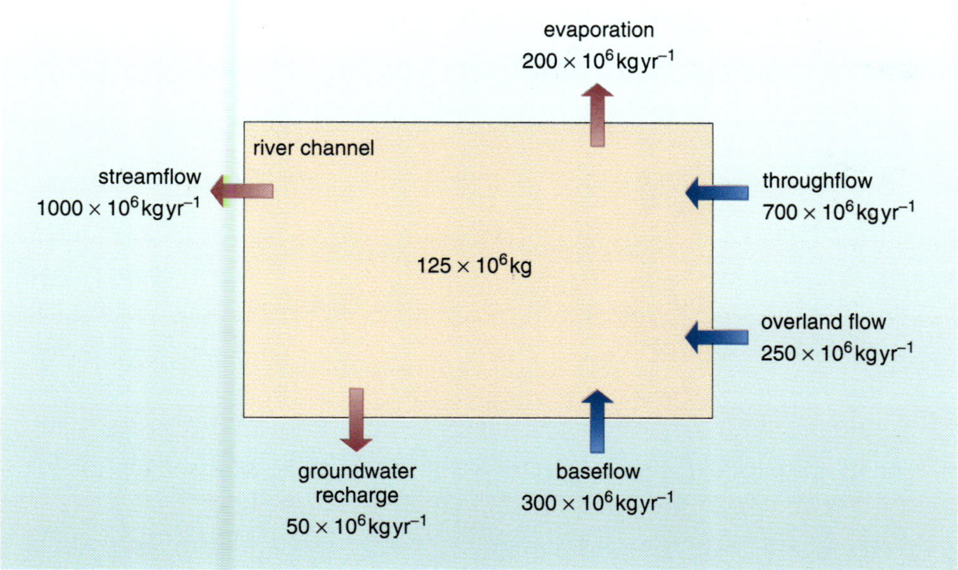

Figure 5.7 Box model representing inputs and outputs of water to a river channel. Input and output fluxes are in kg yr^{-1}; the mass of water in the river channel is 125×10^6 kg.

Learning outcomes for Section 5

When you have completed this section, you should be able to:

5.1 Explain the hydrological cycle in terms of reservoirs of water, and movement of water between them.

5.2 Calculate the residence time of water within a reservoir.

5.3 Demonstrate how to set up a water balance equation in terms of outputs, inputs and storages of water.

5.4 Use a simple hydrological model to both simulate and forecast runoff.

Comments on activities

Activity 3.1

Task 1

Completed Table 3.3 Numbers of parent and daughter atoms during radioactive decay

Number of elapsed half-lives, n	Number of parent atoms, P	Number of daughter atoms, D	D/P
0	1024	0	0
1	512	512	1
2	256	768	3
3	128	896	7
4	64	960	15
5	32	992	31
6	16	1008	63
7	8	1016	127
8	4	1020	255
9	2	1022	511
10	1	1023	1023

Task 2 (a) See Figure 3.50.

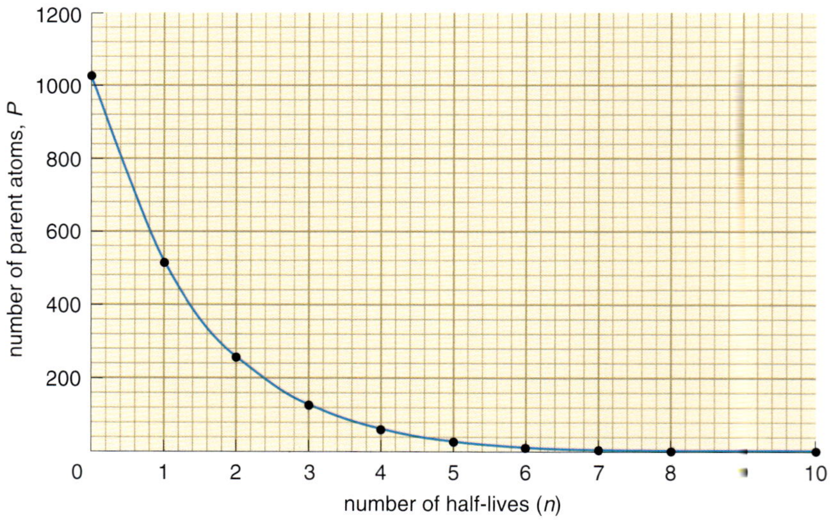

Figure 3.50 Plot of the number of parent atoms, P, versus the number of half-lives, n; data are from the completed Table 3.3 (above).

(b) See Figure 3.51.

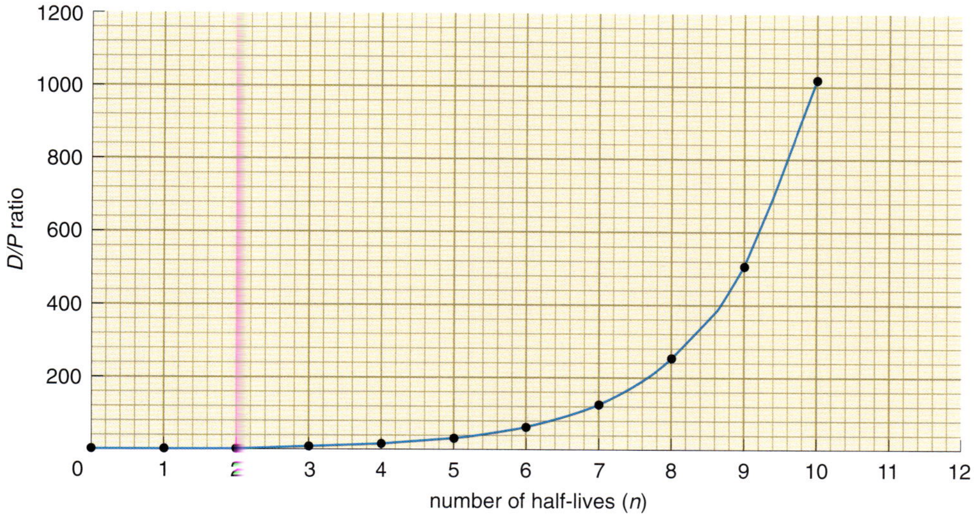

Figure 3.51 Plot of the *D/P* ratio versus the number of half-lives, *n*; data are from the completed Table 3.3 (above).

Activity 3.2

(a) The cation and anion levels in $mmol_c\ l^{-1}$ are:

	Ca^{2+}	Mg^{2+}	Na^+	K^+	Cl^-	SO_4^{2-}	CO_3^{2-}	HCO_3^-
$mmol_c\ l^{-1}$	0.80	0.20	1.90	0.02	1.50	0.18	0.24	1.00

(b) The relative percentage of the cations and anions in $mmol_c\ l^{-1}$ are:

Cations	$mmol_c\ l^{-1}$	% of Total	Anions	$mmol_c\ l^{-1}$	% of Total
Ca^{2+}	0.80	27.4	Cl^-	1.50	51.4
Mg^{2+}	0.20	6.8	SO_4^{2-}	0.18	6.2
$Na^+ + K^+$	1.92	65.8	$CO_3^{2-} + HCO_3^-$	1.24	42.4
Total	2.92	100	Total	2.92	100

(c) See Figure 3.52, which is the completed Figure 3.38.

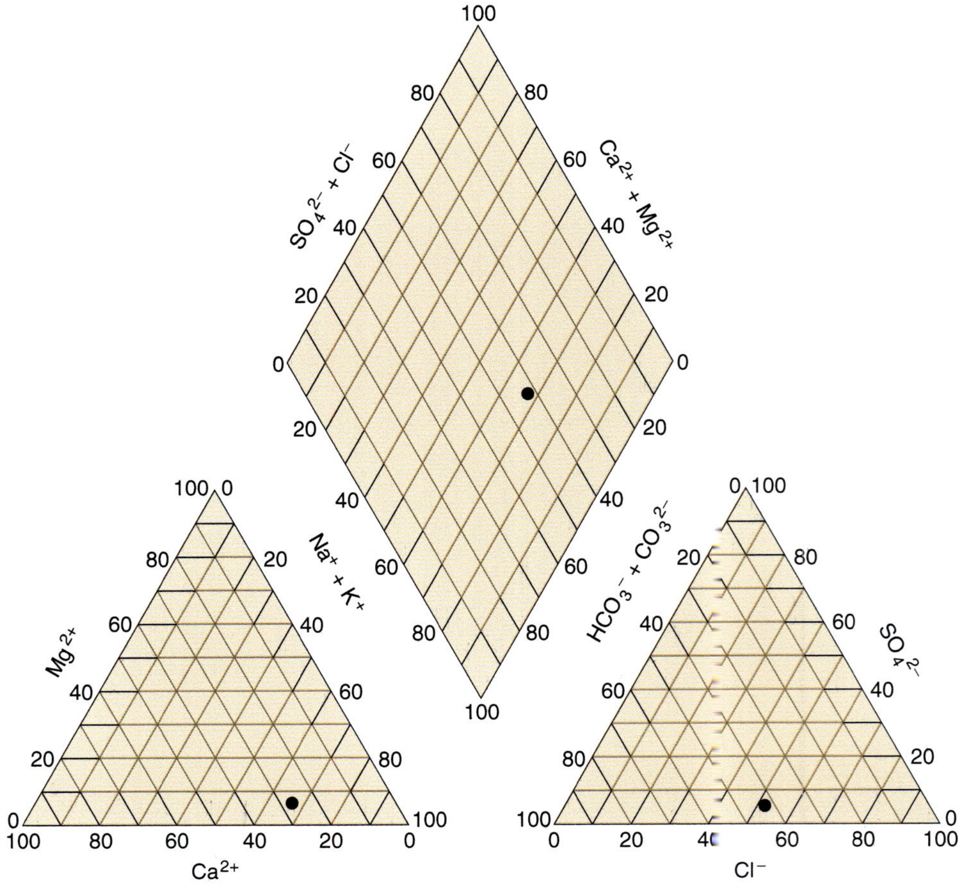

Figure 3.52 Completed Figure 3.38.

Answers to questions

Question 2.1

The total area covered by the raingauge network is 77 km^2. The proportion of the area measured by each gauge is given in the completed Table 2.3 below.

Completed Table 2.3

Raingauge identifier	Rainfall/ mm yr^{-1}	Proportion of total area	Area/ km^2
1	327	0.229	17.6
2	268	0.151	11.6
3	333	0.173	13.3
4	125	0.235	18.1
5	222	0.213	16.4

The mean rainfall over the area of the map = (0.229 × 327 mm) + (0.151 × 268 mm) + (0.173 × 333 mm) + (0.235 × 126 mm) + (0.213 × 222 mm) = 250 mm.

Note that a simple mean of the raingauge readings is 255 mm.

Question 2.2

Wind generally results in an undercatch of rain, an effect accentuated where there are nearby obstructions. The effect can be reduced by using a gauge with a wind deflector. Smaller raindrops are more affected by wind compared with larger ones. With less dense snowflakes, the result is a significant undercatch which can be as high as 50%. Alternative methods are often used for snowfall such as monitoring the pressure change on a bag filled with antifreeze solution or by using gamma probes. The measurement of 6.5 mm rainfall equivalent is likely to be a significant underestimate.

Question 2.3

Rain falling over ocean islands will always have come from over the sea, whereas rain falling over coastal America and Europe may, depending on wind direction, have come from across land and have gathered continental material in the process.

Question 2.4

Mass can be converted into moles by dividing the mass by the relative ionic mass.

For Na^+, 1.32 g l^{-1} corresponds to $\frac{1.32}{23.0}$ mol l^{-1} = 5.74×10^{-2} mol l^{-1}

For Al^{3+}, 0.64 g l^{-1} corresponds to $\frac{0.64}{27.0}$ mol l^{-1} = 2.37×10^{-2} mol l^{-1}

For Cl^-, 2.06 g l^{-1} corresponds to $\frac{2.06}{35.5}$ mol l^{-1} = 5.80×10^{-2} mol l^{-1}

For SO_4^{2-}, 3.38 g l^{-1} corresponds to $\frac{3.38}{96.1}$ mol l^{-1} = 3.52×10^{-2} mol l^{-1}

Question 2.5

Data set A shows a much higher nitrate ion concentration than set B suggesting an urban input. A likely contributor to these enhanced levels is from nitrogen oxides, NO_x, from motor vehicles. In both areas, nitrate is being removed from the canopy as throughfall levels are higher than the concentrations in above-canopy precipitation. However, the urban site shows higher levels in above-canopy precipitation and evidence for nitrate deposition on leaf surfaces.

Question 2.6

Volume of water lost per day = 0.25×10^4 m^2 × 0.47 l m^{-2} d^{-1} = 1.175×10^3 l d^{-1}.

We shall assume that 1 litre of water has a mass of 1 kg. (The relationship is not exact in that there is a small variation with temperature.)

Energy to vaporize 1 kg of water at 25 °C is 2.44×10^6 J.

Energy required for volume of $1.175 \times 10^{3}\,\mathrm{l\,d^{-1}}$

$$= 1.175 \times 10^{3}\,\mathrm{l\,d^{-1}} \times 2.44 \times 10^{6}\,\mathrm{J\,l^{-1}}$$

$$= 2.87 \times 10^{9}\,\mathrm{J\,d^{-1}}$$

This is equivalent to having about 33 one-kilowatt electric fires in operation continuously.

Question 2.7

Although the amounts of rain that fell were identical, the intensity of the rain in case (a) was much higher than the intensity in (b).

As a result, the hydrograph in case (a) would show a high peak discharge and discharge would increase above baseflow over a short period of time. The hydrograph in case (b) would show a relatively low peak discharge, but discharge would remain higher than baseflow for an extended period of time.

Question 2.8

Compare your polygons with those constructed on the map shown in Figure 2.31. The grid helps with the calculation of the areas of the polygons but do note that the question addresses the area of land (excluding the ocean). The data are shown in Table 2.14.

Figure 2.31 For answer to Question 2.8.

Table 2.14 Rainfall data for gauges 1–15.

Raingauge identifier	Land associated with raingauge /km^2	Rainfall depth over one month/mm
1	72.5	48.6
2	63.5	61.1
3	40.0	55.4
4	86.0	63.4
5	87.0	82.9
6	45.5	120.0
7	17.5	117.0
8	35.5	131.0
9	46.5	59.7
10	47.5	43.9
11	56.0	98.6
12	53.0	101.0
13	73.0	48.0
14	29.0	61.3
15	41.5	52.6
total land area	794	

The mean rainfall is calculated in the same way as was done in Question 2.1. The rainfall datum for each gauge is multiplied by the fraction of the land area represented by the appropriate polygon. Then these values for each gauge are added to give the mean rainfall.

So for raingauge 1, the rainfall measured was 48.6 mm and the polygon occupies $\frac{71.6}{790}$ of the land area. Data for all the gauges are shown in Table 2.15.

Table 2.15 Theissen polygon data for raingauges 1–15.

Raingauge identifier	Fraction of land occupied by Thiessen polygon	Contribution to mean rainfall/mm
1	0.0913	4.44
2	0.0800	4.89
3	0.0504	2.79
4	0.108	6.87
5	0.110	9.08
6	0.0573	6.88
7	0.0204	2.58
8	0.0447	5.88
9	0.0586	3.50
10	0.0593	2.62
11	0.0705	6.95
12	0.0668	6.74
13	0.0919	4.41
14	0.0352	2.24
15	0.0523	2.75

The mean rainfall for the month over the land area is 72.6 mm.

Question 3.1

See completed Figure 3.14.

Groundwater is likely to flow more rapidly on the east side of the river as the groundwater elevation contours are closer together.

The tributary in the southwest section of the river is gaining water from groundwater; the water elevation contours form a V-shape pointing upstream as they cross the tributary.

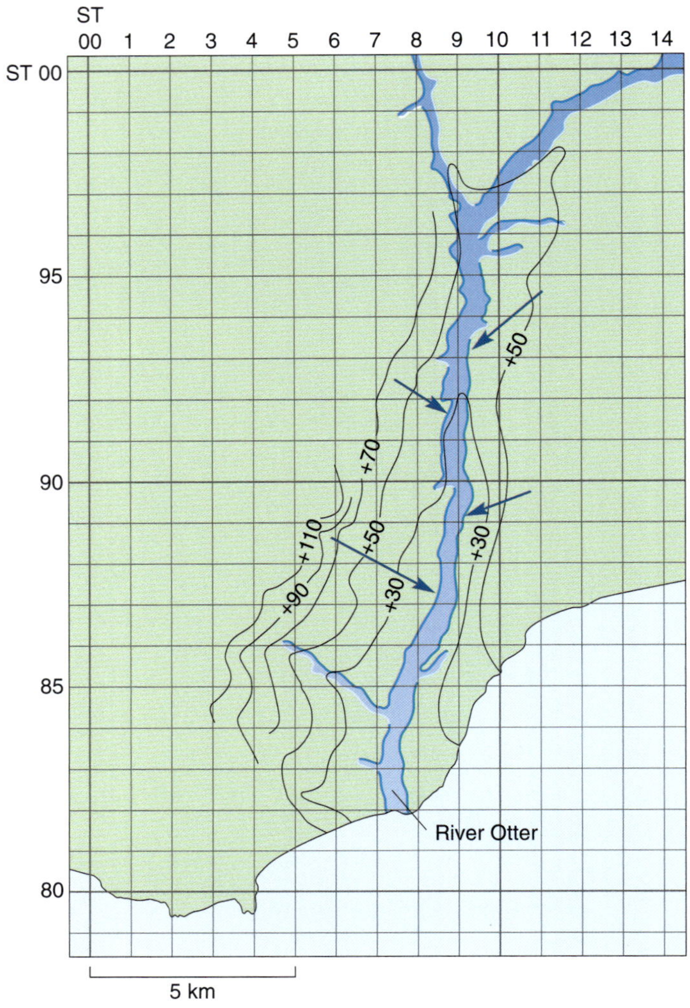

Completed Figure 3.14

Question 3.2

(a) The aquifer is regarded as permeable because its hydraulic conductivity is sufficiently large (aquifers are generally regarded as permeable if $K \geq 1\,\text{m}\,\text{d}^{-1}$.

(b) The fluid flow rate, Q, is given by $Q = v \times b \times w$ (Equation 3.5). Thus $Q = 5 \times 10^{-2}\,\text{m}\,\text{d}^{-1} \times 3\,\text{m} \times 8\,\text{m} = 1.2\,\text{m}^3\,\text{d}^{-1}$.

(c) According to Equation 3.3, $v = K(h/l)$. Rearranging to make h the subject of the equation gives $h = (vl)/K$. Thus $h = (5 \times 10^{-2}\,\text{m}\,\text{d}^{-1} \times 10\,\text{m}) / (1.2\,\text{m}\,\text{d}^{-1}) = 0.42\,\text{m}$.

(d) The transmissivity (T) of the aquifer is given by $T = Kb$. Thus, $T = 1.2\,\text{m}\,\text{d}^{-1} \times 3\,\text{m} = 3.6\,\text{m}^2\,\text{d}^{-1}$.

Question 3.3

The poorly cemented fine sandstone at the top of the sequence has highest porosity because it has the lowest electrical resistivity. The anhydrite layer has lowest porosity as it has the highest resistivity. An increase in the amount of water in a rock increases its ability to conduct electricity, and therefore it has a lower resistivity. As a result, rocks with high porosity will have a lower resistivity than low-porosity rocks.

Question 3.4

(a) In order to convert units of $\text{mmol}\,\text{l}^{-1}$ to units of $\text{mmol}_\text{c}\,\text{l}^{-1}$, the data in Table 3.8 must be multiplied by the number of units of charge on that species.

Table 3.8 (completed) Concentration of chemical species in groundwaters from the London Basin in units of $mmol_c\ l^{-1}$.

	Ca^{2+}	Mg^{2+}	Na^+	K^+	$HCO_3^- + CO_3^{2-}$	SO_4^{2-}	Cl^-
0693	0.28	0.22	6.60	0.10	2.60	2.10	2.50
1171	5.60	0.42	1.21	0.17	6.20	1.00	0.20
1406	4.80	0.40	0.35	0.04	5.20	0.00	0.39
1512	1.00	0.60	2.80	0.26	1.90	2.20	0.56

(b) Relative percentages of cations and anions for groundwaters collected from each site are given below.

Table 3.9 (completed) Relative percentage of $mmol\ l^{-1}$ of cation and anion charge in groundwater.

Site 1171

Cations	$mmol_c\ l^{-1}$	% of Total	Anions	$mmol_c\ l^{-1}$	% of Total
Ca^{2+}	5.60	75.7	Cl^-	0.20	2.7
Mg^{2+}	0.42	5.7	SO_4^{2-}	1.00	13.5
$Na^+ + K^+$	1.38	18.6	$CO_3^{2-} + HCO_3^-$	6.20	83.8
total	7.40	100	total	7.40	100

Site 1406

Cations	$mmol_c\ l^{-1}$	% of Total	Anions	$mmol_c\ l^{-1}$	% of Total
Ca^{2+}	4.80	85.9	Cl^-	0.39	7.0
Mg^{2+}	0.40	7.1	SO_4^{2-}	0.00	0.0
$Na^+ + K^+$	0.39	7.0	$CO_3^{2-} + HCO_3^-$	5.20	93.0
total	5.59	100	total	5.59	100

Site 1512

Cations	$mmol_c\ l^{-1}$	% of Total	Anions	$mmol_c\ l^{-1}$	% of Total
Ca^{2+}	1.00	21.4	Cl^-	0.56	12.0
Mg^{2+}	0.60	12.9	SO_4^{2-}	2.20	47.2
$Na^+ + K^+$	3.06	65.7	$CO_3^{2-} + HCO_3^-$	1.90	40.8
total	4.66	100	total	4.66	100

(c) The completed Piper diagram (Figure 3.45) is given below.

(d) Samples 1406 and 1171 are likely to be from the unconfined part of the aquifer

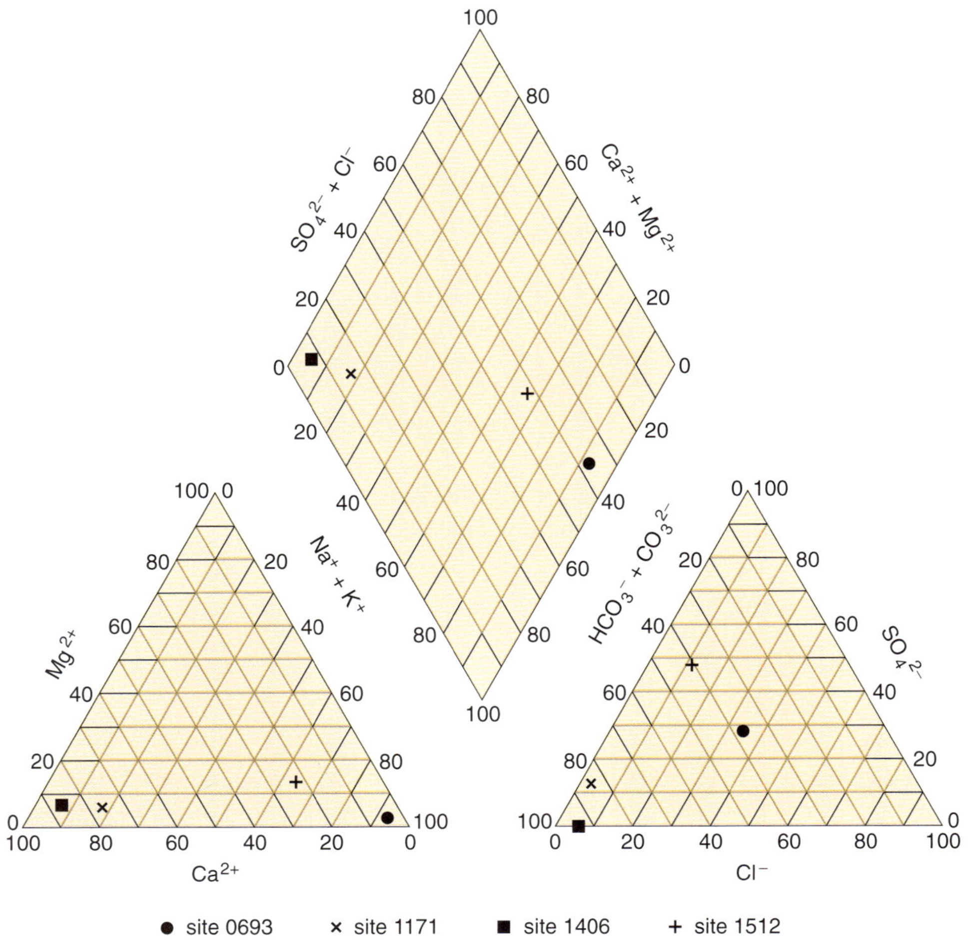

Completed Figure 3.45

as their chemistry is dominated by Ca^{2+} and carbonate species. This reflects simple dissolution of carbonate rocks. Samples 0693 and 1512 are likely to be from the confined part of the aquifer. They have higher levels of sulfate, reflecting reactions with pyrite, higher levels of Na^+ and K^+ due to ion exchange of Ca^{2+} and Mg^{2+} with Na^+ and K^+ located on the surface of clay minerals, and higher levels of Cl^-. In accordance with the Chebotarev sequence, sample 0693 is likely to be the oldest of these two samples because it has slightly lower levels of sulfate (due to its conversion into hydrogen sulfide) and slightly higher levels of Cl^-.

Question 3.5

(a) Groundwater levels are highest between February and June as levels of infiltration are highest in these months in the UK and levels of evaporation are low. Groundwater levels fall throughout the summer as groundwaters are not recharged by precipitation and evaporation is high. Lowest mean groundwater levels are usually recorded in November.

(b) Groundwater levels are close to average for the first six months in 1988, but groundwater levels in the latter part of the year are below average. In early 1989, some of the lowest groundwater levels since measurements began are recorded. Highest groundwater levels in 1989 were about 23 metres below the highest level recorded in the previous year. Groundwater levels are also far lower than average throughout 1990, but they increase to values close to average in the first six months of 1991. In the latter months of 1991, groundwater levels fall to levels that are close to the minimum recorded for that period. As groundwater levels are frequently close to minimum values recorded over the entire measurement period for the period 1988–1991, then they cannot be typical for this aquifer.

Question 3.6

(a) From Equation 3.7:

$$\text{pore space volume} = \frac{\text{mass rock wet} - \text{mass rock dry}}{\text{density of fluid}}$$

$$\text{Therefore pore space volume} = \frac{205\ \text{g} - 200\ \text{g}}{1\ \text{g cm}^{-3}} = 5\ \text{cm}^3$$

for the shale

$$\text{and pore space volume} = \frac{270\ \text{g} - 250\ \text{g}}{1\ \text{g cm}^{-3}} = 20\ \text{cm}^3 \text{ for}$$

the sandstone.

From Equation 3.1:

$$\text{porosity (\%)} = \frac{V_p}{V_{tot}} \times 100$$

Both the sandstone and the shale have a total volume (V_{tot}) of 100 cm^3. Hence:

porosity = 5/100 × 100 = 5% for the shale,

and porosity = 20/100 × 100 = 20% for the sandstone.

(b) Although the individual shale layers may be homogenous, it will be easier for water to flow between the layers, i.e., in the horizontal dimension, than it would be for water to flow through the layers (in the vertical dimension), so the shale is anisotropic. In order for the sandstone to be homogenous, the grain size distribution and degree of cementation must be variable only within small limits. Figure 3.47 shows some degree of grain size variation, so the sandstone is unlikely to be homogeneous. Although the sandstone may have variable permeability on a small-scale (e.g. over short distances), over longer distances these will be smoothed out so the sandstone is probably isotropic on a larger scale.

(c) Although the sandstone has highest porosity, the pores rarely link up so it is unlikely to be highly permeable. In addition, the photograph does not show any evidence for cracks or fissures in the rock. Although the shale is not very porous, groundwater is likely to be able to flow between the individual layers, so it is probably more permeable than the sandstone.

(d) Darcy's law assumes that the rock is homogenous and isotropic. As discussed above, the shale is not isotropic because it is more permeable in the horizontal direction than it is in the vertical direction. For this reason, there may be difficulties in applying Darcy's law to the study of water movement within shale.

Question 3.7

(a) The average hydraulic gradient is given by:

$$\frac{\text{change in height of water-table}}{\text{distance between contaminant source and borehole}} = \frac{15\ \text{m}}{5000\ \text{m}}$$

= 0.003

(b) The speed at which the pollutant is travelling towards the borehole is given by Equation 3.3:

speed = K × hydraulic gradient = 30 m d^{-1} × 0.003 = 0.09 m d^{-1}.

$$\text{Hence it will take } \frac{5000\ \text{m}}{0.09\ \text{m d}^{-1}} = 56\,000 \text{ days,}$$

or about 150 years for the contaminant to reach the borehole. Note that this assumes that the contaminant is travelling at the same speed as the water molecule; this is not necessarily true because effects such as adsorption may mean that the pollutant travels more slowly.

(c) For the fissured limestone, the speed at which groundwater travels is:

5000 m day^{-1} × 0.003 = 15 m d^{-1}.

$$\text{Therefore it will take } \frac{5000\ \text{m}}{15\ \text{m d}^{-1}} = 330\ \text{days}$$

(just less than a year) for the contaminant to reach the borehole.

(d) If the pollutant decomposes into a harmless by-product within five years, then it will not pose a threat to the water supply in the sandstone aquifer as it will have decomposed by the time it reaches the borehole. However, it will not have time to decompose before it reaches the borehole in the limestone aquifer, so it could contaminate the drinking water supply.

Question 3.8

(a) The total number of atoms must always be equal to P_0, i.e. 0.88 Bq m^{-3} units. Thus a sample with 0.75 Bq m^{-3} units ^{85}Kr must have 0.88–0.75 = 0.13

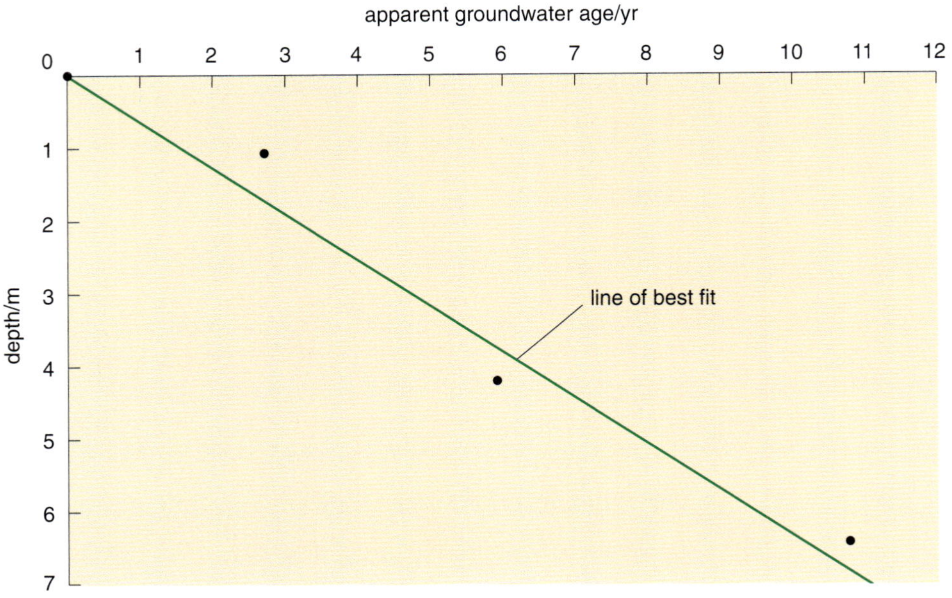

Figure 3.49 Answer to Question 3.3 part (a).

daughter units, hence $D/P = 0.13/0.75 = 0.17$. The remainder of the D/P values given in the completed Table 3.12 are calculated in the same way.

(b) The number of elapsed half-lives, n, corresponding to each D/P value can be read from Figure 3.28. The age of the groundwater samples is $n \times 10.76$ yr (the half-life of ^{85}Kr).

Completed Table 3.12

Depth/m	^{85}Kr /Bq m^{-3}	D/P	No. of elapsed half-lives, n	Age/yr
0 (soil gas/ atmosphere)	0.88	0	0	0
1.1	0.75	0.17	0.25	2.7
4.2	0.60	0.47	0.55	5.9
6.4	0.45	0.96	1	10.8

(c) The age of the groundwater increases approximately linearly with depth (see line of best fit on Figure 3.49). This means that there is unlikely to be any major change in lithology over this depth interval, as this would almost certainly change the hydraulic properties of the aquifer (e.g. change in permeability, hydraulic conductivity). This would affect the speed of fluid flow through the aquifer, and would lead to a non-linear relationship between fluid age and depth.

Question 4.1

We need to use Equation 4.1

$$Re = \frac{vd}{\nu}$$

At 30 °C, $\nu = 0.798 \times 10^{-2}\,\text{cm}^2\,\text{s}^{-1}$. Rearranging the equation and putting $Re = 500$ at the laminar/turbulent boundary criterion,

$$d = \frac{500 \times 0.798 \times 10^{-2}\,\text{cm}^2\,\text{s}^{-1}}{3.2\,\text{cm}\,\text{s}^{-1}}$$

Question 4.2

For the first river,

area of cross-section = 12.0 m × 3.20 m = 38.4 m^2

wetted perimeter = 3.20 m + 12.0 m + 3.20 m = 18.4 m

$$\text{hydraulic radius} = \frac{38.4\,\text{m}^2}{18.4\,\text{m}} = 2.09\,\text{m}$$

For the second river,

area of cross-section = 20.0 m × 1.92 m = 38.4 m^2

wetted perimeter = 1.92m + 20.0 m + 1.92 m = 23.8 m

$$\text{hydraulic radius} = \frac{38.4\,\text{m}^2}{23.8\,\text{m}} = 1.61\,\text{m} = 1.61\,\text{m}$$

Again note that, although both rivers have the same cross-sectional area, the hydraulic radii are quite different.

Question 4.3

To solve the Manning equation, we need values for the hydraulic radius of the stream and the appropriate Manning coefficient.

The stream is quite small (width 2.42 m) and weed clogged so the Manning coefficient will be in the range 0.075–0.150 $\text{m}^{-\frac{1}{3}}$ s. We could use as our value the mean of the extremes of the range, $n = 0.1125\ \text{m}^{-\frac{1}{3}}\ \text{s}$.

To calculate the hydraulic radius R, we need the

area of cross-section =

$$\tfrac{1}{2}\pi r^2 = \tfrac{1}{2}\pi \times (1.21\ \text{m})^2 = 2.30\ \text{m}^2$$

wetted perimeter $= \pi r = \pi \times 1.21\ \text{m} = 3.80\ \text{m}$

$$\text{hydraulic radius} = \frac{2.30\ \text{m}^2}{3.80\ \text{m}} = 0.605\ \text{m}$$

The Manning equation gives

$$\text{speed of water, } v = \frac{(0.605\ \text{m})^{\frac{2}{3}} \times (1.22 \times 10^{-3})^{\frac{1}{2}}}{1.125 \times 10^{-1}\ \text{m}^{-\frac{1}{3}}\ \text{s}}$$

$$= 0.222\ \text{m s}^{-1}$$

Question 4.4

Figure 4.19 shows that the hydrogen carbonate anion represents about 66% of the total dissolved inorganic carbon.

The mass of carbon is:

$$2.0 \times 10^{-4}\ \text{g l}^{-1} \times 0.66 = 1.3 \times 10^{-4}\ \text{g l}^{-1}$$

This represents: $1.1 \times 10^{-5}\ \text{mol l}^{-1}$

As one mole of the hydrogen carbonate anion contains one mole of carbon, the molar concentration of hydrogen carbonate is $1.1 \times 10^{-5}\ \text{mol l}^{-1}$.

Question 4.5

Sample B has a higher Ω value, so it contains the higher concentration of dissolved carbon dioxide. Both samples, however have lower dissolved carbon dioxide concentrations than water that is in equilibrium with the atmosphere (as Ω is less than 1).

Question 5.1

(a) Sources of water to the groundwater reservoir are percolation through soil and recharge from water stored in rivers. (b) Groundwater is lost by capillary rise to soils, and as base flow to rivers.

Question 5.2

(a) If there is no change in the amount of water stored in the river, then the input of water to the river must be equivalent to the output of water from the river. Hence:

throughflow + overland flow + baseflow = groundwater recharge + streamflow + evaporation

or: throughflow + overland flow + baseflow − groundwater recharge − streamflow − evaporation = 0

For the river, over the year long period:

inputs $= (700 \times 10^6) + (250 \times 10^6) + (300 \times 10^6) = 1250 \times 10^6$ kg

and outputs $= (1000 \times 10^6) - (200 \times 10^6) - (50 \times 10^6) = 1250 \times 10^6$ kg

Thus over this period, the input of water to the river does balance the output of water from the river.

(b) The residence time of water in the river is given by:

$$\frac{\text{mass of water in river}}{\text{rate of transfer of water to or from the river}}$$

$$= \frac{125 \times 10^6\ \text{kg}}{(700 \times 10^6) + (250 \times 10^6) + (300 \times 10^6)\ \text{kg yr}^{-1}} \text{ or}$$

$$= \frac{125 \times 10^6\ \text{kg}}{(1000 \times 10^6) + (200 \times 10^6) + (50 \times 10^6)\ \text{kg yr}^{-1}}$$

= 0.1 years (or 5.2 weeks)

(c) The assumption that there is no change in water storage in the river is less likely to be valid for time periods shorter than one year. This is because there is likely to be significant variation in water storage throughout the year, with overall loss of water from the river in the summer months due to excess evaporation over precipitation, and overall addition of water to the river in the winter because precipitation exceeds evaporation.

(d) Other inputs of water to the river system could be direct channel precipitation, other outputs of water include flooding, and use for water supply or irrigation.

Acknowledgements

Grateful acknowledgement is made to the following sources for permission to reproduce material in this book:

Cover illustration: © Mike Dodd/Open University; *Figure 1.1*: © NASA; *Figure 1.2*: © All rights reserved, Beagle2; *Figure 1.3*: © Reuters; *Figure 1.4, Figure 2.18, Figure 4.3, Figure 4.20*: Reproduced from the Ordnance Survey 1:1 500 000 map with the permission of Ordnance Survey on behalf of The Controller of Her Majesty's Stationery Office, © Crown copyright, licence number ED 100020607; *Figure 1.5, Figure 1.6*: Adapted from 2000 EUMETSAT satellite image; *Figure 2.3, 2.31*: Reproduced from the Ordnance Survey 1:1 200 000 map with the permission of Ordnance Survey on behalf of The Controller of Her Majesty's Stationery Office, © Crown copyright, licence number ED 100020607; *Figure 2.4, 4.2, 4.6*: © Stuart Bennett/Open University; *Figure 2.15*: Andrew Syred/Science Photo Library; *Figure 3.2, Figure 3.3, Figure 3.12, Figure 3.46*: Courtesy of the Centre for Ecology and Hydrology © NERC; *Figure 3.4a*: Reproduced by permission of the Director, British Geological Survey: NERC copyright reserved; *Figure 3.4b*: © Peter Skelton/Open University; *Figure 3.6*: Reproduced from the Ordnance Survey 1:3 000 000 map with the permission of Ordnance Survey on behalf of The Controller of Her Majesty's Stationery Office, © Crown copyright, licence number ED 100020607; *Figure 3.7*: © Geoscience Features Picture Library; *Figure 3.8*: © Kevin Church/Open University; *Figure 3.10, Figure 4.22*: Sinclair Stammers/Science Photo Library; *Figure 3.14*: Adapted from Hydrogeological map of the permo-trias and associated minor aquifers of South-West Endland, Institute of Geological Science and South-West Water Authority, Scale 1:100 000, 1982. *Figure 3.33*:© Geomicrobiology Laboratory, Earth Sciences Department, University of Bristol; *Figure 3.37*: © Courtesy of Victoria Art Gallery, Bath; *Figure 3.40*: © The Environment Agency 2000; *Figure 3.44*: Reproduced from the Ordnance Survey 1:1 250 000 map with the permission of Ordnance Survey on behalf of The Controller of Her Majesty's Stationery Office, © Crown copyright, licence number ED 100020607; *Figure 3.47*: © Andy Tindle/Open University; *Figure 4.12*: © Colin Neal/Centre for Ecology and Hydrology; *Figure 5.2*: © Mark Sephton/Open University; *Figure 5.5*: Adapted from catchment map in 'Land use and water issues in the Uplands: The Plynlimon Study', Institute of Hydrology 1999.

The figures listed below have been adapted from the following sources:

Figure 2.7: Figure adapted from Schlesinger W. H., 'Biogeochemistry: An analysis of Global Change', 2nd edn, Academic Press Publ (1991) p. 84, Figure 3.15; *Figure 2.16*: Adapted from Fetter, C. W., 'Applied Hydrogeology', 3rd edn, Prentice-Hall Inc Publ., p. 34, Figure 2.3; *Figure 2.27*: Adapted from Fetter, C. W., 'Applied Hydrogeology', 3rd edn, Prentice-Hall Inc Publ., p. 69, Figure 3.17; *Figure 3.24*: Cook, P. G., Solomon, D. K., *Journal of Hydrology*, **191** (1997), p. 254, Figure 4; *Figure 3.25*: Cook, P. G., Solomon, D. K., *Journal of Hydrology*, **191** (1997), p. 255, Figure 5.

Every effort has been made to trace all the copyright owners, but if any has been inadvertently overlooked, the publishers will be pleased to make the necessary arrangements at the first opportunity.

PART 2

LIFE

David Gowing and Hilary Denny

1 Introduction

Living organisms are systems that:

- require a constant input of energy;
- can grow and reproduce;
- can respond to their environment.

They are found spread throughout the atmosphere, oceans, land surfaces and even at considerable depths within the Earth's crust. The range of environments they inhabit is known as the **biosphere**. This part of Block 3 focuses on how the various types of living organism are distributed within the biosphere, and on the environmental factors that influence their performance. The aim is to introduce you to the concepts and processes by which we can interpret the distributions of organisms and by which we can predict how a change in the environment may impact on them.

There is a vast array of different types of living organism. This variety is often referred to as **biodiversity**. An understanding of environmental science is necessary if we are to conserve our current stock of biodiversity and safeguard it for the future.

In order to embark on this rather imposing task, we return to the Teign catchment to begin in familiar surroundings. A number of general patterns and processes are described such that in Blocks 5–7 they can be applied to very different environments indeed.

Vegetation patterns

2

As we saw in Block 1, the vegetation of Britain is very varied, and different parts of the land are dominated by different types of vegetation: woodland, moors, heaths, pastures and fields full of arable crops. Why is there so much variety and why are there regional variations? The sections that follow attempt to answer these questions — Sections 2.1–2.3 focus on observing and describing the vegetation and Sections 2.4–2.6 consider the factors that influence the distribution of vegetation.

2.1 Example: vegetation of the Teign catchment

It has been suggested that the natural vegetation of most of Britain is 'temperate deciduous forest.' However, when you went on your virtual field trip to the Teign catchment, you visited a number of different habitats, and examined a wide variety of organisms.

Recall the different types of vegetation you encountered on the field trip, and how particular species tended to be associated with particular types.

You should remember that the distribution of species is not uniform and that certain groups of organisms tend to occur together, forming characteristic assemblages. Sometimes the boundaries between the different assemblages are very sharp, but at other times they are more gradual. Some of the assemblages are composed of many species arranged in an elaborate three-dimensional structure, while others have less complex structures containing fewer species.

Figure 2.1 Freshwater shrimp (*Gammarus pulex*), a common invertebrate animal of streams in Britain.

A particular example of a characteristic assemblage of species is called a **community** by biologists, which is defined as all the organisms that occur in a particular location at the same time. The word is often used to refer to organisms of a particular kind, such as the plant community on a heath or the insect community inside a cowpat. Communities are made up of a number of populations of different species. A **population** is defined as all the members of the *same* species that occur in the same place at the same time, like the freshwater shrimps (*Gammarus pulex*, Figure 2.1) in a stream or the oak trees (*Quercus* spp.) in a wood (Figure 2.2). The boundary between two communities is called an **ecotone**. However, ecotones do not have to be abrupt; they can also be more diffuse, where one community grades into another.

Figure 2.2 An oak tree (*Quercus robur*), common throughout Britain both in woodlands and hedgerows.

2.2 Importance of vegetation in the environment

When we look about us we see different communities of plants and animals that fall into recognizable classes with predictable characteristics, and you have now met quite a few of them, e.g. broadleaf woodland, unimproved grassland, moorland, etc. However, the animals are just as characteristic of the communities as the plants. An experienced naturalist knows that you should go to an upland

(a)

(b)

Figure 2.3 (a) Wheatear (*Oenanthe oenanthe*), a bird of upland moors. (b) Dipper (*Cinclus cinclus*), a bird of riverside habitats.

Figure 2.4 White admiral butterfly (*Limenitis camilla*), which lives predominantly in woodland.

site, such as Dartmoor, to see a wheatear (*Oenanthe oenanthe*, Figure 2.3a), you should go to the Teign Valley itself and make your way along the riverbank if you want to see a dipper (*Cinclus cinclus*, Figure 2.3b) but, if it's a white admiral butterfly (*Limenitis camilla*, Figure 2.4) you want to see, you would be better off going to Yarner Wood. So why are the communities named in terms of their dominant or characteristic plant species?

Putting it simply, the answer is that plants underpin all other terrestrial life forms, both through their role as food producers (Section 2.2.1) and through their role as providers of the structural environment of habitats (Section 2.2.2). Much of this part of Block 3 will therefore focus on vegetation and the plants that comprise it.

2.2.1 Plants as primary producers

Plants are fundamentally different from animals and fungi, in that they are capable of building their tissues using light energy from the Sun and simple inorganic substances from the environment, such as water, carbon dioxide and mineral salts. Animals cannot do this: they have to obtain their energy and materials for building their bodies in a ready-to-use, prefabricated form, i.e. they have to get them by consuming other living things, either plants or other animals. Plants are termed **autotrophs** (from the Greek *auto* meaning 'self' and *troph* meaning 'feed'). In contrast, animals are termed **heterotrophs** (*hetero*, from the Greek for 'different'). Therefore, all animals ultimately depend on plants or other autotrophs for the energy and materials with which to build and maintain their bodies.

The fundamental process that plants can carry out, which animals cannot, is called **photosynthesis**. Plants take carbon dioxide (CO_2), water (H_2O) and light energy from the Sun and, through a series of chemical reactions, turn them into the sugar glucose ($C_6H_{12}O_6$). Oxygen gas (O_2) is produced as a by-product of photosynthesis. The process is summarized by the equation:

$$\text{light energy} + \text{carbon dioxide} + \text{water} \longrightarrow \text{glucose} + \text{oxygen}$$

which is summarized in chemical notation as:

$$\text{light energy} + 6CO_2(g) + 6H_2O(l) = C_6H_{12}O_6(s) + 6O_2(g) \qquad (2.1)$$

The glucose can then undergo a multitude of chemical reactions in which it can be combined with other elements such as nitrogen (N), sulfur (S), phosphorus (P) and iron (Fe) and/or converted into many different organic (i.e. carbon-based) substances, such as proteins, fats, and storage carbohydrates such as starch.

The carbon dioxide is usually obtained from the air (aquatic plants obtain it in the form of hydrogen carbonate, HCO_3^-, from the water); water is obtained from the soil, normally by the roots (aquatic plants and simple plants like mosses absorb it over their entire surface). Light is harvested by the special green pigment, **chlorophyll**, which is present in the leaves and young stems. All the other elements, such as nitrogen, phosphorus, sulfur and iron, are obtained from the soil solution by the roots. The growth of plants that results from these processes is termed **primary production**, in that it is the first step in the production of complex, living matter from the simple inorganic environment. The animal tissue that results from the consumption of plants is called **secondary production**. Almost all animals (and most microbes) ultimately depend on photosynthetic organisms for the energy and materials to build and maintain their bodies. Animals also depend on plants for another essential resource.

- ❍ What is the other product of photosynthesis besides energy-rich sugar?
- ● Oxygen.

Virtually all of the oxygen in the Earth's atmosphere is a product of photosynthesis. Without plants to constantly remove carbon dioxide and renew the supply of oxygen in the atmosphere, most other forms of life would eventually cease. In this way, it can be seen that plants are essential for the existence of life as we know it on Earth.

- ❍ Why do all multicellular organisms require oxygen?
- ● They use oxygen in respiration.

In common parlance, the term respiration is used to imply breathing, but in biology **respiration** is more precisely used to describe the chemical reactions that release energy and carbon dioxide from simple carbohydrates such as glucose. It is *aerobic* respiration, which is carried out by all the multicellular organisms you are likely to encounter, that uses oxygen, and this process can be summarized as:

$$C_6H_{12}O_6(s) + 6O_2(g) = 6CO_2(g) + 6H_2O(g) + \text{energy} \qquad (2.2)$$

- ❍ How does this equation compare with the one for photosynthesis (Equation 2.1)?
- ● It is very similar to that for photosynthesis, except the equation operates in the opposite direction

Some organisms (mainly microbes) are capable of respiration without using oxygen (anaerobic respiration), but all plants, most animals, and many microbes use aerobic respiration to generate energy for reactions that maintain their structures and synthesize new substances. This subject is dealt with in more detail in Sections 3.1–3.3.

2.2.2 Primary production and habitat structure

As we have seen, plants exist in a multitude of different shapes and sizes: just think of the contrasts there are between a mighty oak tree (Figure 2.2), a sinuous, clinging ivy plant (*Hedera helix*, Figure 2.5) and a single tuft of grass.

Figure 2.5 An ivy plant (*Hedera helix*) using a tree to support its growth.

Basic structures, such as leaves and petals, have evolved into a staggering variety of different forms and arrangements (in fact petals are just a fancy kind of leaf). Furthermore, when you look at different plant communities, you can see that the plants are arranged in a variety of different ways, each characteristic of the particular community. Why should this be so?

The short answer is so that the available resources of light, moisture and soil nutrients are shared out and used efficiently. This should not be taken to mean that there is any degree of altruism involved. It is just that, if there are resources available, something will usually evolve that makes use of them. Systems that are inherently wasteful, in that resources go unused or are lost, tend to be unsustainable and ultimately disappear.

Case study: A temperate wood

The overwhelming impression one gets on entering a wood is of vertical height. Large trees such as oak, lime and ash are very dominant. Their canopies of leaves cast a lot of shade, and the massive trunks and branches are physically impressive. However, if one looks more carefully, it is soon apparent that there are many other plants that make up the woodland community. There are smaller trees and shrubs, which are particularly prominent around the edges of clearings and the margins of the wood.

On the floor of the wood, many small flowering plants such as bluebells (*Hyacinthoides non-scriptus*, Figure 2.6a), wood anemones (*Anemone nemorosa*, Figure 2.6b) and wood sorrel (*Oxalis acetosella*) abound. If you get down and look very closely, you find that these plants are underlain by a further layer of plants, such as mosses. In other words, woods exhibit strong vertical stratification (Figure 2.7a).

The largest trees, e.g. oak (*Quercus* spp.), lime (*Tilia* spp.) and ash (*Fraxinus excelsior*), form the **canopy** or uppermost layer. Below this is a layer of small trees, e.g. holly, and immature trees. The third layer is the **shrub layer**, which includes species such as blackthorn (*Prunus spinosa*) and hawthorn (*Crataegus* spp.), then comes the **herb layer** of small flowering plants and finally the **ground layer** comprising mosses and liverworts — well not quite finally, because the **climbers**, such as honeysuckle and old man's beard, use the other woody plants for support and drape themselves through and over all the layers.

(a)

(b)

Figure 2.6 (a) Bluebells (*Hyacinthoides non-scriptus*) and (b) wood anemones (*Anemone nemorosa*), common members of the ground flora in older woodlands.

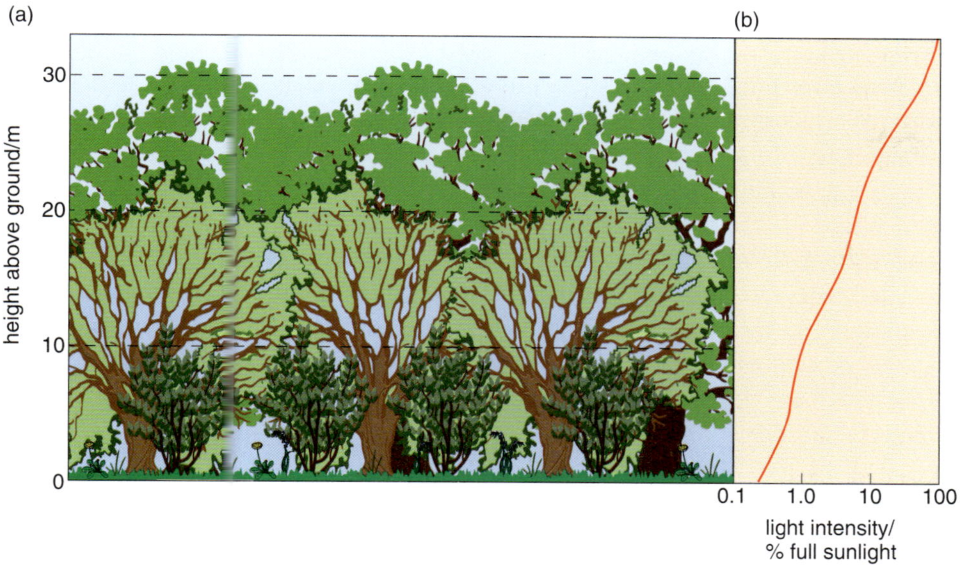

Figure 2.7 Stratification and light extinction in a wood. (a) Different species of trees, shrubs, and herbs bear foliage at different heights above the ground. (b) Different light intensities result from the absorption of sunlight by the foliage.

❍ The availability of which essential raw material for primary production: light, carbon dioxide or water, varies most strongly through the vertical layers of a wood?

● Light.

The leaves and branch surfaces of the canopy trees may absorb and scatter more than half of the sunlight energy (Figure 2.7b). The intensity of light reaching the shrub layer is likely to be a small fraction of that at the canopy. Shrub layer species are adapted to photosynthesize using this weaker light within the forest.

❍ Can you think of other habitats in which the organisms are likely to be stratified vertically and which vary greatly in terms of the amount of light they receive?

● Major examples are aquatic habitats such as deep lakes and the sea.

Plants inhabiting the various layers of a wood (or water body) are adapted to function most efficiently under the conditions they experience there. For woodland herbs such as wood sorrel (Figure 2.8a), **shade tolerance** affects many aspects of the plant's biology.

Shade-tolerant plants usually have intrinsically slow growth rates and can maximize photosynthesis at low light intensities. Some of the woodland plants are evergreen, e.g. holly (*Ilex aquifolium*, Figure 2.8b). Features like these enable the plants to make best use of the light available during the parts of the year when the canopy trees are leafless. Evergreen leaves can photosynthesize slowly throughout the winter, and bulbs allow flowering plants like bluebells to get a quick start at the first signs of spring. Bluebells (Figure 2.6a) have usually flowered, set seed and their leaves have started to wither away, before the leaves of the canopy trees are fully formed. This example highlights the central importance of the form and location of overwintering structures to plant adaptations to particular environments.

(a)

(b)

Figure 2.8 Typical inhabitants of the lower stories of mature deciduous woodland. (a) Wood sorrel (*Oxalis acetosella*) is adapted to grow in the herb layer at low light intensities. (b) Holly (*Ilex aquifolium*) is usually found in the shrub layer and has evergreen leaves.

2.2.3 Classification of plant forms

Botanists have long recognized the importance of plant form. It is primarily by form and structure that living things are classified, evolutionary relationships are deduced and, of most relevance in this case, adaptations to environment are recognized. A Danish botanist, C. Raunkiaer, used a single principal characteristic — the relation of the perennating tissue to the ground surface — to classify plant growth patterns (**life-forms**). **Perennating tissue** is that which is **meristematic**, i.e. capable of growth in terms of cell division and subsequent elongation, but which remains inactive during the winter or dry season, and then resumes growth when favourable conditions return. Perennating tissues include **buds**, which are condensed, embryonic shoots in a protective covering of bud scales (Figure 2.9).

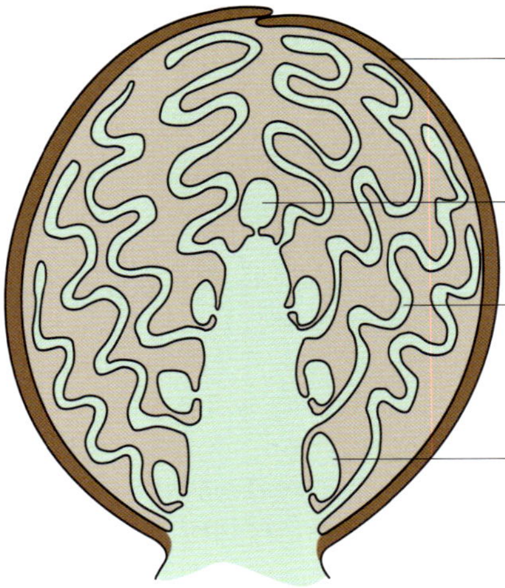

Figure 2.9 The structure of a leaf bud. The protective bud scales enclose and protect the miniature shoot, complete with unexpanded leaves and branches.

Since the perennating tissue makes it possible for the plant to survive periods of unfavourable conditions, such as low light or extremes of climate, its location is a central feature of the plant's adaptation to its environment. The harsher the conditions, the less likely it is that plants have buds far above the ground surface, where they are fully exposed to the cold or drying power of the atmosphere. Five major types of life-forms in land plants are defined by the position of the perennating tissues (Figure 2.10).

Phanerophytes (*phanero* = exposed, *phyte* = plant) are woody plants that have their buds well above the ground surface, fully exposed to the atmosphere. Phanerophytes include trees and shrubs that are predominantly taller than an arbitrary minimum height of 25 cm, and those plants that grow upon them, such as the climber honeysuckle (*Lonicera periclymenum*) and the fern (*Polypodium interjectum*), which often grows on the trunks and branches of established trees as well as on the ground (Figure 2.11).

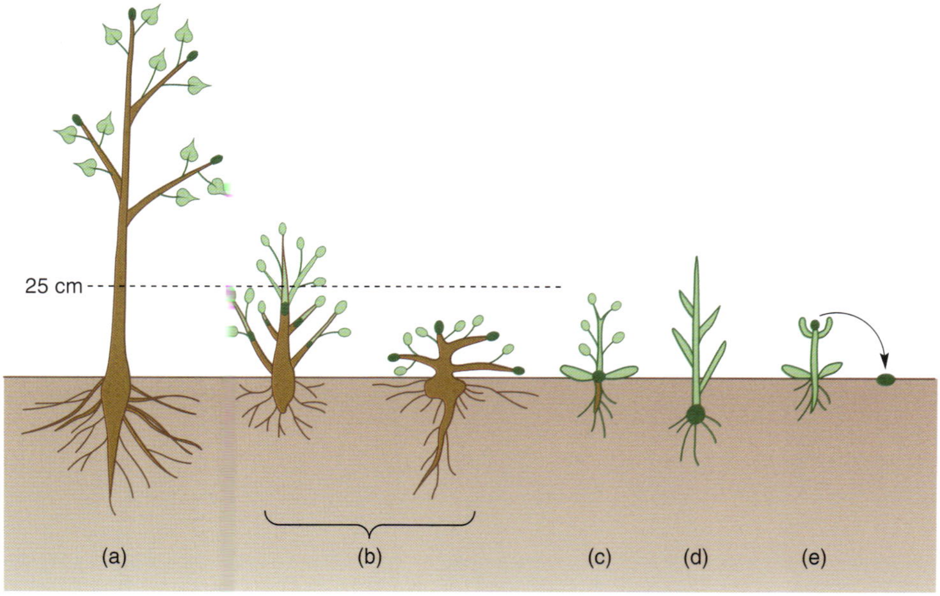

Figure 2.10 Raunkiaer's plant life-forms: perennating tissues are shown in dark green, woody tissues in brown, and deciduous tissues pale green. (a) Phanerophyte, tree or tall shrub with buds more than 25 cm above the ground. (b) Chamaephyte, low shrubs with buds less than 25 cm above the ground. (c) Hemicryptophyte, perennial herb with its bud at the ground surface. (d) Geophyte, perennial herb with a bulb or other perennating organ below the ground surface. (e) Therophyte, annual plant surviving unfavourable periods only as a seed.

(a)

(b)

Figure 2.11 Phanerophytes that use other plants for support: (a) honeysuckle (*Lonicera periclymenum*), which is a climber; (b) the fern, western polypode (*Polypodium interjectum*), which often grows on trees, rooting in the leaf litter and debris accumulated on the surface of branches.

Chamaephytes are plants with their buds above ground but predominantly below the key height of 25 cm (Figure 2.12). The buds are less exposed to cold or drying winds than those of phanerophytes and may be protected by a covering of snow in winter. Chamaephytes include dwarf shrubs, small succulents and rosette shrubs, and can be extended to include ground-dwelling mosses. Many of the popular rock-garden plants and culinary herbs are chamaephytes, e.g. dwarf willows, thyme, marjoram and saxifrages.

(a)

(b)

Figure 2.12 Two examples of the chamaephyte growth form: (a) cowberry (*Vaccinium vitis-idea*), a common plant of peat moorlands in Britain; (b) yellow saxifrage (*Saxifraga aizoides*), a small plant that grows on wet rocks near streams.

Hemicryptophytes (*hemi* = partly, *crypto* = hidden) are **perennial** herbs with their perennating tissues at the soil surface. (Perennial plants live for several years.) At the soil surface they can be protected by leaf litter or dead plant remains as well as snow. Hemicryptophytes include many of the garden plants to be found in the herbaceous border, such as delphiniums, lupins and the like (Figure 2.13).

Figure 2.13 Delphinium (*Delphinium* sp.), a common garden plant that is a typical hemicryptophyte.

Geophytes (*geo* = ground) are perennial herbs with underground perennating tissues that are more fully protected from the above-ground climate, such as **bulbs**, **corms**, **tubers** and **rhizomes**. This group includes many of the popular spring flowers. The bluebell, mentioned above, overwinters as a bulb, which is formed from swollen leaf bases (Figure 2.14a). Crocuses grow from corms, which look a bit like bulbs, but are formed from a swollen stem base (Figure 2.14b). The iris (Figure 2.15) survives the winter as a rhizome which is derived from a horizontal stem (Figure 2.14c). Potatoes, of course, are the classic example of a plant that perennates as a tuber, which is also anatomically derived from a swollen underground stem (Figure 2.14d). In each case the tissues are swollen with food reserves, which enable the plants to grow very rapidly as soon as conditions permit, allowing them to flower and complete their growth and reproduction within a short growing season.

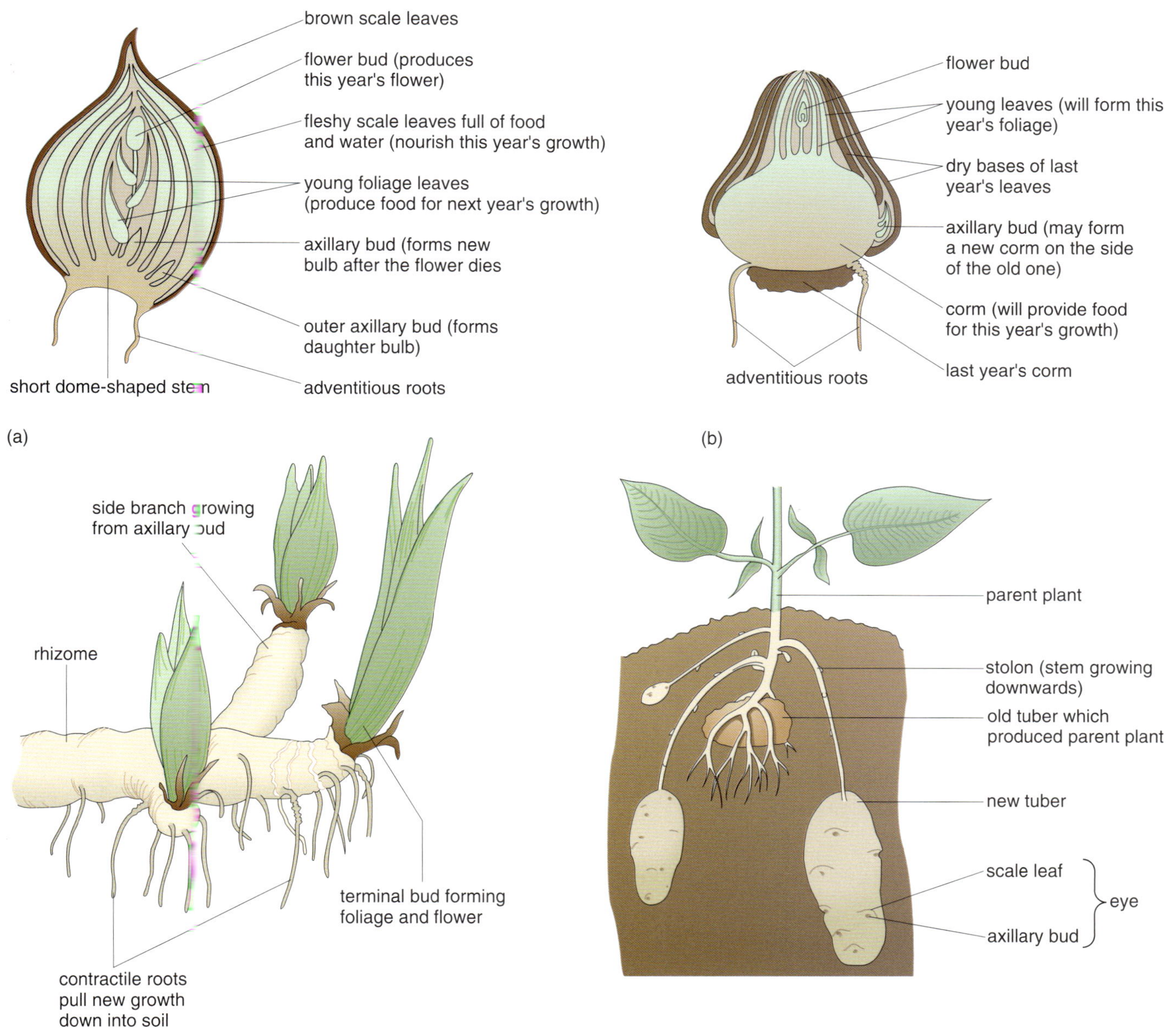

Figure 2.14 Underground perennating structures: (a) bulb, formed from swollen leaf bases; (b) corm, formed from a swollen stem base; (c) rhizome, formed from a swollen, horizontal stem; (d) tuber, formed from an expanded, underground stem.

Figure 2.15 Yellow flag iris (*Iris pseudacorus*), a wetland species in Britain, which — like the garden iris — relies on a modified stem, or rhizome, to survive the winter.

Figure 2.16 Groundsel (*Senecio vulgaris*), a typical therophyte. This species is a common garden weed and is able to complete its life cycle in just 6 weeks.

Therophytes are **annual** or short-lived herbs that survive unfavourable seasons (or, in some cases, several unfavourable years) as seeds, the shoots and roots dying after seed production. (As the name suggests, annual plants complete their life cycle within a single year.) Examples of therophytes include many popular garden flowers, such as candytuft, petunias and marigolds, as well as many weeds, e.g. groundsel (*Senecio vulgaris*, Figure 2.16), chickweed (*Stellaria media*) and fat-hen (*Chenopodium album*).

Life-forms are used to characterize community composition. The species in a particular community or geographic area can be classified by life-form, which, when expressed as a percentage, can be used to generate a **life-form spectrum** (Table 2.1).

Differences in this spectrum reflect the effects of environment, especially climate, on plant adaptation in communities. For example, tropical rainforest is characterized by a predominance of phanerophyte species.

- Which is the most abundant class in (a) tundra and (b) deserts?
- Hemicryptophytes are the most abundant class in tundra, while in desert climates therophytes are most abundant.

Table 2.1 Life-form spectra for various types of vegetation from around the world. Figures are expressed as percentages.

Vegetation type	Climate	Phanero-phytes	Chamae-phytes	Hemicrypto-phytes	Geo-phytes	Thero-phytes
worldwide average		46	9	26	6	13
oak woodland (e.g. as seen in the Teign catchment).	temperate, moderately moist	30	23	36	5	6
tropical rainforest	hot and very wet	96	2		2	
cool temperate forest	cool temperate and moist	10	17	54	12	7
tundra	very cold and wet, but water often frozen	1	22	60	15	2
dry grassland	temperate/tropical, dry	1	12	63	10	14
desert	hot, dry		4	17	6	73

2.2.4 Taxonomy and the evolutionary tree

Life-forms are not the only way to classify organisms. Individual plants are named and sorted into an hierarchical system of groupings that is intended to reflect evolutionary relationships, like a huge 'family tree' stretching back millions of years. The evidence on which we base our classification is very incomplete: much of the original **taxonomy** (naming and classification) was done using the sparse fossil record and comparative morphology (i.e. shape).

Much of this donkey-work was performed by generations of dedicated, botanizing vicars, who spent long hours poring over their pressed flower collections during the 19th century. However, with the advance of technology, new techniques have become available, such as immunology and DNA sequencing, which generate additional evidence. This is used by taxonomists (the scientists that study relationships between, and assign names to organisms) to continually confirm or improve our classification of organisms.

Organisms are grouped into a hierarchy of categories that are thought to have increasingly distant relationships. Thus the smallest, least inclusive category is the **species**, which includes organisms that are so closely related they can freely interbreed and produce fertile offspring. The next level is the **genus** (pl. **genera**), which includes species that have many features in common and can on occasion interbreed to form **hybrids**, which have variable levels of fertility. Hybridization among plants is quite common and has been the mainstay of plant breeders for millennia, but it can occur amongst animals, e.g. the mule is the (infertile) product of a cross between a male donkey and a female horse. Genera with similar traits are grouped into **families**, which are themselves grouped into **orders**, orders into **classes**, and classes into **phyla** (sing. **phylum**). Finally, at the top the phyla are grouped into **kingdoms**, which include organisms that share only really fundamental characteristics (Figure 2.17)

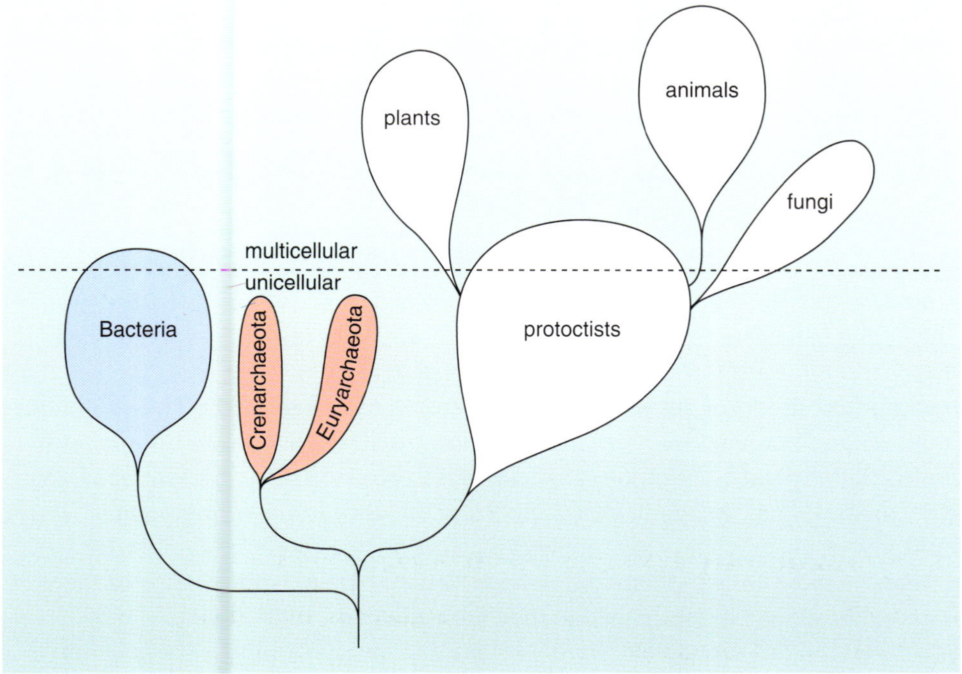

Figure 2.17 Diagram showing the seven kingdoms of organisms, arranged to reflect their evolutionary relationships. Above the dashed line, all the organisms are multicellular, below it they are unicellular.

(a)

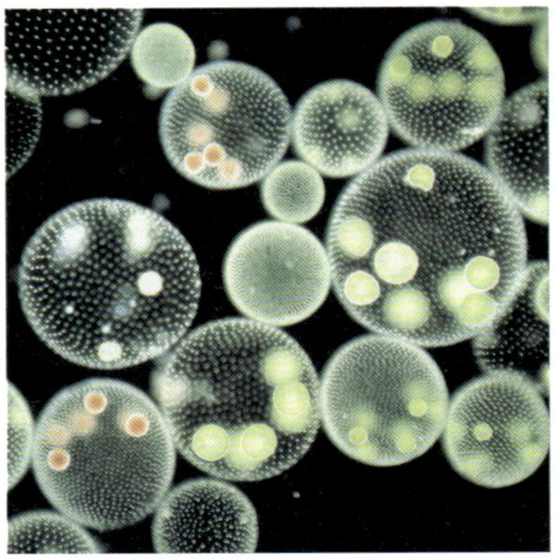

(b)

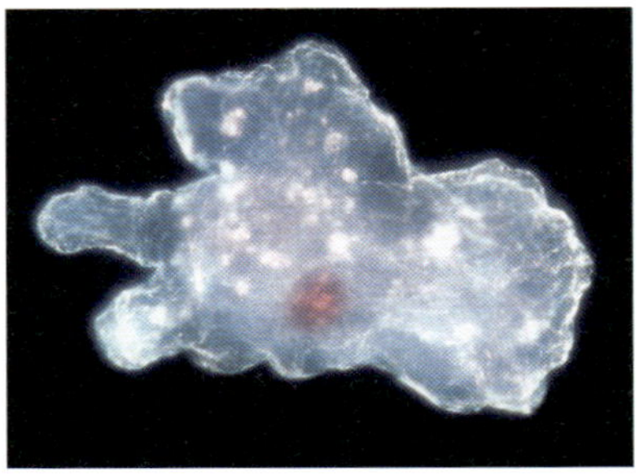

(c)

Figure 2.18 Members of the kingdom Protoctista: (a) the seaweed, knotted wrack (*Ascophyllum nodosum*); (b) a freshwater alga (*Volvox* sp.); (c) a protozoan (*Amoeba* sp.).

Three of the kingdoms, Bacteria, Crenarchaeota and Euryarchaeota, comprise organisms superficially similar to bacteria that can exist as single cells and have a very simple cellular structure. The other four kingdoms comprise organisms with a much more elaborate cellular structure. The plant, animal and fungal kingdoms are instantly recognizable, and are all derived from the 'rag-bag' kingdom known as the protoctists, which includes many of the groups of the simpler organisms, such as the algae and protozoa (Figure 2.18a, b and c), that were previously tacked onto the plant and animal kingdoms in the days when everything had to belong to one or other of these two.

Theoretically, all of the organisms within a particular grouping should share a common ancestor, although clearly this is impossible to prove. In practice, there are no really hard-and-fast rules for deciding how closely related individuals need to be to belong to intermediate categories, so there is an element of arbitrariness about them.

Each species is known by a latinized binomial name, printed in italics, e.g. *Ranunculus repens* (commonly known as the creeping buttercup). The first part, which always starts with a capital letter, denotes the genus the plant belongs to, and the second part, which never starts with a capital, denotes the species. This system was first devised by the Swedish botanist Linnaeus, who published his book *Species Plantarum* in 1753, and it has stood the test of time; most of the names he assigned to plants are still used worldwide 250 years later.

On an evolutionary timescale, plants originated relatively recently, about 500 million years ago, from ancestors that were green algae living in freshwater. Most plants share a common structure because of their mode of nutrition.

- ❍ Recall the raw materials plants need to build and maintain their tissues.
- ● Carbon dioxide, water and light for photosynthesis, and mineral nutrients (such as nitrogen, phosphorus and potassium).

Consequently, virtually all land plants have a green, photosynthetic shoot, which must be exposed to light, and a non-photosynthetic part, which provides anchorage and absorbs mineral nutrients and water. In recently evolved plants, these structures are the roots, but in more primitive plants, e.g. mosses, they may be modified shoots or simple hair-like processes called **rhizoids**. The evolution of an efficient transport system in plants was essential for the development of multicellular plants, with tissues specialized to perform different functions, that were also rigid enough to be self-supporting on land. In flowering plants, transport of water, sugars and mineral ions occurs via more or less tubular vascular tissues called xylem and phloem, which extend from the tips of the roots to the smallest leaves, where they are visible as 'veins'.

Xylem vessels form hollow tubes with walls that are stiffened by a tough chemical called lignin, which forms spiral bands or a net-like pattern pierced by many pits (Figure 2.19a). They conduct water and dissolved mineral nutrients from the roots to all parts of the shoots. Phloem vessels, or seive tubes, are not hollow structures. They still contain active cytoplasm and are separated from one another by perforated end walls. All phloem sieve tubes are associated with companion cells, which provide the energy for transport (Figure 2.19b). Phloem seive tubes transport sugars from the leaves, where they are formed during photosynthesis, to all other parts of the plant.

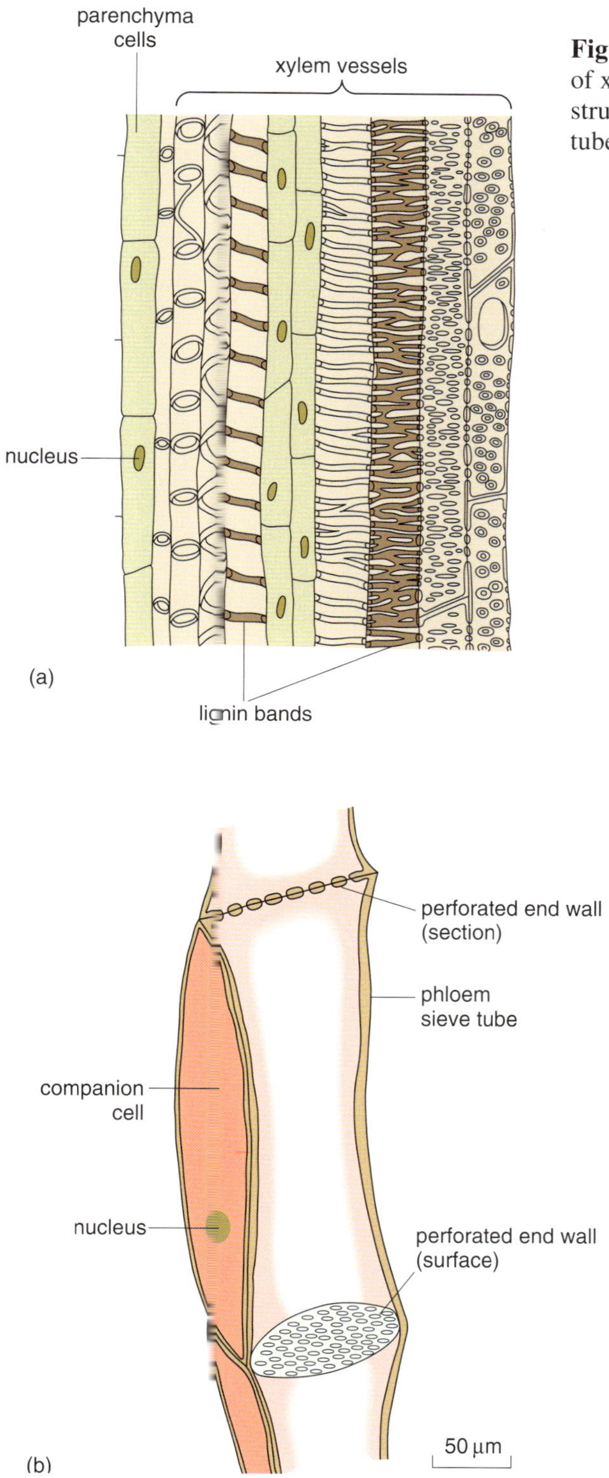

Figure 2.19 Structure of plant vascular tissues. (a) Longitudinal section of xylem vessels, showing a variety of patterns of strengthening by the structural compound, lignin. (b) Longitudinal section of a phloem sieve tubes with perforated end walls.

The distribution of xylem and phloem is different in young stems and roots (Figure 2.20, *overleaf*).

- ❍ Describe the distribution of the vascular tissues in shoots and roots.

- ● In shoots, the xylem and phloem are arranged in discrete bundles, in which phloem is always closer to the surface of the stem. In roots, xylem forms a cross-shaped structure in the middle of the root and the phloem is present in discrete bundles between the 'arms' of the cross.

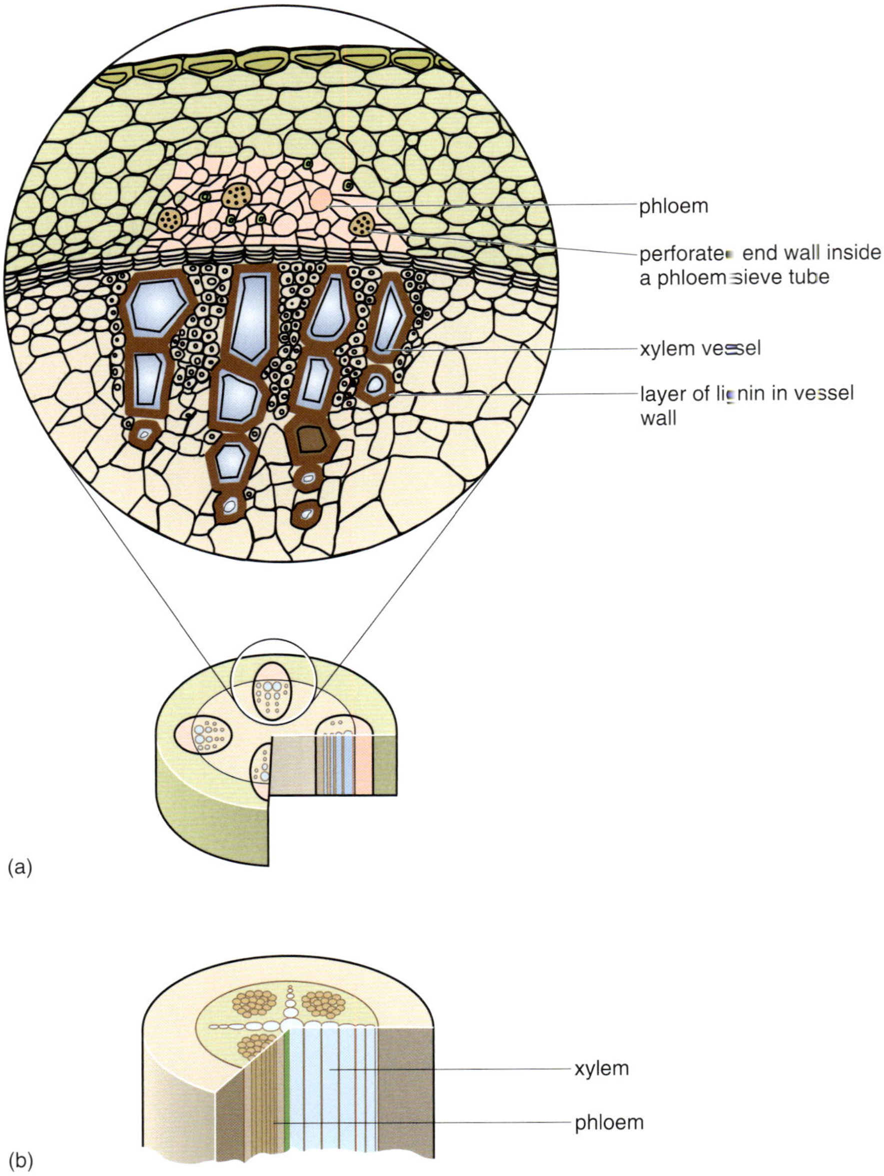

Figure 2.20 Distribution of vascular tissues in stems and roots that are less than one year old. (a) Diagrammatic section of a stem. (b) Diagrammatic section of a root.

As plants grow, the vascular bundles grow sideways to form a complete ring around the stem, and more and more xylem accumulates each year to form wood. However, the basic plant form is retained, even when the shoots become huge and stiffened, as in trees, or even if the plant has changed its mode of nutrition and become parasitic or insectivorous.

Unlike animals, the classification of plants does not relate primarily to overall structure, but to their breathtaking variety of life cycles and modes of reproduction, about which animals are much more conservative. Figure 2.21 shows one evolutionary history and classification system for the most abundant and commonly encountered plant groups.

Figure 2.21 A simplified evolutionary history of the plant kingdom, showing groups with living representatives or descendants. The vertical axis represents time and the width of different lineages indicates the diversity of species within the group, with number of living species given for each. Names, in lower-case print, are either the common name for a phylum (e.g. ferns) or the name of a representative genus (e.g. *Ginkgo*).

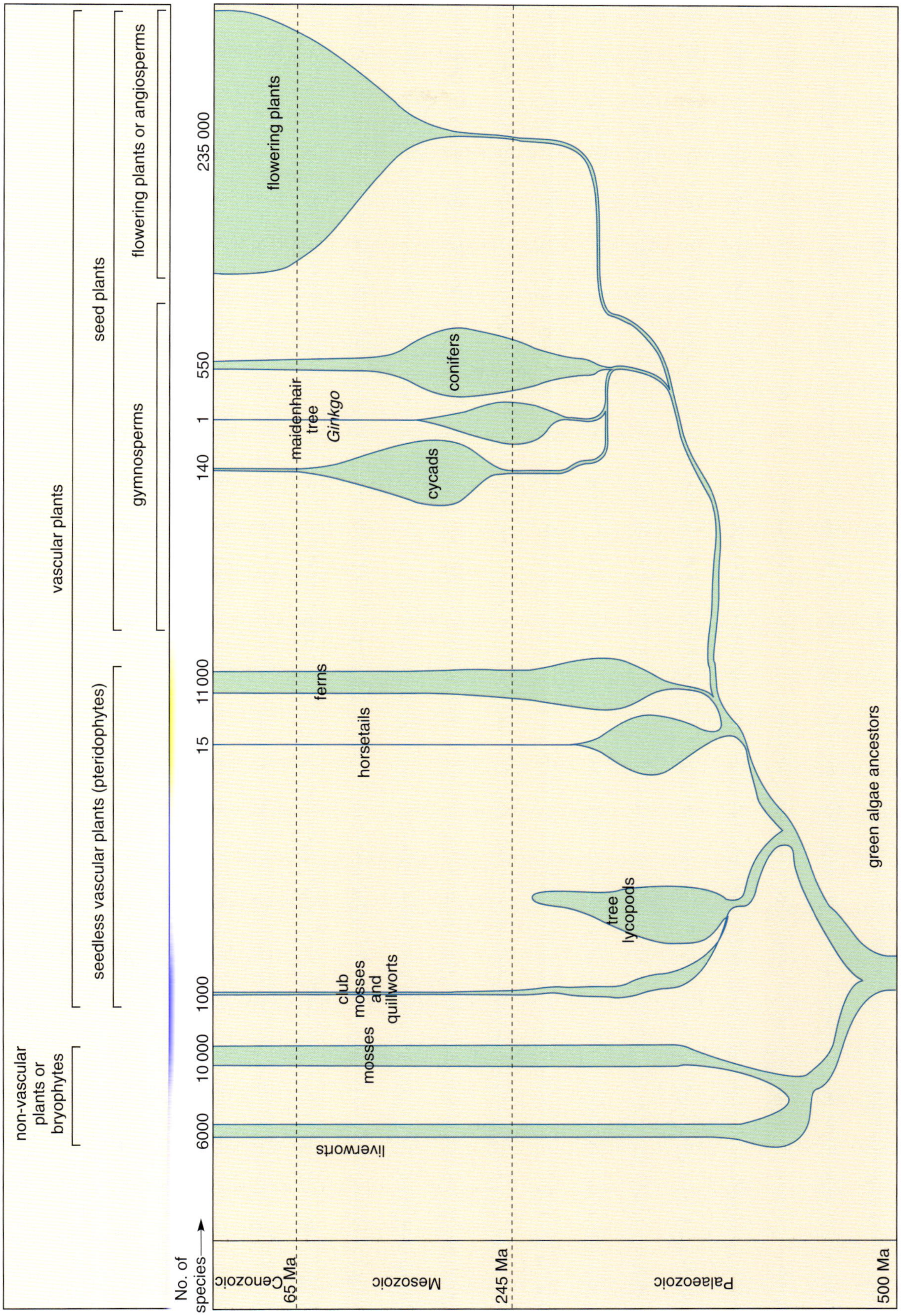
non-vascular plants or bryophytes
vascular plants
seedless vascular plants (pteridophytes)
seed plants
gymnosperms
flowering plants or angiosperms
No. of species
6000
10 000
1000
15
11 000
140
1
550
235 000
liverworts
mosses
club mosses and quillworts
tree lycopods
horsetails
ferns
cycads
maidenhair tree Girkgo
conifers
flowering plants
green algae ancestors
Cenozoic
65 Ma
Mesozoic
245 Ma
Palaeozoic
500 Ma

❍ Which of the groups currently shows the greatest species diversity?

● The flowering plants or **angiosperms** have more species (235 000) than all the other groups put together.

Most of the plants you see today, from oak trees to buttercups, belong to this group, which dominates the land flora. These plants displaced most other groups between 65–142 million years ago. They were able to do this because they possessed a number of key features:

- **Flowers**: structures that aid sexual reproduction and encourage animals, particularly insects, to transport pollen between them.
- Well-protected **seeds** containing food reserves, surrounded by maternal tissues that form a **fruit**. These features enable the angiosperms to reproduce in the absence of free water, and make for more effective dispersal of the seeds than in other plant groups.
- Chemical defences that discourage grazers. Flowering plants produce a huge array of chemicals, including many familiar and often highly toxic substances, e.g. nicotine, caffeine, cyanide, opiates, etc.

❍ Identify the next two most species-rich groups.

● Ferns, with 11 000 species, and mosses, with about 10 000 species.

Mosses and ferns are most abundant in the moist tropics and they occur mainly in habitats that are permanently or periodically damp, because both of these groups require wet surfaces for successful reproduction and growth.

Bryophytes (mosses and liverworts) were the first to evolve from the aquatic, algal ancestors. They produce male sex cells that swim over moist surfaces to fuse with much larger female sex cells, which remain embedded in parent tissue. Bryophytes have no woody cells; therefore they are all low-growing and have small leafy shoots, or a flat pad of tissue with root-like structures (rhizoids) for attachment (Figure 2.22). Bryophytes also have no vascular tissues; water and mineral nutrients are absorbed over the whole surface of the plant. Therefore the surfaces of the tissues are not waterproofed, and the plants are susceptible to drying out.

(a)

(b)

Figure 2.22 Bryophytes: (a) moss (*Leucobryum glaucum*) growing on leaf litter in a beech wood; (b) liverwort (*Lophocolea* sp.) growing over a log.

Mosses produce naked spores (not seeds) in spore capsules that extend high above the main leafy plant. Spores do not have a thick protective coat and a food supply like seeds, so they are more susceptible to drying out and early death than seeds. Mosses and liverworts can withstand low light levels, and can grow on steeply sloping surfaces, such as rocks and tree trunks, which other types of plants could not colonize. They have been largely displaced from more favourable areas by the larger angiosperms and tend only to be dominant in areas where survival is difficult due to poor nutrient supply, such as bogs.

Pteridophytes, i.e. ferns and their relatives, horsetails (Figure 2.23), evolved after bryophytes. The major structural innovation they possess is vascular tissue. The water-conducting vessels are thickened, so they can support more weight. Consequently, ferns and horsetails can grow tall (some were the size of forest trees in days gone by). They can also protect themselves from drying out by having a waterproofed surface to the shoots. However, they have retained a requirement for water for reproduction, having swimming male sex cells like bryophytes, and they produce spores, not seeds.

(a)

(b)

Figure 2.23 Pteridophytes: (a) a typical fern is the lemon-scented fern (*Oreopteris limbosperma*); (b) a horsetail native to Britain is the great horsetail (*Equisetum telmateia*), which can reach 2 m in height.

Ferns can withstand low light conditions, and can grow in poor soils, such as north-facing rocky outcrops and the nutrient-poor layer of pine needles in coniferous woodland.

The conifers are the next most diverse plant group around today. Along with ginkgos and cycads (Figure 2.24), they make up the **gymnosperms**, which are characterized by features that make them less dependent on water.

Gymnosperms have seeds, which have waterproof coats and a food store for the growing embryo. Their male sex cells are wind-carried pollen grains rather than swimming sperm. The pollen grains have thick waterproof coats, and having been blown onto a receptive female organ, they grow through the parental tissues to reach the female sex cells. In conifers, the seeds are borne on cones, as the name implies. Angiosperm trees, such as oak and ash, generally dominate on fertile soils in mild conditions; therefore conifers tend to be restricted to less productive areas than broadleaved trees. Thus conifers are found growing naturally in mountainous regions, such as the Black Forest and the Alps in Europe, and at northern latitudes, e.g. in Scandinavia, Canada and parts of Russia. However, they have also been planted extensively to produce commercial crops of wood, because they often mature more rapidly than many broadleaved trees.

(a)

(b)

Figure 2.24 Gymnosperms largely consist of familiar coniferous trees, but there are some less well-known examples, such as: (a) the maidenhair tree (*Ginkgo biloba*), a species once known only from fossilized leaves and thought to be extinct until it was rediscovered in the 18th century growing in China, and is now widely cultivated as an ornamental tree; (b) the cycad (*Cycas media*), another relict of a once more-abundant group.

2.2.5 Summary of Sections 2.1 and 2.2

1. Species are not distributed evenly; groups of organisms tend to occur together in particular environments to give characteristic assemblages called communities. Boundaries between communities (ecotones) can be sharp or diffuse.
2. Plants are autotrophs. They take carbon dioxide, water and light energy from the Sun and convert them into sugars, by a series of reactions known as photosynthesis. Oxygen is produced as a by-product.
3. Virtually all organisms rely on the carbon and energy from sugars originally produced by photosynthesis for growth and maintenance of their tissues. They obtain their energy by respiration, which is essentially the reverse of photosynthesis.
4. Plants are arranged in a variety of different ways that are characteristic of a particular community. For example, woods are arranged in a series of layers with tall, canopy trees at the top and creeping mosses at the bottom.
5. Plants are differentially adapted to exploit the varying availability of resources, such as light, which occur at contrasting positions within a particular environment.
6. Plants can be classified into a series of life-forms, depending on the position of their perennating tissues in relation to the ground. Different life-forms dominate in different climatic conditions.
7. All organisms can also be classified into hierarchical groups that reflect their evolutionary relationships.

2.3 Describing patterns in vegetation

Block 1 highlighted the range and diversity of vegetation types to be found in a relatively small area. In order to interpret this diversity, a means of describing it is required. We have learnt that plants are usually classified according to the species to which they belong. However, as the Teign catchment supports over 1000 different plant species and any single habitat may contain up to 100 of these, characterizing a habitat by means of a species list is not necessarily the most effective way of describing its vegetation. Table 2.2 provides the species lists for four habitats in the Teign catchment. The actual species names are not important in this context, but note how the length of list varies between habitats and that some species occur in several lists, even though three of the four habitat types differ widely from one another.

- ○ Which habitat contains the highest number and which the lowest number of species recorded only in that habitat?
- ● The woodland glade has the highest number and the semi-intensive pasture the lowest number — in fact it has none!

Table 2.2 Species lists from four different habitat types within the Teign catchment, all of similar area. Species highlighted in red occur in two lists, those in blue occur in three. These habitat types were illustrated in Block 1.

Heathland	Traditional hay meadow	Semi-intensive pasture	Woodland glade	
Agrostis curtisii	*Achillea millefolium*	*Anthoxanthum odoratum*	*Agrostis setacea*	*Luzula multiflora*
Aira caryophyllea	*Agrostis capillaris*	*Bellis perennis*	*Ajuga reptans*	*Luzula pilosa*
Aira praecox	*Ajuga reptans*	*Cerastium fontanum*	*Alnus glutinosa*	*Luzula sylvatica*
Calluna vulgaris	*Anthoxanthum odoratum*	*Cynosurus cristatus*	*Anemone nemorosa*	*Lysimachia nemorum*
Carex binervis	*Bellis perennis*	*Dactylis glomerata*	*Bellis perennis*	*Melampyrum pratense*
Carex laevigata	*Cardamine pratense*	*Holcus lanatus*	*Betula pendula*	*Mentha aquatica*
Carex panicea	*Carex caryophyllea*	*Hypochoeris radicata*	*Betula pubescens*	*Molinia caerulea*
Cuscuta epithymum	*Carex flacca*	*Phleum pratense*	*Calluna vulgaris*	*Myrica gale*
Danthonia decumbens	*Carex nigra*	*Plantago lanceolata*	*Cardamine flexuosa*	*Oenanthe crocata*
Erica cinerea	*Carex panicea*	*Poa trivialis*	*Carex nigra*	*Oxalis acetosella*
Erica tetralix	*Carex pulicaria*	*Ranunculus acris*	*Carex pendula*	*Plantago lanceolata*
Euphrasia anglica	*Centaurea nigra*	*Ranunculus repens*	*Carex pilulifera*	*Poa trivialis*
Galium saxatile	*Centaurium erythraea*	*Taraxacum officinale*	*Carex remota*	*Polygola serpyllifolia*
Hypericum humifusum	*Cerastium fontanum*	*Trifolium dubium*	*Carex sylvatica*	*Potentilla erecta*
Hypochoeris radicata	*Cirsium palustre*	*Trifolium pratense*	*Chamerion angustifolium*	*Potomogeton polygonifolius*
Isolepis setacea	*Cynosurus cristatus*	*Trifolium repens*	*Chrysoplenium oppositifolium*	*Primula vulgaris*
Leontodon autumnalis	*Dactylis glomerata*		*Circaea lutetiana*	*Quercus petraea*
Leontodon hispidus	*Dactylorhiza fuchsii*		*Conopodium majus*	*Ranunculus flammula*
Linum catharticum	*Dactylorhiza praetermissa*		*Coryllus avellana*	*Ranunculus repens*
Molinia caerulea	*Danthonia decumbens*		*Crataegus monogyna*	*Ribes rubrum*
Pinus sylvestris	*Festuca pratensis*		*Dactylis glomerata*	*Rubus fruticosus*
Plantago coronopus	*Festuca rubra*		*Deschampsia cespitosa*	*Salix caprea*
Potentilla erecta	*Filipendula ulmaria*		*Digitalis purpurea*	*Sambucus nigra*
Pulicaria dysenterica	*Holcus lanatus*		*Hyacinthoides non-scriptus*	*Scrophularia nodosa*
Sonchus asper	*Hypericum pulchrum*		*Epilobium palustre*	*Silene dioica*
Trichophorum caespitosum	*Isolepis setacea*		*Erica cinerea*	*Solidago vigaurea*
Trifolium dubium	*Juncus acutiflorus*		*Erica tetralix*	*Sorbus aucuparia*
Trifolium micranthum	*Juncus articulatus*		*Festuca ovina*	*Stellaria neglecta*
Trifolium repens	*Leontodon autumnalis*		*Fragaria vesca*	*Taraxacum officinale*
Vicia angustifolia	*Leontodon hispidus*		*Frangula alnus*	*Teucrium scorodonia*
	Leucanthemum vulgare		*Fraxinus excelsior*	*Ulex europaeus*
	Lotus corniculatus		*Galium palustre*	*Ulex gallii*
	Lotus pedunculatus		*Galium saxatile*	*Urtica dioica*
	Luzula campestris		*Geranium robertianum*	*Vaccinium myrtillis*
	Oenanthe pimpinelloides		*Geum urbanum*	*Veronica chamaedrys*
	Orchis morio		*Hedera helix*	*Veronica montana*
	Phleum pratense		*Holcus lanatus*	*Veronica officinalis*
	Plantago lanceolata		*Holcus mollis*	*Viola palustris*
	Potentilla erecta		*Hypericum androsaemum*	*Viola riviniana*
	Pulicaria dysenterica		*Hypericum pulchrum*	
	Ranunculus acris		*Isolepis setacea*	
	Ranunculus bulbosus		*Juncus acutiflorus*	
	Rumex acetosa		*Juncus articulatus*	
	Senecio jacobea		*Juncus effusus*	
	Succisa pratensis		*Lonicera perclymenum*	
	Taraxacum officinale		*Lotus corniculatus*	
	Trifolium pratense		*Lotus pedunculatus*	
	Veronica chamaedrys			

How can we summarize the information contained in these lists in a form that can be communicated easily? The concept of community, which was introduced earlier, is useful in this respect. The same groups of species often occur together in many different locations within a landscape, as a result of their all being adapted to survive in a particular type of environment. For example, field poppies (*Papaver rhoeas*), chickweed (*Stellaria media*) and black-grass (*Alopecurus myosuroides*) are often found growing with one another because they are all adapted to environments experiencing annual soil disturbance (Figure 2.25). At the other extreme, lime trees (*Tilia cordata*) are often found in association with oak trees (*Quercus* spp.) and hazel bushes (*Corylus avellana*) in environments where soil disturbance has not occurred for centuries (Figure 2.26). We are able to use this phenomenon of co-occurrence to provide a labelling system for vegetation, which refers to the plant community type. The exact status of a community is a subject of debate (see Box 2.1), but their classification is a useful concept, whichever view you take. A limitation to the mapping technique is that the boundary between two communities is rarely a sharp one and it is often necessary to draw somewhat 'fuzzy' lines.

Figure 2.25 A 'community' of cornfield weeds comprising species that are adapted to regular soil disturbance. Numerous species are flowering: the red flowers are of poppy (*Papaver rhoeas*), the yellow of sow-thistle (*Sonchus arvensis*) and the white of mayweed (*Tripleurospermum inodorum*).

Figure 2.26 A small-leaved lime tree (*Tilia cordata*) in an ancient wood. This is the nearest surviving vegetation to a remnant of the English wildwood. The species here are native to Britain but their structure and relative abundance have been heavily influenced by human activity. This type of vegetation is described as being **semi-natural**.

Box 2.1 What is a community?

This is a question that has exercised plant ecologists for most of the last century and is still the subject of debate. The argument was perhaps begun by a book entitled *Plant Succession*, published in 1916 by the American Frederic Clements. His view was that a community of organisms living together behaved in a coordinated way, which made their sum more than their individual parts. In other words the community itself had evolved properties over and above those of its constituent species. The community was therefore a natural entity in itself and communities could be classified in much the same way as species had been. The great Swiss botanist, Josias Braun-Blanquet, was a supporter of this view. He set about constructing a classification of plant communities in Europe, which he published in 1932.

This view of the community as a super-organism came under attack from a number of ecologists, the first and most notable of whom was another American, Henry Gleason. He argued that communities were not distinct entities, but just collections of individual species. He pointed out that the composition of a community is not fixed, but tends to vary across its range. He argued that the classification of communities was for the convenience of the observer describing the assemblage, not a reflection of a natural organization.

One piece of evidence favouring the Gleason school of thought is that the boundary between communities is rarely sharp and there are often large zones of transition (ecotones). This suggests that communities are simply collections of species requiring similar environmental conditions rather than an interdependent society. Modern ecologists tend to favour Gleason's 'individualistic' idea of species behaving as independent units, but some new findings continue to add support to the views of Clements.

- ❍ Some people argue that a plant community can act in a way similar to a single organism, in which the different parts of it (species) work together. Is this view consistent with the thinking of Clements or Gleason?
- ● Clements, because he regarded community members as interdependent.

2.3.1 The system for naming plant communities in the UK

In Britain, it is only comparatively recently that a comprehensive system for describing plant communities has been introduced. It is presented as a series of five volumes edited by John Rodwell and entitled *British Plant Communities*, which were published between 1991 and 2000. In continental Europe an equivalent system, developed by Josias Braun-Blanquet, was published 60 years earlier. Other equivalent systems have been or are being developed for many areas around the world. The British system, known as the National Vegetation Classification (NVC), has been widely adopted by Government agencies and non-governmental organizations involved in nature conservation. The classification covers all natural, semi-natural and major artificial habitats throughout Great Britain. It ranges from remnants of the wildwood (Figure 2.26) to the vegetation you would find between the cracks of pavements in towns (Figure 2.27).

Figure 2.27 Oxford ragwort (*Senecio squalidus*) growing in an urban environment. The community in which it is often found is classified as open vegetation type OV19 by the NVC. Many alien (non-native) species have colonized Britain, taking advantage of the entirely new environments created by people.

The assigning of labels to blocks of vegetation based on their community type is not an exact science, however. Plant species can occur together in a huge number of different combinations. It is not sensible to have a unique community label for each of the possible combinations. Instead, the system identifies groups of species that regularly occur together. At the coarsest level, the classification divides vegetation up into the broad types listed in Table 2.3. Within these types, a number of distinct communities are then described. In some cases these communities are further subdivided into subcommunities, which have many species in common, but which represent a particular pattern of occurrence.

Table 2.3 The broad types of vegetation defined by the National Vegetation Classification.

Vegetation type	Code	Number of communities	Description of typical habitat
aquatic communities	A	24	permanent water bodies
calcicolous grassland	CG	14	soils based on chalk or limestone, maintained by grazing
calcifugous grasslands and montane communities	U	21	acidic soils with low nutrient availability often at high altitude
heaths	H	22	nutrient-poor soils, often acidic and managed by light grazing or burning
maritime cliff communities	MC	12	sea coasts with steep or rocky cliffs
mesotrophic grasslands	MG	13	soils of neutral pH with moderate-to-high nutrient availability, where vegetation is managed by cutting or grazing
mires	M	38	wet soils with poor nutrient availability
saltmarshes	SM	28	inter-tidal areas with fine-textured sediments
shingle, strandline and sand-dune communities	SD	19	sea coasts with coarse-textured sediments
swamps and tall-herb fens	S	28	wet soils with relatively high nutrient availability
vegetation of open habitats	OV	42	areas with bare ground, usually experiencing some disturbance
woodlands and scrub	W	25	area with at least some soil development in which the management allows at least a component of the vegetation to grow to a height of several metres

❍ From Table 2.3, which aspect of the environment appears to be the most important in determining the broad type of vegetation growing there?

● Soil type, particularly its nutrient availability.

In order to classify an area of vegetation, it is first necessary to identify which plant species are present there and how frequently they occur. When defining the area of interest, botanists use the term **stand** to describe a block of vegetation that appears to be uniform in its species composition. It is within a stand that the vegetation should be sampled and it is to that stand that a community label is assigned. It is necessary to take at least five independent samples from the stand and record all the species present in each. The proportion of samples in which a given species occurs is called its **frequency**. The reliability of the frequency estimate depends upon the number of samples taken (Figure 2.28).

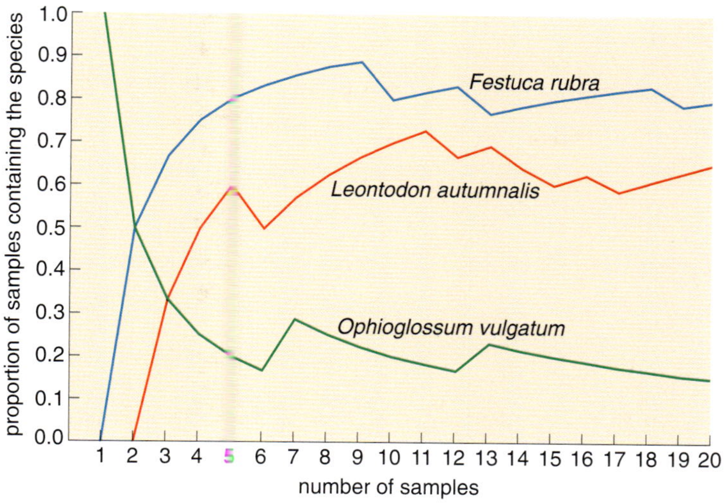

Figure 2.28 A plot of how the estimated frequency of three species: red fescue (*Festuca rubra*), autumn hawkbit (*Leontodon autumnalis*) and adder's-tongue fern (*Ophioglossum vulgatum*), growing together in an area of semi-natural grassland, altered as the number of samples increased.

- ❍ With reference to Figure 2.28, why does the NVC methodology recommend at least five samples are taken from a given stand of vegetation?
- ● The reliability of the estimate is poor if fewer samples are taken. It shows that the estimate may continue to improve if more than five samples are taken.

Some species are more important than others in determining the community label to be assigned. It is possible to distinguish three categories of species based on their importance:

- **Constants** are those species that almost always occur within a stand of the particular community type and usually are found frequently within it.
- **Preferentials** are the species that define one of the subcommunities within the broader community type. These species occur frequently within a particular subcommunity, but rather infrequently, if at all, in the other subcommunities.
- **Associates** are species that are found within stands of the community, but at a lower frequency; they may not be particularly characteristic of that community, but occur more commonly in others.

Table 2.4 An extract from a floristic table for a wet grassland community. Frequencies are conventionally expressed on a five-point scale denoted by the Roman numerals I–V, where I represents a frequency of < 20%, II denotes 20–40%, etc., up to V representing a frequency of > 80%.

Species name	Subcommunity 'a'	Subcommunity 'b'
Deschampsia cespitosa	V	V
Holcus lanatus	IV	IV
Poa trivialis	IV	II
Arrhenatherum elatius	I	V
Alopecurus pratensis	II	II
Carex hirta	I	I

- ❍ Divide the species in Table 2.4 into constants, preferentials and associates.
- ● *Deschampsia cespitosa* and *Holcus lanatus* are constants. *Poa trivialis* and *Arrhenatherum elatius* are preferentials and the remaining two are associates.

In summary, some species are very faithful to a particular community type, whilst others occur across a broad range of different communities. It is the particular combination of species that defines a community, not simply the presence of two or three characteristic ones.

Question 2.1

Table 2.5 reproduces a floristic table from the series *British Plant Communities*. Study the frequency values from the species listed, then (i) identify the habitat type in which you expect to find this community; (ii) give the number of community constants in S2; (iii) list the seven species that may best help you to distinguish subcommunity 'a' from subcommunity 'b'; and (iv) explain why *Juncus subnodulosus* occurs below the lowest horizontal line, when it is much more frequent within the community than the species listed above it.

2.3.2 Uses of the National Vegetation Classification

The NVC provides a method for mapping vegetation in detail. Each community type has been allocated a short alpha-numeric code such as (A1 or CG2), which helps in the presentation of maps. To ensure consistency between different surveyors, the method for collecting data from the field and for interpreting it in order to assign a community label to each distinct stand of vegetation, is clearly set out in a field manual. Activity 2.1 allows you to follow this procedure as part of a virtual field exercise. The information provided by an NVC map can be of considerable help in the management of sites for nature conservation. Each of the 286 communities covered by the NVC has a published account of what is known about its habitat requirements in the *British Plant Communities* series. This forms a valuable reference source for countryside managers. The ability to identify the community type is necessary in order to access this information.

Activity 2.1 Plant community mapping

The DVD activity *Using the National Vegetation Classification* allows you to create a map of field vegetation and then to draw conclusions on processes underlying the pattern displayed by the map. It is based on a site known as Hollington Basin in Bedfordshire. The difficult bit of identifying species has been done for you! Your role is to delineate a stand and interpret the data.

Since vegetation is dynamic, the composition of a stand will change from one year to the next. This may be in response to weather patterns: for example, warm springs promote the growth of some species, whilst wet summers benefit others. Alternatively, a temporary change in management, such as not allowing an area of grassland to be grazed for a year, or deliberately burning an area of heath, will have considerable effects in the short term. Whilst altering the relative abundances of the species present, this type of short-term change will tend not to alter the classification of the community. If the alteration in weather pattern or in the management of the site were sustained over a number of years, however, the community type would begin to change, reflecting the new environment.

Table 2.5 The floristic table for community 'S2' as defined by the NVC and published in the series *British Plant Communities*. ('a' = *Cladium mariscus* subcommunity; 'b' = *Menyanthes trifoliata* subcommunity.)

Species name	Subcommunity 'a'	Subcommunity 'b'
Cladium mariscus	V	V
Calliergon cuspidatum	II	
Solanum dulcamara	II	
Salix cinerea sapling	I	
Fissidens adianthoides	I	
Equisetum palustre	I	
Phragmites australis	II	V
Menyanthes trifoliata	I	IV
Potentilla palustris		II
Carex lasiocarpa	I	II
Scorpidium scorpioides	I	II
Utricularia vulgaris		II
Mentha aquatica		II
Utricularia minor		I
Utricularia intermedia		I
Riccardia multifida		I
Campylium stellatum		I
Hippuris vulgaris		I
Sphagnum subnitens		I
Carex panicea		I
Caltha palustris		I
Carex rostrata		I
Juncus subnodulosus	II	III
Galium palustre	I	II
Carex elata	I	II
Myrica gale	I	II
Lythrum salicaria	I	I

Another important point to note is that there can often be a delay, or lag, in the response of vegetation to an environmental change. Plant populations can display a form of 'inertia'. That is to say that if a species is well established at a site before an environmental perturbation, the species may maintain its place in the community, even though the habitat conditions are no longer optimal for it.

Figure 2.29 Meadow-rue, a once common plant of British wet grasslands and fens, but now in decline.

An example is a species found in wet grasslands known as meadow-rue (*Thalictrum flavum*, Figure 2.29). It grows best in soils that are saturated with water for part of the year. If a formerly wet site is drained, this species often persists, perhaps because it has large underground storage reserves, which can sustain it through unfavourable years. However, it is no longer able to compete as effectively against other species, many of which may benefit from the better-drained conditions. The old plants persist for decades, but no new individuals are recruited and the population will eventually decline and disappear. This is an example of an individual species' response, but the same is true of communities. If one were to classify and map the original community containing the meadow-rue before the drainage operation occurred, and then to repeat the exercise each year thereafter, the effect of the environmental change may not be picked up for several years.

Vegetation maps drawn using the NVC methodology may not, therefore, be an ideal way of monitoring the environment when considering changes that occur from year to year. Their utility becomes apparent when looking at longer-term trends (perhaps over a 5 or 10-year period). In contrast, the abundance of individual species, especially those with an annual life cycle, is often a function of specific weather or management events. A plot of their population size over time (Figure 2.30) may give us information about year-to-year variation in the environment, but the plot often shows too much variation to be useful as an indicator of longer-term environmental change.

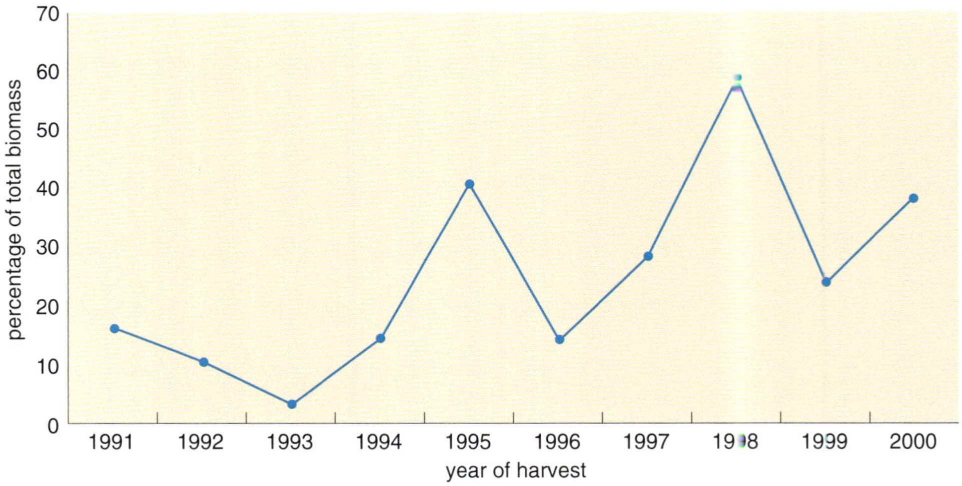

Figure 2.30 The abundance of the meadow foxtail (*Alopecurus pratensis*) within a grassland plot at the Park Grass experiment, Rothamsted, England.

- ❍ If the abundance of *A. pratensis* had only been monitored since 1998, what would your conclusions about its trend be?
- ● That it has declined.

In fact, over the 10-year period, *Alopecurus* seems to be increasing its abundance within the stand. However, there is no reliable evidence that the frequency of the species within the stand is increasing. The variation in abundance is due to this species producing more biomass in warm, wet springs. This illustrates the need for long-term monitoring to characterize a vegetation change and the danger of relying on the abundance of a single species to indicate change within a community.

2.3.3 What is a habitat?

The term **habitat** as applied to wildlife can be defined in a number of ways, but it essentially means the place where an organism lives. Definitions that are in common usage are discussed below.

The range of environments in which a given species occurs. This definition is perhaps the truest to the original derivation of the term, as the place where something lives. When applied in practice, however, the set of environments defined can be very varied, depending on the species chosen. If we take one that inhabits quite a narrow range of environments, such as a wild member of the lily family, known as the snakes-head fritillary (*Fritillaria meleagris*, Figure 2.31), one can describe its habitat as traditionally managed hay meadows on alluvial soils in the floodplains of lowland rivers. In this case, the current definition of 'habitat' seems clear and useful for classifying or mapping areas of wildlife interest.

On the other hand, the same definition can give rise to a very broad set of environments if a species such as the common eel (*Anguilla anguilla*, Figure 2.32) or the grass, red fescue (*Festuca rubra*, Figure 2.33) is considered. Eels occur in a great variety of different environments at different stages of their life cycle: reed-swamps, ponds, wet grasslands, streams, slow-flowing lowland rivers, estuaries, temperate coastal waters, the open ocean and tropical waters full of giant marine algae. Figures 2.34 and 2.35 illustrate two environments that could both be described as 'eel habitat,' yet they have little other similarity.

Figure 2.31 The snake's-head fritillary, an indicator species of floodplain meadows in England.

Figure 2.32 The common eel (*Anguilla anguilla*), a widely travelled animal that exploits many different environments during its life cycle.

Figure 2.33 Red fescue (*Festuca rubra*), a grass species that is genetically variable and able to thrive in numerous different environments.

Figure 2.34 A small stream in which eels may be found.

Figure 2.35 The Sargasso Sea in the western Atlantic Ocean, showing free-floating giant algae (*Sargassum* spp.), which is also an eel habitat.

Similarly, red fescue occurs in almost every type of grassland plant community within the UK, most of continental Europe and increasingly round the globe as it is transported in agricultural seed mixtures. For neither of these species is it useful to combine the multitude of environments in which they occur under single headings of 'eel habitat' or 'red fescue habitat.' Instead, one would say that these species occur in a number of habitats. In the case of the eel, the species moves between habitats during the course of its life cycle. In the case of red fescue, although it is classified as a single species, there is evidence that the strains growing in chalk grasslands (Figure 2.36) differ from those in wet grasslands, which differ from the ones in acid, dry grassland, which in turn are different from the coastal marsh strains (Figure 2.37). There is a degree of genetic adaptation to each of the various environments in which it occurs. To lump these environments together under the label of 'red fescue habitat' is potentially misleading. The genetic variants of the species *Festuca rubra* are denoted by adding a third latinized name to the usual binomial one. Thus *Festuca rubra rubra* is the typical subspecies found, for example, in a chalk grassland, whilst *Festuca rubra pruinosa* is a specialized subspecies found growing in saline environments.

An alternative and probably more practical definition assumes a habitat to be *a distinct unit in the landscape, supporting a distinct community of organisms*. Using this definition of the term, one is forced to make rather arbitrary choices about what constitutes a 'habitat.' Some commentators will regard grassland as one 'habitat', woodland another and wetland a third. Another authority may regard grassland as a broad habitat type, which they would subdivide into a number of distinct 'habitats,' such as:

- agriculturally improved grassland
- chalk grassland
- upland acidic grassland
- grassy heath
- damp hay meadow
- fen meadow
- grazing marsh
- rush pasture.

Figure 2.36 An area of dry chalk grassland containing the subspecies of red fescue called *Festuca rubra rubra*.

Figure 2.37 An area of upper saltmarsh containing the subspecies *Festuca rubra pruinosa*.

The subdivisions can continue to form smaller and smaller units, basing the definition of 'habitat' on the dominant plant species present or the major plant community. Although this definition is rather less precise than the one above, because one cannot tie the habitat definition to one particular species, it is a more utilitarian approach if mapping of habitats is the aim. It was the one used in the Phase 1 habitat survey in Block 1.

To remove the arbitrary aspect of categorization, publications such as the *Handbook for Phase 1 Habitat Survey* (published in 1990 by the Nature Conservancy Council) define a set number of habitat types for the British vegetation. If more detailed mapping is required, then the NVC field methodology can be used to create a map of plant communities.

'Habitat management,' 'habitat restoration' and 'habitat creation' are increasingly used terms in the field of wildlife conservation. They are used to describe the scientific principles and practical methods employed to maintain, enhance or create areas for the benefit of indigenous species.

Question 2.2

From what you saw in Block 1, sort the following species according to whether they are indicative of a particular habitat or whether they should be regarded as being non-specific with respect to habitat type: (a) bracken; (b) swimming mayfly; (c) wheatear; (d) brown hairstreak butterfly; (e) humans; (f) sea lavender.

2.3.4 Summary of Section 2.3

1 Groups of plant species consistently co-occur, allowing them to be labelled as communities.

2 Published accounts of plant community classification exist. In Great Britain, the National Vegetation Classification is the most relevant.

3 Plant communities can be mapped and labelled. Such maps are useful tools for monitoring environmental change. They are better used for long-term monitoring than studying annual variation in plant populations.

4 The term 'habitat' strictly refers to the range of environments in which a given species occurs, but in practice, it is usually used to describe a distinct unit in the landscape which supports a particular group of organisms.

2.4 Environmental factors affecting distribution of species

You already know from Block 1 that the distribution of organisms is not uniform. The common rock-rose (*Helianthemum nummularium*, Figure 2.38) is found only on sites overlying **calcareous** (calcium carbonate-containing) rocks, because it performs best when on soils with a high pH. The dipper (*Cinclus cinclus,* Figure 2.3b) has to live near streams because it eats freshwater invertebrates and builds its nest in holes near water. The white admiral butterfly (*Limenitis camilla,* Figure 2.4) inhabits dense woodland, because it lays its eggs on shady, honeysuckle plants, which the larvae eat, and the adults need bramble plants to provide them with nectar.

Figure 2.38 Common rock-rose (*Helianthemum nummularium*), a small chamaephyte typical of chalk grassland.

Figure 2.39 Greater horseshoe bat (*Rhinolophus ferrumequinum*) is now a very rare mammal in Britain.

The greater horseshoe bat (*Rhinolophus ferrumequinum,* Figure 2.39) has very complex requirements. Its habitat must provide: cattle pasture to provide prey in the form of dung beetles, cockchafers and daddy-long-legs; woodland to provide prey in the form of moths; shaded rides and perches, as well as old buildings, trees, caves, etc. to provide summer and winter roosts. If any of these elements are missing from a particular location, then the species cannot survive there.

2.4.1 Describing an organism's environment

From an organism's point of view, essential requirements can be divided into: (1) resources, something that an organism uses and which can be in short supply, e.g. food, nest sites, etc.; and (2) conditions, which influence the external environment in ways that affect how organisms function, but are not actually used by it, e.g. temperature, shade.

The environment itself can be divided into **abiotic** (non-living) components and **biotic** components, i.e. living organisms or substances derived from them. (Think back to the humble cowpat!) Abiotic factors can be further subdivided into those that are *physical*, e.g. light, temperature, humidity, space, wind, water movements (waves and currents), soil texture and fire; and those that are *chemical*, e.g. mineral nutrients, oxygen, carbon dioxide, water, salinity, toxins and pH. In some cases the abiotic factors can be readily identified as either a resource or a condition, but the relationship is not always straightforward.

❍ Can you identify one physical factor that is both a resource and a condition?

● Light; it is a resource for photosynthesis, and a condition where it affects day length (because this influences growing season, breeding season, etc.).

Water and soil are also factors that can be considered from more than one point of view; they certainly both have physical and chemical aspects to them. However, one thing is clear: many of these factors are intimately related to climate and geology.

2.4.2 Climatic factors affecting distribution

On a global scale, it is climate that determines the dominant, terrestrial vegetation types, or **biomes**. (Strictly speaking, the term 'biome' includes all the characteristic organisms within the community, but only in as much as the other organisms are associated with particular vegetation types.) Figure 2.40 shows how the climatic characteristics of temperature and precipitation relate to produce some of the major biomes.

❍ What types of vegetation can be found with a mean annual temperature of 20 °C?

● A mean annual temperature of 20 °C may be found in desert, grassland or deciduous forest, depending on annual precipitation.

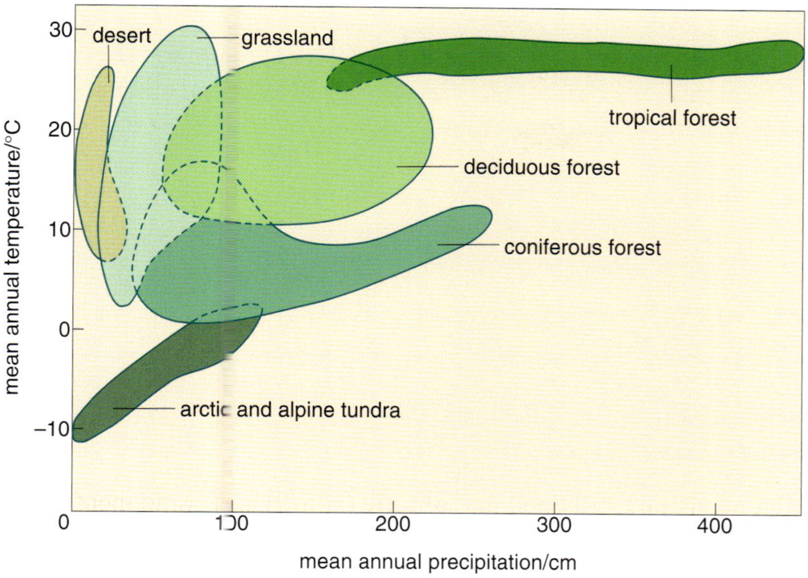

Figure 2.40 Six major biomes related to their environmental tolerances in terms of mean annual temperatures and mean annual precipitation.

Plant growth is extremely sensitive to temperature. Often a difference of a few degrees leads to a noticeable change in growth rate. Temperature is a key determinant of rates of reaction; with increases in temperature, the increased kinetic energy of the reacting molecules results in an increased rate of reaction, but increasing temperature also results in an increased rate of damage to **enzymes**, which are protein molecules that catalyse specific biochemical reactions. The upshot of this is that enzymes have an **optimum temperature** at which their function is maximal, and they only work at all over a defined temperature range, as Figure 2.41 shows.

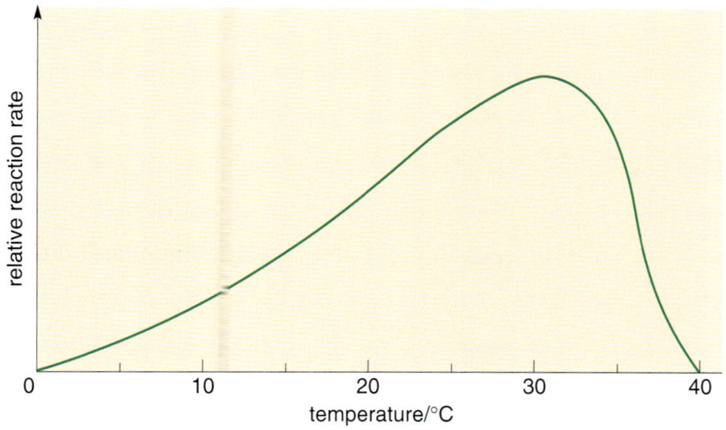

Figure 2.41 The activity of a typical enzyme as a function of temperature.

- ❍ What is the optimum temperature for the activity of the enzyme shown in Figure 2.40?
- ● The optimum temperature is about 30 °C.

Most of the enzymes in a single organism tend to have broadly the same temperature optimum, and if an organism is consistently exposed to temperatures away from this optimum, the enzymes can be damaged to the extent that the organism functions inefficiently, or not at all.

- ❍ What do you think would happen if the plant from which the enzyme in Figure 2.41 came, was maintained at a temperature above 40 °C?
- ● The plant would die if kept above 40 °C for very long, because its enzymes would be damaged or destroyed.

Plants, in common with many animals, are described as **ectothermic**, i.e. the temperature of the organism's tissues is determined by the temperature of the environment. Birds, mammals and a few specialized plant tissues are **endothermic**, i.e. they can maintain an internal temperature that is different from that of the environment. Both ectothermic and endothermic organisms can have structural, or behavioural/life-form adaptations that enable them to *tolerate* some variations in environmental temperature, and to *avoid* periods of extreme heat or cold, but only endotherms are able to use metabolic energy to control the temperature of their tissues to some degree.

Figure 2.42 The swallow (*Hirundo rustica*) may spend the winter in Africa but breeds in Northern Europe.

- ❍ Given the temperature tolerances of enzymes, what implications does this difference in temperature regulation have for the global distribution of ectotherms and endotherms?
- ● Generally, ectotherms are restricted to a narrower range of habitats than endotherms.

For example, many migratory birds such as the swallow (*Hirundo rustica,* Figure 2.42) move between climatic zones at different times of the year. Humans are an example of a species that has adapted to live on all continents. In contrast, most invertebrate species and plants are restricted to quite narrow climatic zones and often to very narrowly defined habitats within them. For example, the cockchafer (*Melolontha melolontha,* Figure 2.43) is restricted to pastureland in temperate Europe.

Precipitation (rain, snow and hail) is no less important than temperature to living things. All organisms are composed of at least 50% water and almost all bodily reactions take place in solution (i.e. the reactants are dissolved in water). Furthermore, water is a key resource for both plants and animals. It is required by all plants as a raw material for photosynthesis, and by many plants and animals for reproduction. Water also has an important role in supporting non-woody tissues in plants.

Figure 2.43 The cockchafer (*Melolontha melolontha*) feeds in grassland as a larva and upon trees as an adult.

- ❍ What happens to herbaceous plants when they become very short of water, but before they die?
- ● They wilt.

See Box 2.2 for details.

Box 2.2 Water and plant support

Unlike animal cells, plant cells are bounded by a fairly rigid cell wall (Figure 2.44). Plant cells differ from animals cells in a number of ways. They possess a large fluid-filled, membrane-bound space (vacuole). Freshwater tends to enter all cells, which makes them swell. Under these conditions, having no cell wall, an animal cell may eventually burst. In contrast, a plant cell starts by swelling, but it does not burst because expansion is restricted by the cell wall. The cell becomes quite stiff (like a fully inflated football). *En masse*, plant tissues can be rigid enough to stand erect, like a fully inflated bouncy castle, and, like a bouncy castle, the whole structure sags if the inflation is not complete.

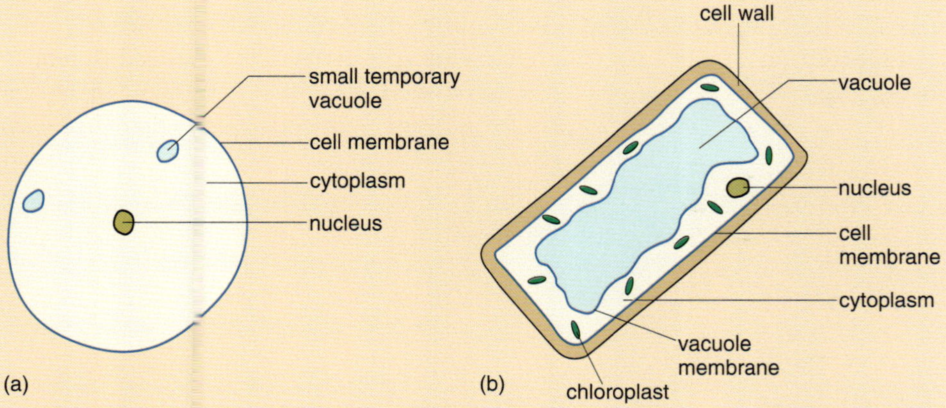

Figure 2.44 (a) Structure of a typical animal cell. (b) Structure of a typical plant cell.

While deciduous forest may be the natural vegetation type for much of Europe, including Britain, it does not mean that all the plants that grow in the forest are adapted to exactly the same set of conditions, or indeed, that forest grows everywhere. Conditions of temperature and precipitation vary on a smaller scale, and a number of other factors come into play, to produce a variety of **microhabitats**.

Plants such as wood anemone (*Anemone nemorosa*) that grow on the woodland floor receive less light, and experience a narrower range of temperatures than bell heather plants (*Erica cinerea*) on a open, upland heath (Figure 2.45).

Figure 2.45 Bell heather plants (*Erica cinerea*) grow on exposed moorlands with no shelter from Sun or wind.

- ❍ What other features of the landscape could influence local temperatures?
- ● Aspect, exposure and altitude can all affect temperature.

All gardeners know that in the Northern Hemisphere you get less frost and a longer growing season in south-facing, lowland gardens that are sheltered from strong winds, than in north-facing gardens that cling to windswept hillsides. Aspect, exposure and altitude can also affect plants' water relations. High temperatures and winds dry out both the soil and the plant, but topography can also affect soil moisture significantly. The rainshadow effect, in which clouds tend to deposit their burden of moisture as they start to rise over hills and mountains, means that land on the leeward side of high land tends to be relatively drier than that on the windward side (Figure 2.46).

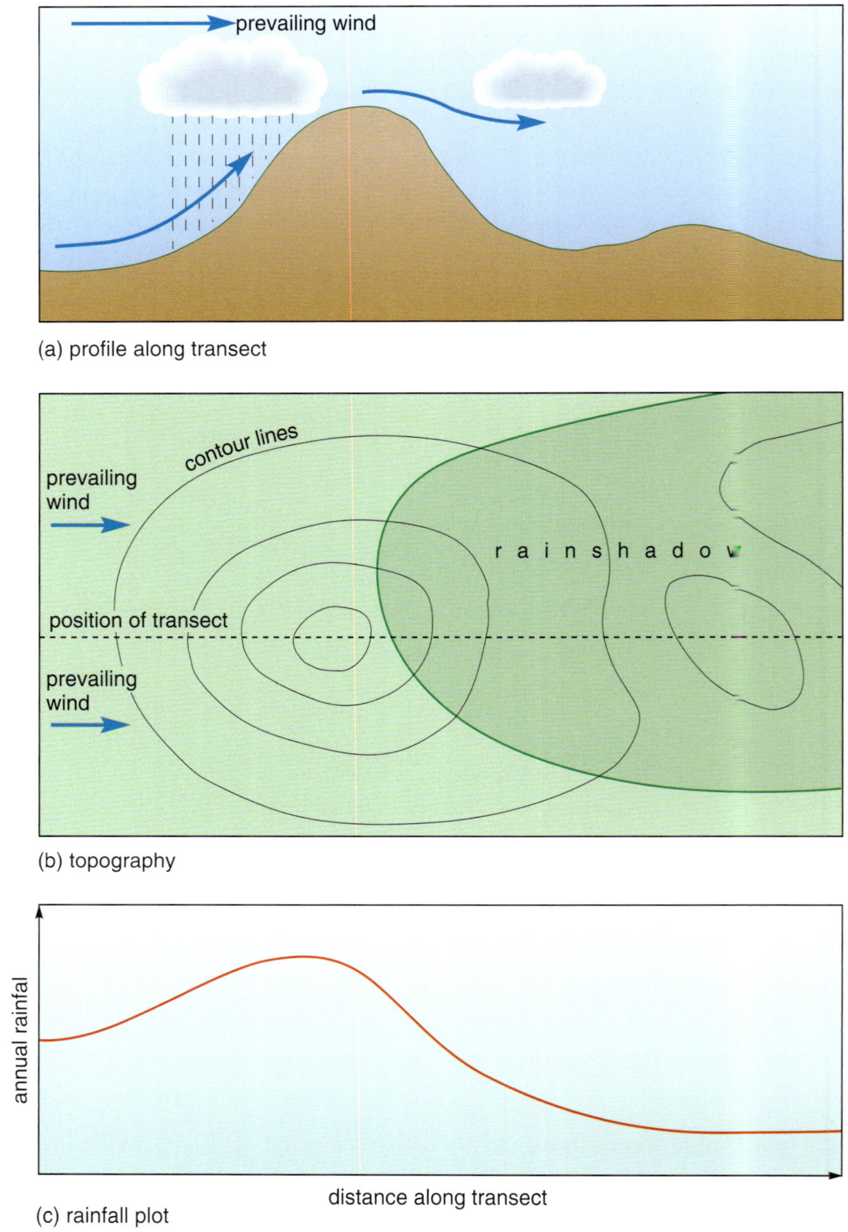

Figure 2.46 Diagrams to illustrate how topography can influence local precipitation through the rainshadow effect: (a) rain falls most on the windward side of high ground, depleting the air of moisture, so the area in the lee of the hill (b) receives less rain, creating a local gradient in moisture availability (c).

Environmental conditions can vary significantly over very short distances, so microhabitats can be very small. Organisms that are ectotherms (e.g. most plants and invertebrates) tend to have very narrow tolerances. As a consequence of these two facts, it is often the case that a number of closely related species have evolved, each of which is adapted to a different set of environmental conditions. Among the British flora, there are at least 13 separate species of pansies and violets (*Viola* spp.). Amongst these species, the field pansy (*Viola arvensis*) is found on low-lying arable and wasteland, whilst the mountain pansy (*Viola lutea*) is found on upland grasslands and moors (Figure 2.47). Differential tolerance to ambient temperatures is a key part of their adaptation.

(a) (b)

Figure 2.47 (a) The field pansy (*Viola arvensis*) is less adapted to growth at low temperatures than its relative (b) the mountain pansy (*Viola lutea*).

- ❍ On average, how are the temperatures experienced by the two species likely to differ?
- ● The mountain pansy is likely to be exposed to generally lower temperatures than the field pansy, because temperature decreases with altitude.

Topography can also influence soil moisture in that water tends to flow downhill and accumulate in valley bottoms, a phenomenon that can have important consequences for soil fertility too (see Section 2.4.3).

2.4.3 Geological factors affecting distribution

After temperature and precipitation, soil type is one of the major factors affecting plant distribution. The reasons for this are obvious.

- ❍ Name the resources that plants obtain from soils.
- ● Water and mineral nutrients.

Mineral elements are acquired from the soil, and all except nitrogen (which is derived from the air) are ultimately derived from the underlying rocks. Because rocks differ in their chemical composition, the proportions of the different mineral elements in the soils derived from them also vary. Furthermore, rocks that are rich in aluminosilicates (e.g. granite) contain a lot of aluminium, which can cause problems of aluminium toxicity in acid conditions. Soil texture is determined mostly by the size of the rock particles. The bigger the average size of the particles, the coarser the texture and the more free-draining a soil tends to be. The smaller the size, the finer the texture and the more water-retentive the soil tends to be.

- ❍ From a plant's point of view, would a sandy or a clayey soil be better in a dry climate?
- ● The clayey soil would be better because it would be more likely to retain moisture during dry periods.

Geology also has an important influence on soil pH. Soils derived from base-rich and, especially, calcareous rocks tend to produce soils of high pH. In contrast, soils derived from base-poor rocks tend to be more acidic. In turn, pH has an enormous influence on plant distribution, because the solubility, and therefore the availability, of soil mineral nutrients varies with pH, as Figure 2.48 shows.

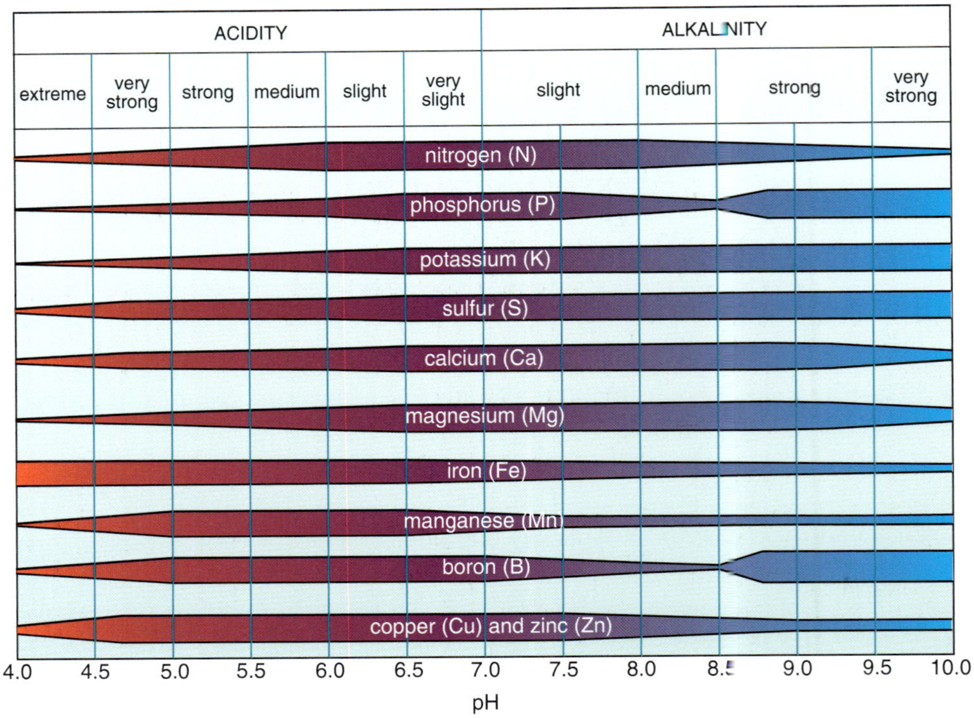

Figure 2.48 The relationship between the pH of the soil and the availability of mineral nutrients: the wider the bar, the greater the availability.

- ❍ Which elements are likely to be in short supply to plants at low pH values (around pH 5)?
- ● Potassium and sulfur may be somewhat limiting, but calcium, magnesium and especially phosphorus are most commonly deficient at low pH.

- ❍ Which elements are likely to be in short supply to plants at high pH (around pH 8.5)?
- ● Copper and zinc may be somewhat limiting, but iron, manganese and boron plus phosphorus again are most commonly deficient at high pH.

Most plants are adapted to thrive in soils around pH 6.5, at which pH all of the nutrients are readily available to plants. However, some plants have evolved to specialize in growing in soils at either high or low pH. Lime-lovers, or **calcicoles**, are adapted to grow well only on calcium-rich soils of alkaline pH through their enhanced ability to take up Fe, Mn, B, Cu and Zn when present at very low levels, and to manage with low levels of P. The poor growth of calcicoles on acid soils may be either because of nutrient shortages — sometimes the major nutrients are not readily available under these conditions — or because of mineral toxicity.

- ❍ Which ions are most likely to cause problems of toxicity to calcicoles in soils of low pH?

- Fe, Mn, B, Cu and Zn are all readily available in acid soils — the same elements that calcicoles are adapted to take up particularly efficiently.

Iron toxicity is the most commonly encountered, but high levels of dissolved aluminium may also cause toxicity in acid soils. The limited agricultural potential of acid soils is said to be due largely to the presence of aluminium in solution, a situation that is exacerbated by acid precipitation. Typical calcicoles include common rock-rose (*Helianthemum nummularium*), cowslip (*Primula veris*) and the pasque flower (*Pulsatilla vulgaris*, Figure 2.49). In contrast, lime-haters, or **calcifuges**, are usually restricted to acid soils. They include species such as cross-leaved heath (*Erica tetralix*, Figure 2.50), heath bent grass (*Agrostis curtisii*) and bilberry (*Vaccinium myrtillus*). If such plants are grown on alkaline soils, they show severe yellowing, which is caused by a deficiency of iron. They may also be affected by calcium toxicity.

Figure 2.49 The pasque flower (*Pulsatilla vulgaris*), a calcicole, which is restricted to chalk grassland habitats.

2.4.4 Water as a condition: aquatic habitats

Most of the discussion above has concentrated on abiotic, environmental factors that are particularly important in determining the distribution of *terrestrial* organisms. However, the principal abiotic factors affecting species distribution in *aquatic* environments are quite different. Water has a greater capacity for storing thermal energy than air; only small changes in temperature occur in water bodies when comparatively large amounts of heat are absorbed or lost. Consequently, temperature variations affecting the thermostability of enzymes (the tendency of enzymes to break down as temperature rises) are less important than on land. Water as a vital *resource* is clearly not an issue, except perhaps in the case of watercourses that dry up in summer, but water as a *condition* is a very important factor. Water currents have a huge impact on aquatic organisms: a fast-flowing river is a very different environment from a stagnant pool. Furthermore, both water temperature and water movements have an impact on the amount of oxygen available to aquatic organisms. Therefore, oxygen concentration and water currents are the principal abiotic factors determining the distribution of organisms in aquatic habitats.

Figure 2.50 Cross-leaved heath (*Erica tetralix*), a calcifuge, which is restricted to habitats with acidic, often peaty, soils.

Oxygen concentration

Temperature affects the amount of oxygen dissolved in water as Figure 2.51 illustrates.

- ❍ What effect does increasing the water temperature have on the oxygen content of the water?
- ● It lowers it.

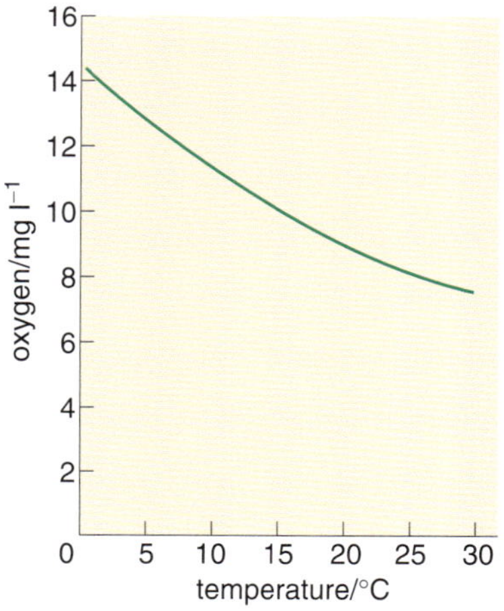

Figure 2.51 Effect of temperature on the oxygen content of pure water.

Whereas the concentration of oxygen in the air is fairly stable, oxygen concentrations in water can vary widely, depending not only temperature but also on turbulence (churned water is able to replenish oxygen from the air more effectively than stagnant water) and the amount of oxygen being removed from or added to the water by chemical and biological processes. Oxygen is removed from the water by chemical reactions such as the formation of iron oxides. Photosynthesis is the only biological process that increases the oxygen concentration in water. Oxygen is removed by many of the other activities of organisms, such as respiration and the chemical reactions involved in decomposition. The conversion of ammonia (NH_3) to nitrates (NO_3^-) and of complex organic molecules to simple salts and carbon dioxide uses up a lot of oxygen. These processes are performed predominantly by microscopic organisms. The sum total of oxygen-requiring activities carried out by organisms is known as the **biological oxygen demand (BOD)**, and it is an important factor in determining the oxygen availability in freshwater habitats such as the River Teign.

❍ What sorts of substances are likely to enter a watercourse that contain significant quantities of ammonia and/or complex organic molecules?

● Dead vegetation such as leaf litter, sewage effluent, farmyard slurry, some fertilizers and waste products of some industrial processes such as food production and leather tanning.

Figure 2.52a and b shows what happens to some of the key abiotic factors downstream of a sewage outfall pipe.

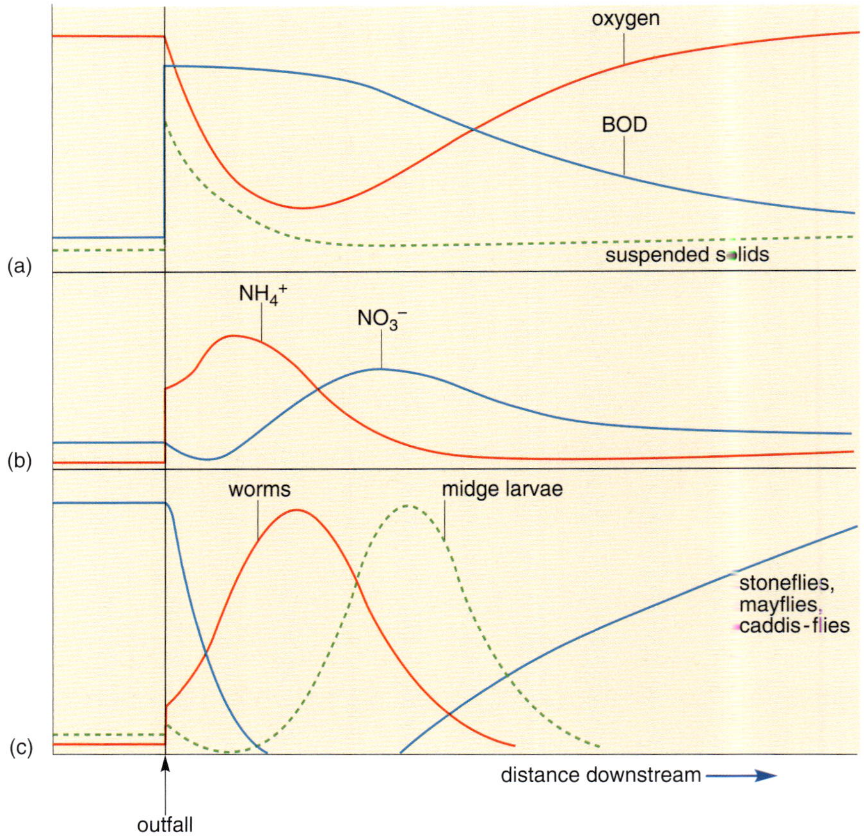

Figure 2.52 Diagrammatic representation of some of the effects of organic pollution on a river and the changes that occur downstream of the outfall: (a) oxygen, BOD and suspended solids; (b) mineral nutrients; (c) invertebrates.

❍ Describe the relationship between (i) oxygen concentration and BOD, and (ii) the concentration of ammonium ions (NH_4^+) and nitrate ions (NO_3^-), downstream of the outfall pipe.

● (i) There is an inverse relationship between oxygen concentration and BOD. (ii) There is an inverse relationship between the concentration of ammonium ions (NH_4^+) and nitrate ions (NO_3^-) downstream of the outfall pipe.

There is an immediate sharp drop in available oxygen below the outfall pipe, which is accompanied by a high BOD. As the BOD gradually falls away downstream, the oxygen concentration rises. A similar inverse relationship exists between ammonium and nitrate ions because sewage has a high concentration of ammonia. As it is oxidized to nitrate, the concentration of ammonium ions falls and the concentration of nitrate rises. The process of oxidation is carried out by microbes in the water and it is this process that causes the high BOD and low oxygen concentration in the water immediately downstream of the outfall pipe.

Freshwater invertebrates vary greatly in their tolerance to the low oxygen concentrations associated with organic pollution. Consequently, there is a predictable series of changes to the invertebrate fauna, which reflects the sequence of changes to the oxygen concentration (Figure 2.52c).

- ❍ Are the data in Figure 2.52c consistent with your findings from the *Freshwater invertebrates* activity in Block 1?
- ● Yes, worms and midges increased in abundance, and mayflies, stoneflies and caddis-flies decreased in abundance immediately downstream of the sewage outfall pipe.

- ❍ What do the graphs in Figure 2.52 suggest would have happened to the abundance of these various groups of organisms if more samples and measurements of abiotic factors had been taken further downstream of the outfall on the River Teign?
- ● There would be a gradual reversal of these changes: oxygen concentrations would rise, BOD would fall, ammonium concentration would fall, and the numbers of mayflies, stoneflies and caddis-flies would increase while the numbers of worms and midges decreased.

Water currents

The speed at which water moves has had a major influence on the evolution of the shape of aquatic creatures, as well as determining their distribution.

Activity 2.2 Effect of flow rate on swimming mayfly abundance

In this DVD exercise you are required to look at the possible relationship between water flow rate and the abundance of swimming mayflies (Baetidae) in a specific stretch of the River Teign. The exercise involves the calculation of a Spearman rank correlation coefficient, which is a very useful statistical test for evaluating the strength of a potential link between two variables such as the abundance of a taxonomic group and the strength of an environmental variable. See the booklet entitled *Using the Spearman Rank Correlation Coefficient* for further details.

- ❍ What relationship did you find between the two variables: water flow rate and numbers of swimming mayflies per sample for the stretch of river you studied in Activity 2.2?
- ● There is a positive correlation between the two factors, i.e. as one variable increased, so did the other variable.

However, at very slow and very fast water flow rates, swimming mayflies are not found. They are replaced at lower flow rates by burrowing mayflies (such as the family Ephemeridae) and at very high flow rates by crawling mayflies (family Ecdyonuridae). (See Figure 2.53.)

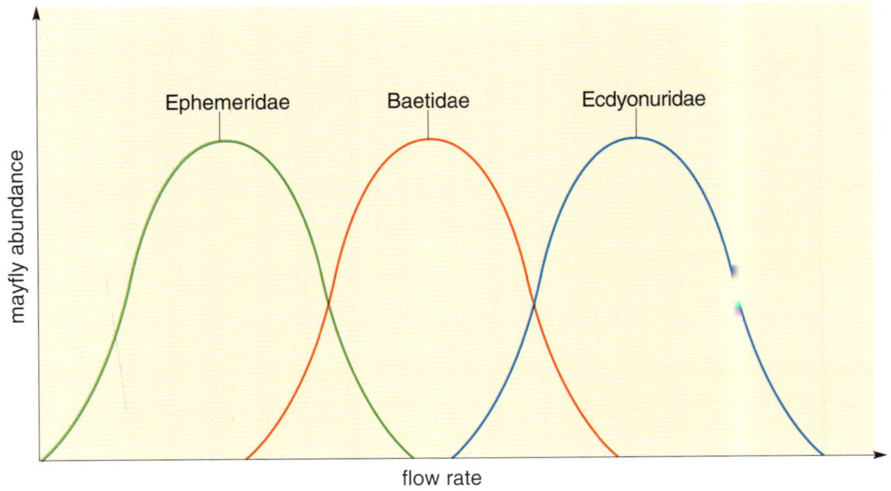

Figure 2.53 Graph showing the relative abundance of three families of mayflies in water of varying flow rates: burrowing mayflies (Ephemeridae) swimming mayflies (Baetidae), and the crawling mayflies (Ecdyonuridae).

In each case, the basic mayfly anatomy has been modified to produce a group of species well adapted for the microhabitats that result from different water flow rates. *Ephemera danica* is large, relatively sedentary and has gills that are raised up vertically. It buries all but its gills in the soft sediments found under slow-flowing waters (Figure 2.54a). The streamlined, teardrop shape of *Baetis rhodani* enables it to swim efficiently in moderately fast-flowing water. Its long tails serve as rudders, keeping the insects facing upstream, at which orientation they are most streamlined (Figure 2.54b). Stones on the bottom of a fast-flowing stream possess a layer of slower-flowing water called the **boundary layer**. The strongly flattened shape of *Ecdyonurus venosus* allows it to keep within this boundary layer as it crawls over stones in fast-flowing water (Figure 2.54c).

Flow rate can vary greatly over very small distances, creating a multitude of microhabitats (Figure 2.55). Therefore, mayflies of all three families may exist within a short stretch of river or stream.

Environmental scientists are often called upon to make judgements about the reasons for changes in the distribution and abundance of particular species, and to make recommendations about how the environment could be managed in order to promote or discourage a particular species or community. They must have reliable means of detecting changes in the abundance of organisms or changes in the composition of communities, and linking this to particular environmental factors that have produced the changes.

Because of their differential sensitivity to organic pollution, freshwater invertebrates can be used as indicators of such pollution. They live in the watercourse for extended periods, often their whole lives, and so reflect the long-term state of the water much better than a one-off measurement of water chemistry. Various methods have been devised for deriving a numerical index indicating the state of the water, from the invertebrate fauna that exist there.

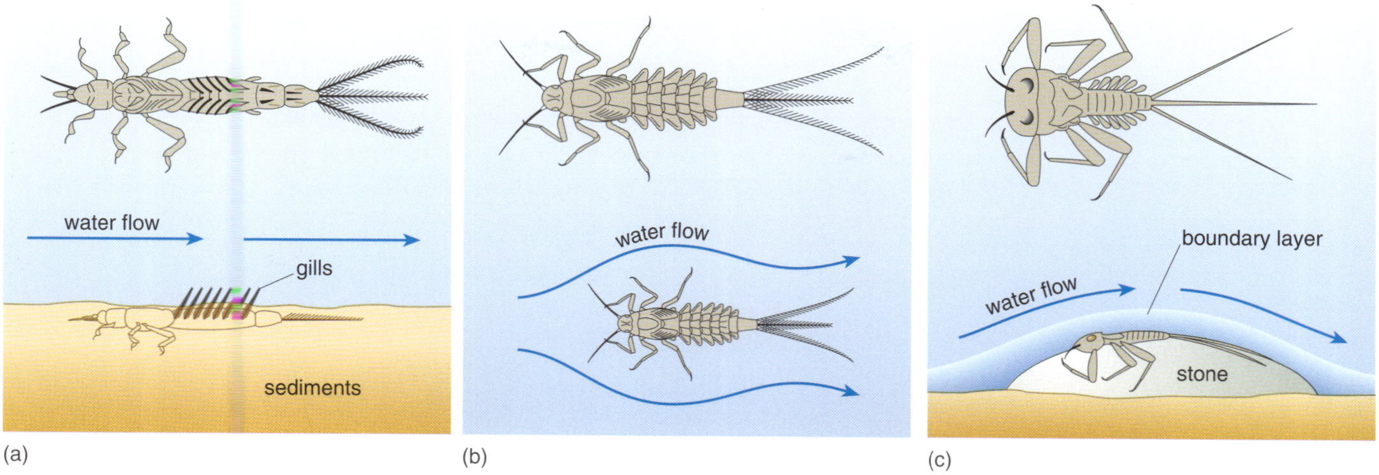

Figure 2.54 Diagram showing: (a) *Ephemera danica* raising its gills above the soft sediments of slow-flowing waters; (b) water flowing smoothly round the teardrop shape of *Baetis rhodani*; (c) the flattened shape of *Ecdyonurus venosus*, which keeps it within the slow-flowing boundary layer of water next to stones in the substratum of fast-flowing rivers and streams.

Figure 2.55 A stretch of the River Teign, showing varying flow rates, which create a variety of aquatic microhabitats.

The Environment Agency monitors the quality of river water by carrying out regular sampling of the invertebrates in watercourses, and calculating the Biological Monitoring Working Party (BMWP) index from the sample data. You calculated the BMWP score for two stretches of river, situated above and below a sewage outfall on the River Teign on your virtual field trip. Therefore, you should recall that the system works by assigning a score for each of the invertebrate families listed in Table 2.6 that are present in the samples taken from a particular stretch of river, etc. The individual scores are then added together to produce the BMWP index value. The index works on the principle that freshwater invertebrates differ in their sensitivity to organic pollution. The more sensitive they are, the higher the score assigned to that family.

Table 2.6 Sample BMWP scoring system for producing a biotic index for monitoring water quality.

Group	Families	Score
mayflies	Siphlonuridae, Heptageniidae, Leptophlebidae, Ephemerellidae, Ecdyonuridae, Potaminthidae, Ephemeridae	10
stoneflies	Taeniopterigidae, Leuctridae, Capniidae, Perlodidae, Perlidae, Chloroperlidae	10
caddis-flies	Phryganeidae, Molannidae, Beraeidae, Odontoceridae, Leptoceridae, Goeridae, Lepidostomatidae, Brachycentridae, Sericostomatidae	10
river bugs	Aphelocheiridae	10
caddis-flies	Philopotamidae, Glossosomatidae	8
dragonflies	Lestidae, Agriidae, Gomphidae, Cordulegasteridae, Aeshnidae, Corduliidae, Libellulidae	8
mayflies	Caenidae	7
stoneflies	Nemouridae	7
caddis-flies	Polycentropidae, Rhyacophilidae, Limnephilidae	7
snails	Neritidae, Viviparidae, Ancylidae	6
caddis-flies	Hydroptilidae	6
mussels	Unionidae	6
shrimps	Corophidae, Gammaridae	6
dragonflies	Platycnemididae, Coenagriidae	6
bugs	Mesovelidae, Hydrometridae, Gerridae, Nepidae Naucoridae, Notonectidae, Pleidae, Corixidae	5
beetles	Hydrophilidae, Clambidae, Elmidae, Dryopidae, Haliplidae, Hygrobiidae, Dytiscidae, Gyrinidae, Helodidae, Chrysomelidae	5
caddis-flies	Hydropsychidae	5
true flies	Tipulidae, Simuliidae	5
flatworms	Planariidae, Dendrocoelidae	5
mayflies	Baetidae	4
alderflies	Sialidae	4
leeches	Piscicolidae	4
snails	Valvatidae, Hydrobidae, Lymnaeidae, Physidae, Planorbidae	3
pea cockles	Sphaeridae	3
leeches	Glossiphonidae, Hirudidae, Erpobdellidae	3
hog louse	Asellidae	3
midges	Chironomidae	2
worms	Tubificidae, Naididae, Lumbricidae	1

Question 2.3

Freshwater invertebrate samples were taken and BMWP scores calculated for a small river downstream of a dairy farm in Devon, over a two-year period. Use the survey data in Table 2.7 to calculate a BMWP score for this site in January 1999 and enter it into the appropriate place in Table 2.8. During the period 1998–99, there was an occasion when slurry from the dairy farm spilled into the watercourse. Comment on the nature of any evidence in Table 2.8 that might indicate when the slurry spill occurred.

Table 2.7 Record sheet of invertebrate families recorded at a site on the River Teign in January 1999, and the associated BMWP scores.

Group	Family within group	BMWP score
caddis-flies	Limnephilidae	7
shrimps	Gammaridae	6
snails	Ancylidae	6
beetles	Dytiscidae	5
	Hydrophilidae	5
caddis-flies	Hydropsychidae	5
true flies	Tipulidae	5
snails	Hydrobiidae	3
	Lymnaeidae	3
leeches	Erpobdellidae	3
hoglouse	Asellidae	3
midges	Chironomidae	2
worms	Lumbricidae	1

Table 2.8 BMWP scores for a stream in Devon, 1998–99.

Year	Jan	Mar	May	Jul	Sep	Nov
1998	125	105	95	110	130	135
1999		58	70	100	120	130

2.4.5 Biotic influences on species distribution

Abiotic factors have an important influence on the distribution of species. They are particularly important in determining the potential range of a species. However, the influence of biotic factors on species distribution should not be underestimated. Biotic factors, i.e. other organisms and their products, can also act as resources and conditions. Think back to the example of the greater horseshoe bat; the moths and dung beetles are clearly resources that the bats needs to eat, and old trees are a resource in that they provide roosts and perches.

- ❍ In what sense are trees also a condition for the bats?
- ● Trees provide shade for flight paths and cover near the roost.
- ❍ What is the most likely reason for bats needing these conditions?
- ● To provide them with protection from potential predators such as barn owls.

This point brings us on to the important influence of biotic factors on species distribution. By the very fact that many biotic factors are living things, they can interact with other organisms in ways that abiotic factors cannot. From an individual organism's point of view, other organisms may be seen not only as potential food items, but also as competitors for a given resource (see Section 2.5), as beneficial partners (see Section 3.7), or as predators (which includes **herbivores**, i.e. grazing animals that 'predate' on plants). Clearly, these sorts of interactions will affect the distribution and abundance of species.

This brings us to a third way that an organism's environment can be described. Most of the resources and conditions you have met so far relate to absolute needs, which effectively determine the maximum number of individuals of a particular species that may live in a particular place, i.e. its **carrying capacity**. These resource-based limits on distribution are known as **bottom-up controls**. In contrast, if the presence of a species is limited or excluded by a predator, this is known as **top-down control**. The distribution of predators and prey are inextricably tied to one another. Predators like bats and dippers are dependent on the existence of suitable numbers of their prey species. But equally, the populations of their prey species, i.e. dung beetles and freshwater invertebrates respectively, can only withstand certain levels of predation. If recruitment to the population through birth or immigration does not keep pace with losses through predation, old age and disease, then the long-term existence of a population is threatened, as the following case study shows.

Case study: the water vole (Arvicola terrestris)

Figure 2.56 The water vole (*Arvicola terrestris*), a once abundant mammal of the British countryside.

The water rat (Ratty in the *Wind in the Willows*) is more properly known as the water vole (*Arvicola terrestris*, Figure 2.56). Inhabiting burrows along the banks of lowland rivers or ponds with abundant low vegetation, the water vole was, until recently, a common sight swimming purposefully across the surface of the water in search of the water plants and bank-side plants that form the bulk of its diet.

However, since the early 1900s there has been a big decline in the distribution and abundance of this species. Only a third of the sites occupied by water voles a century ago were still occupied in the 1990s, and within the inhabited areas, the numbers of voles have fallen. Factors such as: increased afforestation with coniferous species; acidification of waterways due to acid precipitation and the aforementioned afforestation; habitat destruction from engineering works on the rivers causing increasing fragmentation and isolation of the colonies; and water pollution resulting from urbanization and agricultural runoff have all contributed to the species decline.

- ❍ Classify these factors as top-down or bottom-up controls.
- ● They are all bottom-up controls because they all affect the availability of suitable habitat for water voles, which is a resource.

However, the factor most likely to cause the final extinction of the water vole in Britain is predation by the feral American mink (*Mustela vison*, Figure 2.57). The word 'feral' refers to an animal that has escaped from captivity beyond its normal range and thus become an alien species. The mink, a smaller cousin of the otter (*Lutra lutra*), having initially escaped from fur farms, has successfully spread and established itself across much of lowland Britain, occupying many of the sites previously occupied by otters, which are also ideal vole habitats.

Figure 2.57 American mink (*Mustela vison*), a feral species in Britain.

A voracious predator, the mink eats a variety of foods, including small mammals, fish and the young of ground-nesting birds (which include not only game birds such as partridge and grouse, but also species of conservation interest, such as waders, terns, waterfowl, etc.). Researchers have investigated whether there is a link between the spread of mink and the decline of water voles. Figure 2.58 illustrates the relationship between the presence of mink and regional variation in water vole density. Water vole numbers are inferred from the frequency of water vole latrines (each vole habitually uses the same place to deposit its dung).

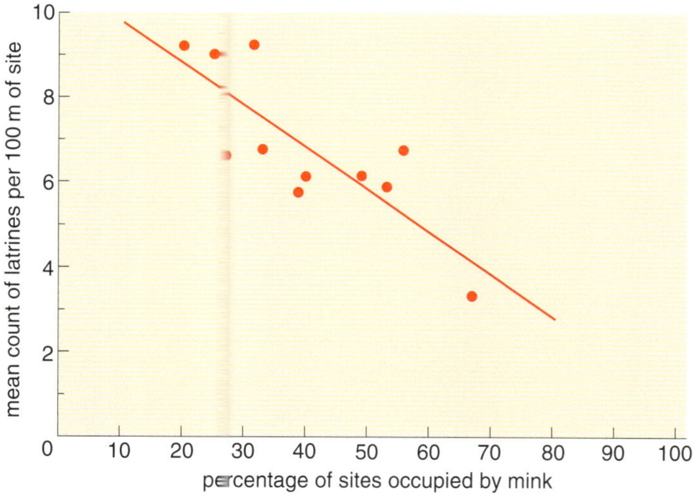

Figure 2.58 Relationship between the presence of mink and the frequency of water vole latrines.

- ❍ From Figure 2.58, what appears to be the relationship between mink abundance and water vole abundance?
- ● As mink abundance increases so water vole numbers decline.

Activity 2.3 Is the case against mink proven?

This exercise uses the Spearman rank correlation coefficient again to determine whether the relationship between two sets of data is likely to be due to chance or not.

The figures in Table 2.9 relate to the points on the graph in Figure 2.58. They illustrate the relationship between water vole abundance (as indicated by mean number of latrines per 100 m of site) and mink abundance (as indicated by percentage of sites occupied by mink).

Table 2.9 Values for mink abundance and water vole abundance at various sites throughout Britain.

Site number	1	2	3	4	5	6	7	8	9	10	11
% sites occupied by mink	67	39	53	49	40	56	33	27	27	20	31
mean vole latrines per 100 m of site	3.3	5.7	5.8	6.2	6.1	6.9	6.8	6.6	9.0	9.2	9.2

(a) Using the information presented in *Using the Spearman Rank Correlation Coefficient*, calculate a correlation coefficient for the two sets of data to test the null hypothesis that there is no correlation between mink abundance and water vole abundance.

(b) State whether the correlation is statistically significant, and if it is, whether the relationship is positive or negative.

(a)

(b)

(c)

Figure 2.59 Water vole predators that are native to Britain:
(a) weasel (*Mustela nivalis*);
(b) stoat (*M. erminea*);
(c) brown rat (*Rattus norvegicus*).

Water voles have always been preyed upon by predators such as weasels (*Mustela nivalis*), stoats (*M. erminea*), rats (*Rattus norvegicus*) (Figure 2.59a, b, and c respectively) and otters.

So why has the mink had such a detrimental effect on water vole numbers? Weasels, stoats, and rats are small enough to enter the vole burrows, but the vole could escape by entering the water, and otters do take voles in the water, but are too big to enter the burrows. However, mink are small enough to enter the vole burrows and are also good swimmers. Therefore, for the vole there is no escape from the mink. At present, Ratty's future looks bleak.

- ❍ On the basis of the evidence presented here, are top-down or bottom-up controls having the greatest impact on water vole abundance?
- ● Top-down controls, in the form of predation by mink.

The impact of one species on the distribution of others is generally limited. However, in some cases the influence of one species can be much more wide-ranging.

- ❍ Identify one species that could be reasonably said to have influenced the distribution and abundance of virtually every species in Britain (if not the world).
- ● Humans (*Homo sapiens*).

Ironically, one of the main reasons for the prevalence of feral mink in Britain is the deliberate release of these animals from fur farms by animal rights activists, and one of the most effective means of controlling mink is through hunting with dogs. Clearly it is very important to have a thorough and objective appreciation of the life history and ecology of an organism, before attempts are made to control or promote their abundance.

2.4.6 Human influences on species distribution

As hunter-gatherers, humans harvested naturally occurring plants and animals. However, since the development of agriculture, human activity, including mining and especially agriculture, has shaped the entire landscape of Britain. It is almost certainly true that there is not a single scrap remaining of the original wildwood that once covered Britain. Traditional activities, such as forest clearance, ploughing, mowing, grazing, hedging and ditching, influence the ability of organisms to complete their life cycles, an essential prerequisite if populations are to be sustainable. More recently, industrial activities have caused substantial changes to the atmosphere, land and watercourses, which have profoundly affected organisms living all over the world. In all these respects, humans have influenced the distribution of species.

- ❍ In terms of top-down and bottom-up controls on other organisms, how has the shift from a hunter-gatherer to an agricultural and then industrial economy changed the way humans influence the distribution of other species?

- ● Hunter-gatherer communities took a proportion of the naturally occurring plants and animals around them, through harvesting or hunting. Therefore they exerted predominantly top-down controls on other organisms; as agriculture and industrial economies have developed, humans have made a much greater impact on the quality of the environment around them, so there has been a shift in emphasis towards bottom-up controls.

Today, it is the pressures of intensification of farming methods going hand-in-hand with diversification into the leisure industry that is changing the landscape. It is erroneous to suggest that the distribution of any species in Britain (or indeed most of the world) is 'natural', implying that it is free from human influence. The entire British flora and fauna are 'unnatural' to some extent, and this is as true for 'set-aside' agricultural land as it is for a golf course or the municipal bowling green. However, it is also erroneous to conclude that human influence is always detrimental to other species, as the following examples illustrate.

Case study 1: White admiral butterfly

You have already established on your virtual field trip (Block 1) that the caterpillar of the white admiral butterfly (*Limenitis camilla*) requires shady conditions and honeysuckle to feed on, while the adults need brambles for nectar. This species has spread since the 1920s.

A crucial causative factor seems to have been the reduced amount of coppicing of woodland that has occurred during the 20th century. Coppicing involves cutting back trees to ground level, which encourages new growth of multiple stems from the remaining 'stool'. These stems are harvested every few years (5–15 yr, depending on the size of wood needed) and can have many uses, including woven fences and baskets. Cessation of coppicing benefits the white admiral because this species needs shade. It lays its eggs on honeysuckle, which flourishes in shady conditions. When coppicing is carried out, shade is reduced and honeysuckle is cut back (foresters consider it a weed). When coppicing declines, as it has done throughout the 20th century, honeysuckle thrives and so does the white admiral. The butterfly has also benefited from the spread of conifer plantations where honeysuckle grows along the shady ride edges, and where bramble patches often develop.

Case study 2: High brown fritillary

The high brown fritillary (*Argynnis adippe*, Figure 2.60) over-winters as eggs, the larvae hatch in early spring and spend long periods basking on dead bracken or in short, sparse vegetation. The temperatures in these microhabitats can be 15–20 °C higher than in surrounding grassy vegetation, which allows the larvae to develop quickly in otherwise cool spring weather. The caterpillars eat violets, mainly the common dog violet (*Viola riviniana*, Figure 2.61), and the adults feed on aphid honeydew in tree canopies and on nectar from thistles and brambles. This species has undergone a great decline since the 1950s (Figure 2.62).

Figure 2.60 A declining butterfly, the high brown fritillary (*Argynnis adippe*).

Figure 2.61 The common dog violet (*Viola riviniana*), a food source of high brown fritillary caterpillars.

In Devon, the species is restricted to the steeper sloping land surrounding the edge of Dartmoor, and in the deciduous woodland of the upper Teign Valley. In fact, this area is now an important national stronghold for the species. The reason for this is as follows. The high brown fritillary used to be found in the open woodland associated with regular coppicing. With the decline in coppicing during the 20th century, this butterfly has disappeared from woods. Today, bracken and violets only occur together in sufficient quantities on south-facing slopes, which are warmer, and where ponies and cattle graze them, as occurs on the fringes of Dartmoor. Cattle and ponies are particularly effective at breaking up the dead bracken litter. This trampling creates a mosaic of grass and bracken where violets and therefore the high brown fritillary can flourish. If grazing ceases, the bracken litter quickly builds up and eventually eliminates violets as well as the butterfly.

Question 2.4

Indicate whether the following changes in abundance result from bottom-up or top-down controls, or a combination of the two. List the principal reason for the change.

(a) Decline of the high brown fritillary butterfly (*Argynnis adippe*); (b) increase of the white admiral butterfly (*Limenitis camilla*); (c) decline of the water vole (*Arvicola terrestris*); (d) decline of the brown argus butterfly (*Aricia agestis*); (e) decline of the otter (*Lutra lutra*); (f) decline of wolf (*Canis lupus*).

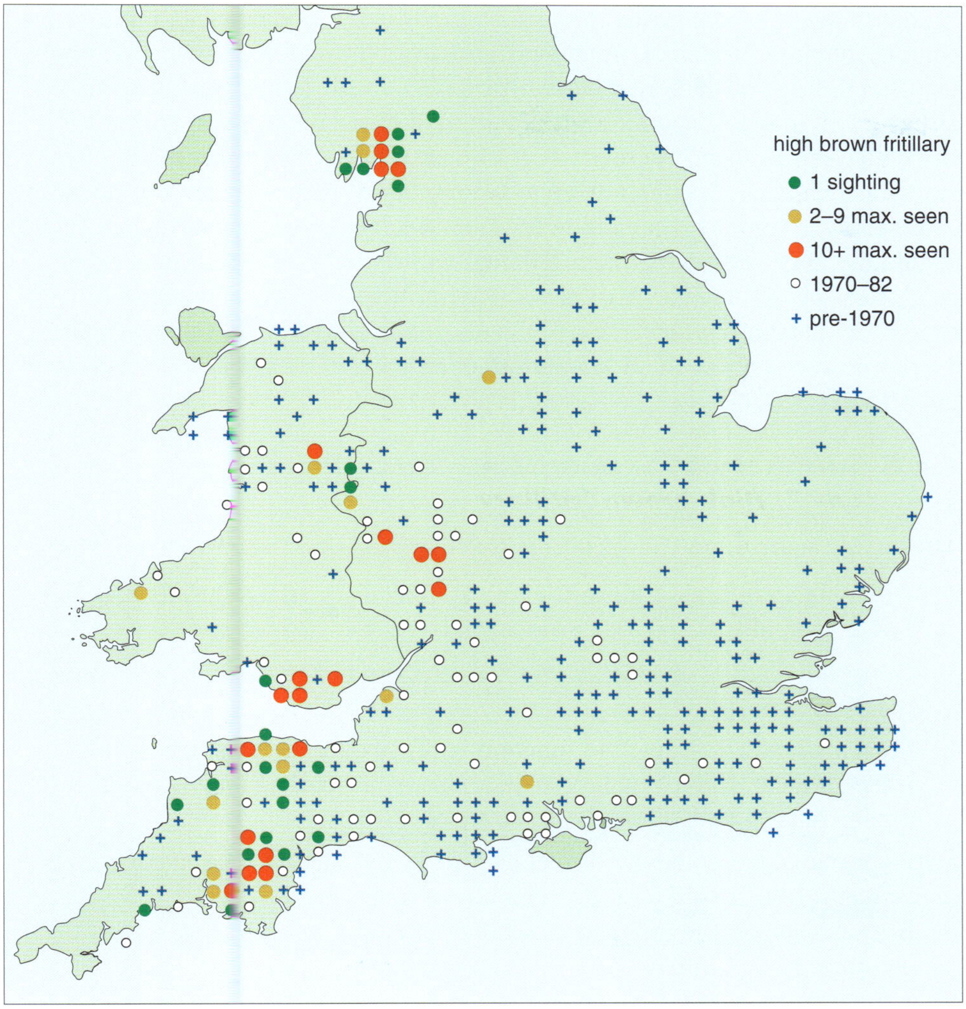

Figure 2.62 Map showing the distribution of the high brown fritillary in 1990s and its former range during the 1970s and earlier.

2.4.7 Conclusions

It should now be clear that organisms are influenced by both abiotic and biotic elements in their environment, and that all organisms have differing resources and conditions that they require. Nevertheless, it is possible to make some broad generalizations in terms of the key factors that determine the distribution of plants and animals. Plants, as autotrophs, need only simple inorganic raw materials from the air and soil. Therefore, plant distributions are determined in the first instance by abiotic factors such as underlying rock type and climatic factors that will determine soil type (including its moisture and pH) and temperature ranges. In contrast, animal distribution is determined primarily by food source and the availability of physical resources such as nest sites and refuges from predators. Therefore, animal distribution tends to be linked closely with the distribution of food species and the geology and topography of the land. These facts explain why animals are so closely linked to vegetation type. Both animal and plant distribution is greatly affected by agriculture, including management activities (hedge laying and cutting, excavation of ditches, mowing, insecticide and herbicide use), the introduction of animals (e.g. mink), and industry; the latter generates pollution such as effluents, which can affect the immediate locality, or acid precipitation, which affects more distant sites.

Humans have affected the distribution of all virtually all species. In times gone by, humans influenced abundance primarily through top-down controls, such as hunting and gathering naturally produced fruits, etc. There are examples of animals becoming extinct through over-hunting, e.g. bear, wolf and wild boar. However, in more recent times, humans affect the distribution of other organisms by influencing the availability of particular resources and conditions through management of the land. Such bottom-up control has potentially very profound effects, although not always to the detriment of an individual species. Human activities are the principal reason why many animals and plants are increasing or decreasing in abundance. Humans have the power to determine the destiny of all other species with which they share the world.

2.4.8 Summary of Section 2.4

1 An organism's essential requirements can be classified as resources, e.g. oxygen availability, or conditions, e.g. temperature.

2 An organism's environment has both abiotic, e.g. rock and water, and biotic components, e.g. other organisms.

3 On a global scale, climate determines the dominant, terrestrial vegetation types (biomes).

4 Each organism is adapted to function within a defined range of temperatures, which may be broad or narrow depending on whether the organism is an endotherm or an ectotherm.

5 Soil type is one of the major factors affecting plant distribution because it influences the availability of most mineral nutrients and water. The rocks from which soils are derived affect their chemical composition, pH and texture.

6 Most plants are adapted to grow on soils with a pH of around 6.5; at this pH all of the essential nutrients are readily available to plants. However, some plants, called calcicoles, thrive at pH above 7.5 and some, called calcifuges, thrive at pH below 5.5. At these more extreme pH values, some of the mineral nutrients may be present at potentially toxic or deficient concentrations.

7 In aquatic habitats, water acts as a condition. Water movements have a huge impact on the distribution of aquatic organisms. A range of adaptations have evolved to adapt individual species to microhabitats with closely defined water movement characteristics.

8 The amount of organic matter, both living and dead, determines the biological oxygen demand (BOD), which affects the availability of oxygen in water. Oxygen availability is a key factor affecting the distribution of aquatic species. The freshwater invertebrate fauna can be used as a biotic index of water quality.

9 A biotic factor may be: a food source, a predator, a beneficial partner or a competitor.

10 The distribution and abundance of a species is affected by the availability of key resources and conditions (bottom-up controls) and the intensity of factors like predation (top-down controls).

11 The water vole has undergone a dramatic decline in the second half of the 20th century. This is partly because its habitat has been destroyed or degraded by human activity, but mainly because of the predatory activity of feral mink.

12 Human activity has influenced the distribution of most species. As hunter-gatherers they exerted mainly top-down controls, but globally they have had a more profound, and frequently catastrophic effect, on the distribution of species through their agricultural and industrial activities, which exert predominantly bottom-up controls.

2.5 Competition and coexistence

2.5.1 Competition

Competition between species (**interspecific competition**) is defined as the interaction that occurs when the abundance of two or more neighbouring species is limited by the same resource. The effect of competition is to reduce the abundance of both species because they are having to share a finite resource. Either species would grow or multiply faster if the other were not there. For plants, the limiting resource is often light, but may also be water, nitrogen, phosphorus or other mineral nutrients. In order to avoid competing with neighbours, each species has tended to develop a slightly different requirement for each resource and a different capacity for capturing each of them. As a result, each species has a particular range of environments that can support it, which might differ from the range of another species, though there is often much overlap. This range is known as a species' **fundamental niche** (in some texts it is referred to as the physiological niche). A fundamental niche can often be very broad, that is to say it can encompass a wide spectrum of environments. Take the daffodil (*Narcissus* spp.) as an example. One can see cultivated varieties growing throughout Britain in all climates and in a wide range of soils and habitats from motorway embankments to window-boxes (Figure 2.63a) — a very broad niche indeed! One of the parents of these cultivated hybrids is a British native plant, known as the wild daffodil (*Narcissus pseudonarcissus*). It is found mainly in the southwestern half of Britain and predominantly in deciduous woodland on freely-draining soils in moderate shade (Figure 2.63b). Its actual distribution, therefore, is only a very small subset of its fundamental niche (Figure 2.64), assuming it could potentially grow wherever its cultivated forms do. The reason a species fails to occupy its full potential niche is that other species compete for the same space or resource. The range of environments a species actually occupies in nature is referred to as its **realized niche** (or ecological niche, Figure 2.65).

(a)

Figure 2.63 (a) Daffodil cultivars (*Narcissus* spp.) growing in an artificial habitat. (b) The native wild daffodil (*Narcissus pseudonarcissus*) in its natural habitat, woodland.

(b)

Question 2.5

In Figure 2.65, the wild daffodil is often found in environments denoted by point A, but not often at those denoted by points B and C. From the information in the diagram, explain why it is not found at (i) B and (ii) C.

Why is the cultivated daffodil found so much more extensively than the wild one? The obvious reason is that it is planted everywhere and so has human-aided dispersal, whilst its wild cousin has to rely on chance distribution of seeds. The less obvious, but equally important, reason is that human activity protects the cultivated daffodil from competition. The grass or weeds amongst which they grow are mown or removed by gardeners. Without this intervention, the cultivated daffodil would largely succumb to competition, whilst the wild relative would tend to continue in its realized niche.

❍ In Figure 2.64, observe the yellow circles, which indicate where wild daffodils have established successfully, following an artificial introduction to the locality. Considering their distribution, do you think the areas in which this species is absent are beyond the fundamental niche of the species or is its distribution limited by its dispersal ability?

● Successful introductions are scattered all over the country, suggesting that the species has a broad fundamental niche but is limited by its ability to reach new habitats.

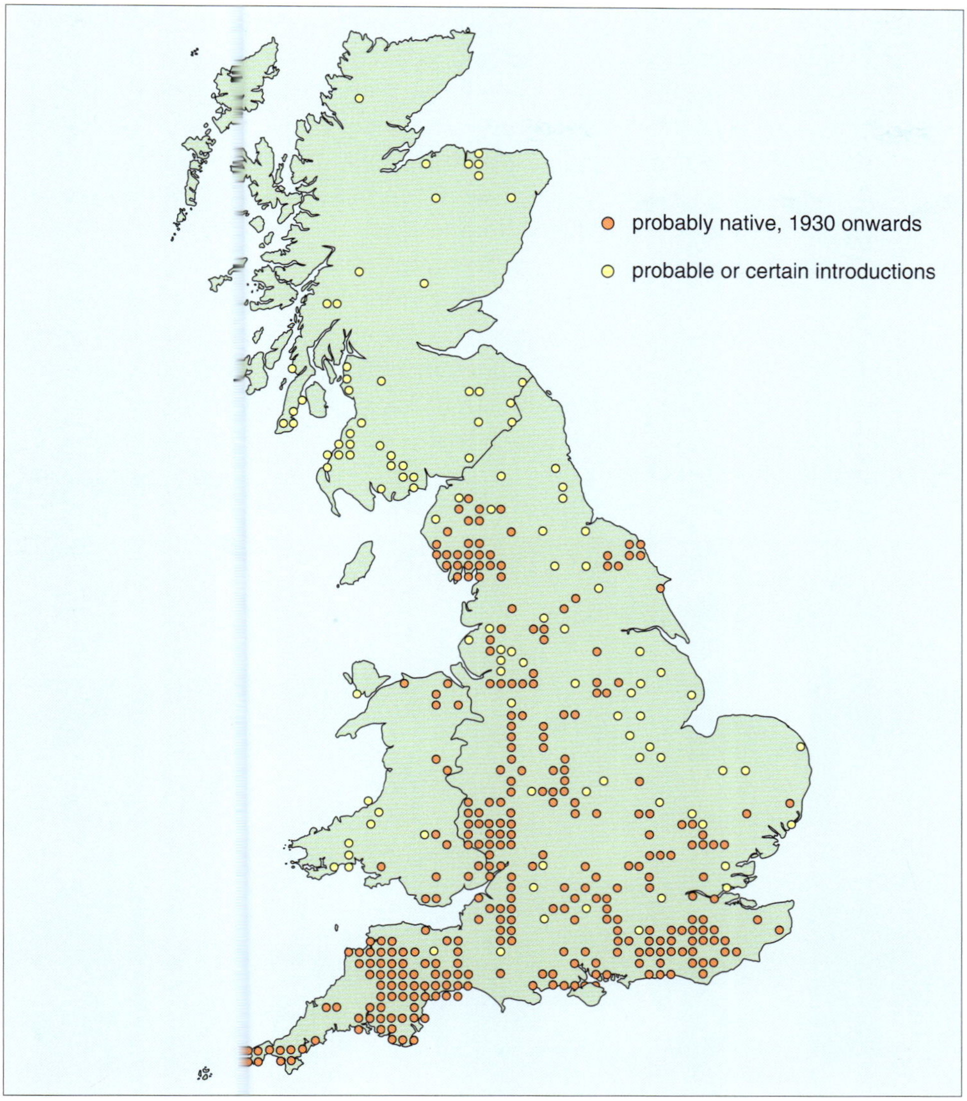

Figure 2.64 The distribution of the wild daffodil in Britain.

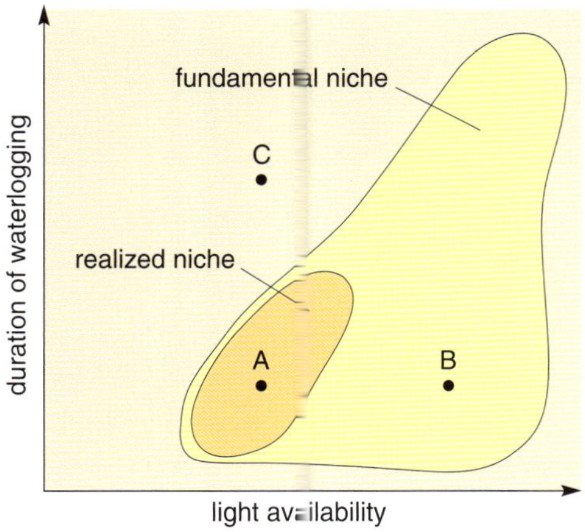

Figure 2.65 An estimate of the wild daffodil's fundamental and realized (ecological) niches against just two (light availability and soil wetness) of many possible axes.

Realized niches tend to be narrow. The range of environments in which a given species is able to outcompete its neighbours is relatively small, compared to the full range of conditions in which it could grow, if allowed to do so without competition. This is of importance when considering changes to an environment. For example, one may not expect an apparently small increase in soil wetness to cause the buttercups in a field to disappear, when it is known that buttercups can grow on wet soils. However, if the additional wetness causes the grass in the meadow to grow faster and taller, then the buttercups will experience a reduction in light intensity due to the increased shading from the grasses (see Figure 2.66). If the buttercups' upward growth is slowed down because they are not receiving enough light to grow at their potential rate, then they receive even less light. In such a way, a small environmental alteration can trigger a large change in the distribution of a species. Competition between species is the process that is responsible for this phenomenon.

Figure 2.66 Meadow buttercups (*Ranunculus acris*) growing in competition with grasses.

This effect was shown elegantly in an experiment by Heinz Ellenberg (a famous German botanist). He grew six grasses along a gradient of water-table depth. Each species was grown either alone (no competition between species) or in a mixture of all species together (species in competition with one another for the same resources). Two of the species used were the meadow foxtail (*Alopecurus pratensis*) and the upright brome (*Bromopsis erecta*). Each species' performance was assessed by measuring the amount of shoot biomass they produced over a season. Figure 2.67 presents a simplified extract of the results relating to these two species, from a much larger and more complex experiment.

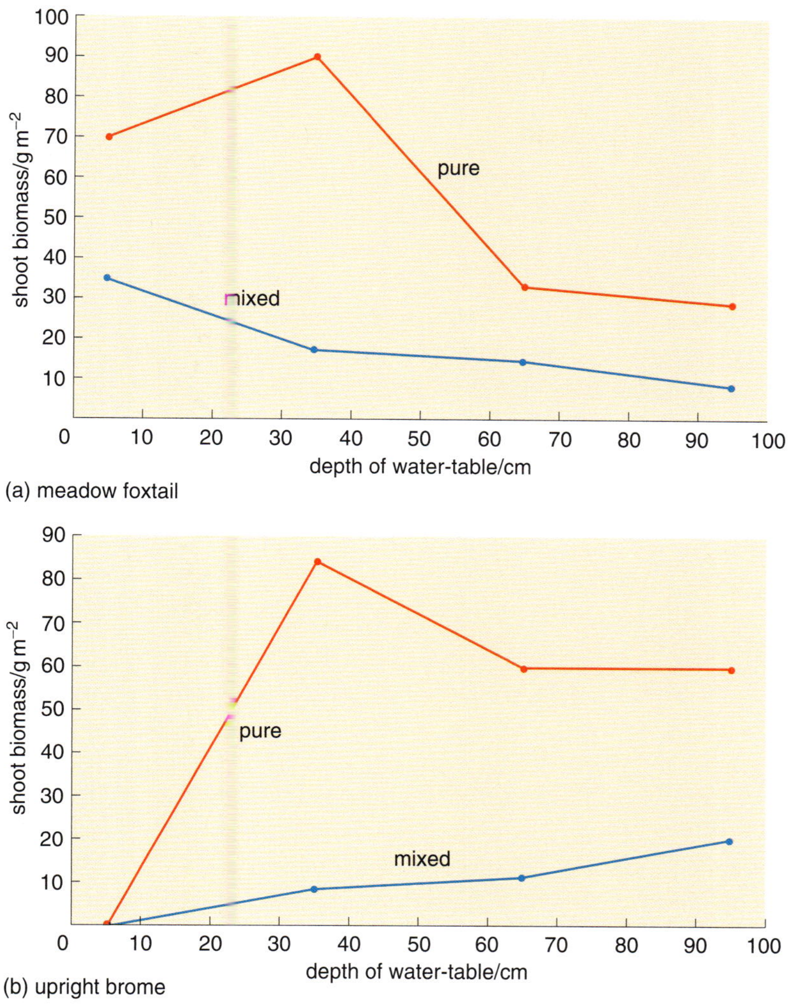

Figure 2.67 Biomass yields for grasses based on the results of an experiment by Heinz Ellenberg: (a) the yield of meadow foxtail (*Alopecurus pratensis*) grown either as a single-species stand (pure) or in combination with five other grass species (mixed); (b) equivalent data for upright brome (*Bromopsis erecta*).

- ❍ Which species has the broader fundamental niche?
- ● Meadow foxtail. Upright brome is absent from the shallowest water-table treatment, even when grown alone.

- ❍ For these two species, how does their growth in mixture compare to their growth when grown alone, and what is the reason for this?
- ● Both species grow less well in mixture. The reason is that competition from one another and from the four other species deprives them of resources for growth.

- ❍ Would you expect to find upright brome in the wild on soils with shallow or deep water-tables?
- ● Deep water-tables. The species grows best in a mixture when the water-table is deep.

2.5.2 Competitive exclusion

If two species, A and B, are competing for exactly the same set of resources and species A is more efficient at gathering them than species B, then one would expect species B to die out. It would have been excluded from that particular habitat by species A. This is a theoretical concept, referred to as **competitive exclusion**. If the earlier example involving buttercups is extended, one could argue that the species able to grow fastest and tallest within the meadow (such as the grasses) should be able to capture most of the sunlight and exclude all other species (such as the buttercup). As plants all have the same basic requirements — light, water, oxygen, carbon dioxide and mineral nutrients — they are likely to be always in competition for these resources with their neighbours. Taking the idea further, we could argue that each habitat should contain only one species. The species that is best-adapted to an environment and able to grow fastest should be able to exclude all others by outcompeting them for the resources available.

How then is it possible for two or more plant species to coexist in the same habitat?

2.5.3 Coexistence

There is no simple answer to the question of coexistence, but there are a number of theories, which we shall now discuss. Each of them has some supporting evidence and it would be possible for all of them to be true and for the different mechanisms all to occur at once, so that in combination they may explain the observed species richness of many habitats. There is not scope within this course to review the evidence for each of the theories, but the ideas are outlined below to illustrate the range of mechanisms that may allow coexistence of species with similar requirements.

1 *Resource variability.* The habitat may vary on quite a fine scale in terms of its resource availability. For example, one patch of soil perhaps only a few centimetres across may be rich in the mineral nutrient nitrogen and favour species A, whilst an adjoining patch may be lower in nitrogen but richer with respect to another nutrient and favour species B. For instance, a grass may dominate the first patch whilst clover, which does not rely on soil nitrogen, may dominate the second. Two species may live in close proximity, yet be exploiting subtly different niches.

2 *Niche separation.* The niche each species exploits may differ in time or space. For example, different rooting depths may allow several species to exploit the same soil profile by extracting water and nutrients from different layers (Figure 2.68). Alternatively, growth spurts in different seasons allow species to occupy the same space without intense competition for light. This can be seen in a hay meadow where early-season grasses (such as the meadow foxtail, which attain greatest leaf area before the hay cut) can coexist with species that develop later in the season (e.g. Yorkshire fog, *Holcus lanatus*). Thereby, plants avoid competing for exactly the same resource at exactly the same time. This is also referred to as **resource partitioning**.

Figure 2.68 Resource partitioning by two plants of the woodland floor (bluebell, *Hyacinthoides non-scripta*, and soft grass, *Holcus mollis*) through variation in rooting depth.

3 *Establishment opportunities*. New individuals require gaps in the vegetation to establish themselves. When a gap appears, the species that colonizes it is likely to have an effective dispersal ability (see later) and/or a bank of dormant seeds in the soil beneath. Some species that are weak competitors may be good dispersers, and thereby maintain their place in the habitat by exploiting gaps before stronger competitors can reach them. The small grass species known as annual meadow grass (*Poa annua*) is too small to maintain itself in a dense grass sward by competing for resources but, by having a high seed production rate and seeds that can survive in the soil for years, it is able to persist in a wide range of vegetation types, wherever and whenever the soil is disturbed to create a gap and to expose its seeds.

4 *Natural enemies*. The success of a given plant species (i.e. the exclusion of most other species of plant through competition and the formation of a dense canopy over an extended area) will favour those parasites, diseases and herbivores that have adapted to feed on it. The destruction of its tissues by these other organisms will tend to reduce its abundance. When a single species does dominate a habitat, over time any worm able to eat its roots, any fungus able to live in its stem, or any insect able to eat its leaves will itself be very successful and multiply in that habitat. The dominant plant species would thus be weakened. A second species of plant that had been relatively infrequent and had not built up large populations of parasites, etc., may then be able to compete with the original species and to coexist with it. In particular, tree seedlings often struggle to establish under the canopy of a mature tree of their own species due to the 'rain' of leaf-eating insects falling from the high canopy with a taste for their leaves.

5 *Climatic variability*. Although under normal conditions the dominant species may be able to exclude others, in years that are perhaps particularly dry or particularly cold, another species will be the better adapted to the habitat and will succeed in establishing a place for itself. If the new species is long-lived, it may then maintain itself in the habitat for a number of years. Temporal variation in environmental conditions can therefore facilitate coexistence. Take, for example, two grasses of the North American prairie, buffalo grass (*Buchloë dactyloides*) and blue grama (*Bouteloua gracilis*), which occupy very similar niches and compete strongly with one another. The buffalo grass is the stronger competitor and in favourable (wet) years it suppresses blue grama. However, blue grama is the more drought-tolerant and maintains its place within the habitat as a result of periodic droughts that occur in the region (Figure 2.69), which allow it to spread out amongst the weakened buffalo grass.

❍ What role do species from outside the plant kingdom play in maintaining the diversity of plant communities?

● Fungi and animals that consume living plant tissue and weaken the competitiveness of dominant species, allow other less dominant species to compete.

❍ Based on Figure 2.69, what might be expected to occur if drought years became the norm in the region, as a result of climate change?

● One would expect buffalo grass to decline and possibly disappear, if the habitat were no longer within its fundamental niche. Blue grama may then become the dominant grass.

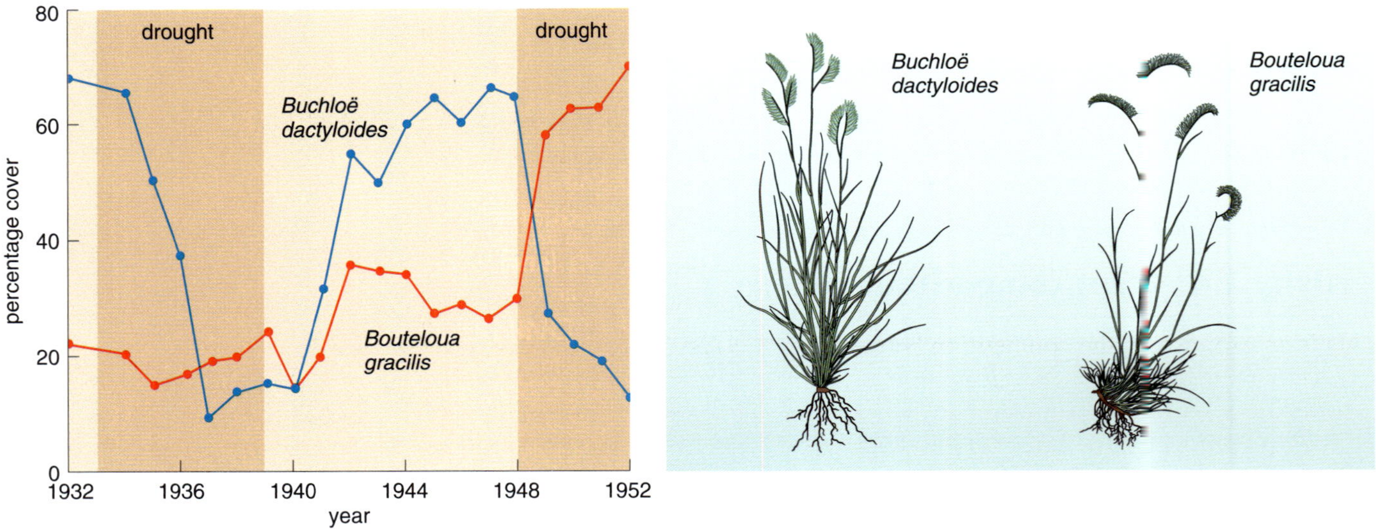

Figure 2.69 Relative abundances of buffalo grass (*Buchloë dactyloides*) and blue grama (*Bouteloua gracilis*) in Kansas in relation to episodes of prolonged drought. The *y*-axis indicates the percentage of ground each species covers. After prolonged droughts there would be a lot of bare ground, which would be exploited by annuals following rain.

The net result of these various mechanisms can be that large numbers of plant species coexist. Some grassland plant communities have 40 different species living within the same square metre! (See Figure 2.70.) Recent evidence suggests that to attain such diversity amongst plant species requires a similar diversity in the soil fungal community. (The subject of mycorrhizal fungi is addressed in Section 3.7.)

Figure 2.70 A diverse plant community in a British floodplain meadow with over 30 species per square metre. Species you can see include the blue devil's-bit scabious (*Succisa pratensis*), the yellow meadow buttercup (*Ranunculus acris*) and the red greater burnet (*Sanguisorba officinalis*), in addition to a variety of grass species.

- ❍ Which plant community would you expect to be most diverse: one growing on a very uniform soil or one growing on a soil whose physical and chemical properties varied spatially on a fine scale, and why?
- ● The one showing fine-scale variation in soil properties, because it may provide a range of different niches and thereby promote coexistence of species with varying requirements.

Activity 2.4 Plant competition on heathland

This is an exercise on the competition between heather and bracken. It should take you about 2 hours to work through.

Question 2.6

The environment in an arable field is deliberately manipulated to promote the growth of a single species, such as wheat (*Triticum aestivum*), which is then managed to maximize its capture of available resources. One may expect all other plant species to be excluded by competition from the wheat plants, yet some of them do manage to grow within the field. Describe three distinct mechanisms by which these other species may be able to access the resources they need for growth in the face of competition from a dense crop of wheat.

2.6 Response of vegetation to environmental change

2.6.1 Environmental indicator values for plant species

In order to predict the effect of an environmental change on the species composition of the vegetation, it would be useful to have information on the environmental tolerances of each species. To do this, the response of a species to many factors, such as light intensity, temperature regime, water and nutrient availability, would all have to be experimentally studied. It is theoretically possible to assign a tolerance range with respect to each of these factors, within which the plant could grow. This would amount to defining its niche. To do this in reality is very difficult, because all the different factors tend to interact and in order to define a realized niche, the plant would need to be studied in the presence of all its possible competitors.

Another approach to the problem is to simply observe where it does grow in the environment. If the species is observed enough times, some indication of its tolerance to the various environmental factors can be deduced. This non-experimental approach cannot yield a truly quantitative description of a species' niche, but it has the advantage of studying the species in the presence of the other species that actually compete with it. The results of this approach are qualitative. It can be observed that species X generally tolerates more deeply shaded habitats than species Y and that species Y tolerates more frequently waterlogged habitats than species Z. In this way, ecological rankings can be derived. Box 2.3 describes an example of this approach.

Box 2.3 Ellenberg's indicator values

The most famous example of ecological rankings being derived from observations is the work done by the German botanist Heinz Ellenberg. From his long experience of observing plants, he subjectively assigned them to categories in terms of their tolerance to seven environmental variables. These **ecological rankings** or **indicator values** are widely used in environmental studies. Here we give a sample of the information available. Ellenberg originally derived his ecological rankings from observations in central Europe, but subsequent work by British botanists has demonstrated them to be generally applicable to the UK as well.

One of the environmental variables described by Ellenberg is the water regime. It is known as his 'water value' or, in German, *Feuchtezahl* (hence many texts refer to it as the Ellenberg F-value.) Species are assigned to one of 12 water-value categories, as listed in Table 2.10.

Table 2.10 Ellenberg's water-value classification of plants.

Water value	Plant description
1	indicators of extreme dryness; often restricted to places that dry out completely
2	intermediate between 1 and 3
3	dry-site indicators; more often found on dry ground than moist places; not found on damp soil
4	intermediate between 3 and 5
5	moist-site indicators, mainly on soils of average dampness; absent from both wet ground and places that may dry out
6	intermediate between 5 and 7
7	damp-site indicators; mainly on constantly damp, but not wet, soils
8	intermediate between 7 and 9
9	wet-site indicators; often in water-saturated, badly aerated soils
10	indicators of sites occasionally flooded but free from surface water for long periods
11	plants rooting under water, but at least for a time exposed above or floating on the surface
12	submerged plants, permanently or almost constantly under water

The water values for four contrasting British species are given in Table 2.11.

Table 2.11 Examples of water values.

Common name	Scientific name	Water value
common rock-rose*	*Helianthemum nummularium**	3
stinging nettle†	*Urtica dioica*†	6
greater tussock sedge	*Carex paniculata*	9
pond water-crowfoot	*Ranunculus peltatus*	12

* Shown in Figure 2.38; † shown in Figure 2.73.

The other two species listed are shown in Figure 2.71.

(a)

(b)

Figure 2.71 (a) Greater tussock sedge (*Carex paniculata*). (b) Pond water-crowfoot (*Ranunculus peltatus*).

The other six variables for which Ellenberg assigned rankings to species are:

- light
- soil pH
- soil nutrient status
- temperature
- continentality of climate
- salinity.

○ A site formerly dominated by the tussock sedge (*Carex paniculata*, Figure 2.71) was invaded by stinging nettles (*Urtica dioica*, Figure 2.73). What does this suggest has happened to the site's water regime?

● It has become drier. *C. paniculata* is ranked as preferring waterlogged sites whilst *U. dioica* prefers sites that are merely moist.

2.6.2 Classifying the response of the vegetation

In earlier examples considering competition between buttercups and grasses and then buffalo grass and blue grama, we saw that plant communities may be finely balanced with respect to the competitive interactions taking place between species. A small perturbation in the environment may give a particular species a competitive advantage over its neighbours and a significant change to the composition of the community could result. An important aspect of environmental science is predicting what this change may be or conversely, having observed a change, interpreting the cause. It can be difficult to analyse the interactions between vegetation and the abiotic environment at a species level. There may be in excess of 100 plant species involved at a particular site. Reliable data on their individual tolerances to a range of environmental variables are not easy to find (though the use of indicator values can be informative). Environmental scientists have therefore searched for a method of grouping species to make the task more manageable. There are two approaches to this problem:

- To analyse the vegetation in terms of its plant community rather than its component species. This was described in Section 2.3. The use of communities reduces the number of units that one needs to describe in terms of their environmental tolerances. Also, it is often easier to generalize about the requirements of a particular community than it is for a single species.
- To assign species to **functional types**, according to their ecological behaviour. Species within a particular type would have a set of characters in common, employ a similar strategy for survival and respond similarly to a given perturbation. By understanding how a typical species within a functional type responds to an environmental change, the response of other species within that type can be inferred. For example red clover (*Trifolium pratense*) is a relatively slow-growing legume found in grasslands. Several other species of slow-growing legume, all of which rely on atmospheric rather than soil nitrogen for their nutrition, also grow in this habitat. Together they can be regarded as a functional group. It has been shown that applying nitrogen fertilizer to a species-rich grassland greatly reduces the abundance of red clover. Using the functional group approach, one would infer that any other slow-growing legumes in the sward would also decline.

○ Why might you expect legumes to decline when nitrogen fertilizer is added?

● Legumes are generally well supplied with nitrogen from the atmosphere and so tend not to benefit from additional nitrogen availability in the soil. Their competitors such as grasses would benefit from the additional nutrient and grow faster, outcompeting the legumes for light.

Both these approaches aim to simplify the actual system under consideration. This is an essential aspect of all science. The use of the community classification route is limited by its geographical constraints. The British system was discussed earlier (Section 2.3). The results can be used to tap an extensive ecological literature on British vegetation, but the system cannot be readily applied elsewhere in the world, because different species around the world assemble themselves into different communities. However, as a counter-example, the British system has been applied with some success in New Zealand, because some of the grassland communities there are so dominated by introduced species of British origin! (See Figure 2.72.)

The functional-type approach has more potential as a universal system. If group membership can be defined on the basis of a set of characteristics, it should be possible to classify any species, wherever it is found in the world.

Figure 2.72 A grassland plant community, which was photographed in New Zealand but is dominated by a number of European species that were introduced by settlers and have spread as aliens. The grassland is so European in character that it could be classified by the British NVC!

2.6.3 The use of functional types in describing vegetation patterns

There has been a wide array of different methodologies proposed for classifying species into functional groups, some using broad categories, others much more detailed. It appears that the more detailed the system, the less universal its application. Furthermore, the classification can consider individual species or entire vegetation types.

- ❍ Which system for classifying plants according to their structure and, by inference, their function have you encountered earlier?
- ● The Raunkaier system (Section 2.2.3), which classifies plants according to the position of their growth points relative to the ground surface.

We shall consider a further two systems that are very different in terms of their objectives. The first is an example of a detailed method, in which individual species are categorized according to the suite of traits they have developed to enable them to survive, referred to as their **survival strategy**. The second is an example of a more broad-brush approach, considering the physiological requirements of the dominant species within a vegetation type.

C–S–R strategies

Figure 2.73 A typical species categorized as a competitor, the stinging nettle (*Urtica dioica*).

A scheme that has detailed rules for assigning species to classes, and that represents the most highly developed system within the British context, is the plant-strategy approach of Philip Grime of Sheffield University, UK. This considers how plants survive and exploit their environment during the mature phase of their life cycle. At its broadest level, the system assigns species to one of three groups. We shall look at each of the three in turn.

Competitors (C) are species that survive by outgrowing their neighbours. To do this, they need rapid growth rates and a tall upright habit. By growing taller than their neighbours, they shade them and inhibit their growth. Competitive species also tend to have deep and extensive root systems that capture effectively the soil resources of water and nutrients. Plants have a limited amount of energy and building materials for new tissues. The cost of adopting this strategy is that reproduction by seed must be delayed because all available resources are required for rapid growth. These plants tend only to reproduce (flower) once they have gained a secure place in the canopy and can afford to divert resources to reproduction. The other drawback of the strategy is the need for a plentiful supply of nutrients to support this rapid growth. Because growth is rapid and a dense shade is cast, older leaves soon become redundant and are discarded. Similarly with roots, because they are extensively exploring and depleting the soil resources, many of them too become redundant. The plant therefore loses a proportion of the mineral resources it has captured, in these discarded tissues, and so needs to capture more. The common stinging nettle (*Urtica dioica*) is a classic competitor (Figure 2.73). It grows rapidly, reaches a height of >1 m in a single growing season (Figure 2.74), casts a dense shade and tends not to flower until towards the end of its second year.

Figure 2.74 The development of a clump of stinging nettles over a growing season.

Figure 2.75 The cowslip (*Primula veris*), a stress-tolerant species.

Stress tolerators (S) are plants that have become adapted to harsh environments. One of the important resources for growth (light, water, mineral nutrients) may be in short supply, or other adverse factors such as temperature extremes, exposure to wind or the presence of toxic chemicals may create stressful conditions for plant growth. The strategy adopted by stress tolerators in these environments is one that uses the resources in short supply in an efficient and conservative manner. These plants have a very low turnover of organs such as leaves and roots. Once they have invested resources in growing a leaf, they often retain it for more than one year and so increase the return on their investment. This contrasts with the stinging nettle, which may make use of an individual leaf for only a few weeks. Stress tolerators therefore have a much lower demand for resources than competitors and are more efficient at obtaining scarce nutrients. They tend to be much more reliant on mycorrhizal relationships with fungi (see later) than do competitors. They can inhabit environments that are very poor in terms of nutrient availability, such as bogs or sand dunes. The price they pay for this ability is a slow growth rate and a small stature in comparison to the competitors. They are therefore unable to compete for resources effectively when conditions permit competitive species to grow. Their only similarity to competitors is that they too delay reproduction until the parent plant is well established. A plant may be two or more years old before it takes the risk of using some of its precious resources to make flowers and fruit. An example of a stress-tolerant species is the cowslip (*Primula veris*, Figure 2.75), which grows in old grasslands that have not been fertilized. This species forms a slow-growing but long-lived, ground-hugging rosette of evergreen leaves, which first flowers in its second or third growing season (Figure 2.76). Relatively little new growth is added each year, but existing tissues are retained from one year to the next.

Figure 2.76 The development of a cowslip rosette over a growing season.

Figure 2.77 An example of a ruderal species, the field poppy (*Papaver rhoeas*).

Ruderal (R) is the label given to short-lived species, which flower just once, then die. Gardeners would refer to them as annuals. The essential difference between these plants and those in the other two categories is that they devote their resources to reproduction as soon as they can. They do not maximize their resource capture as competitors do, nor make most efficient use of resources, like stress tolerators. Ruderals are 'built for speed.' They produce their flowers and seeds in as short a timespan as possible. Some can race through their entire life cycle in six weeks, giving them the potential to have several generations during a single growing season. The drawbacks of this strategy are that they can neither compete effectively with competitors, which outgrow them, nor can they tolerate the harsher environments colonized by stress tolerators. Their niche is in environments that are either regularly physically disturbed, or where the environmental conditions are unpredictable. Physical disturbance, such as soil tillage for arable agriculture, removes all surface vegetation. Long-lived species, which wait to establish themselves before reproducing, are disadvantaged in this environment. Short-lived species, which can return seeds back to the soil before the disturbance occurs, are better adapted to the conditions. The majority of 'weeds' in arable fields or domestic gardens are ruderals. A good example is the field poppy (*Papaver rhoeas*, Figure 2.77), which is a summer annual of arable fields, bearing large flowers throughout much of the summer, with a single plant producing up to half a million seeds before dying. The seeds can lie dormant in the soil for many years and rapidly germinate when they are brought to the surface by cultivations. In non-agricultural environments, ruderals are found in areas such as deserts, with unpredictable rainfall. Again, the seeds lie dormant for long periods until the rains finally arrive. The plant will then grow, flower and set seed in the few weeks before the soil water is completely exhausted. The huge numbers of new seeds produced will then lie dormant, awaiting the next rain, which may not be for several years. In areas of such extreme unpredictability, no other strategy is really viable.

The main characteristics of the three strategies described above are summarized in Table 2.12.

Table 2.12 Lists of characteristics that are used to categorize herbaceous species according to the strategy they display in their established phase.

	Strategy		
Attribute	C	S	R
height	tall	short	varied
growth rate	high	low	high
lifespan	moderate to long	long to very long	short
flowering	late	late	early
examples:	nettle, reed	cowslip, thyme	poppy, groundsel

- ❍ Into which strategy would you place a long-lived chamaephyte?
- ● The S (stress-tolerant) strategy, because chamaephytes are low-growing woody plants according to the Raunkaier system and therefore do not have the tall attribute required to be classed as C, and the long-lived attribute discounts the R strategy. An example of a long-lived chamaephyte is the bilberry you encountered in the Teign Valley field trip.

- ❍ Which strategy opts for early reproduction and what is its analogue under the Raunkaier system?
- ● The R strategy of Grime, which is analogous to Raunkaier's therophyte life-form.

It should be noted that not all species of plant fall neatly into one of these three categories. There is a continuous spectrum of variation between them. This spectrum is often represented as a triangular grid, within which a species can be plotted according to the mixture of traits it displays (see Figure 2.78). It is possible for a species to have traits from all three strategies, in which case it is represented by a point towards the centre of the triangle.

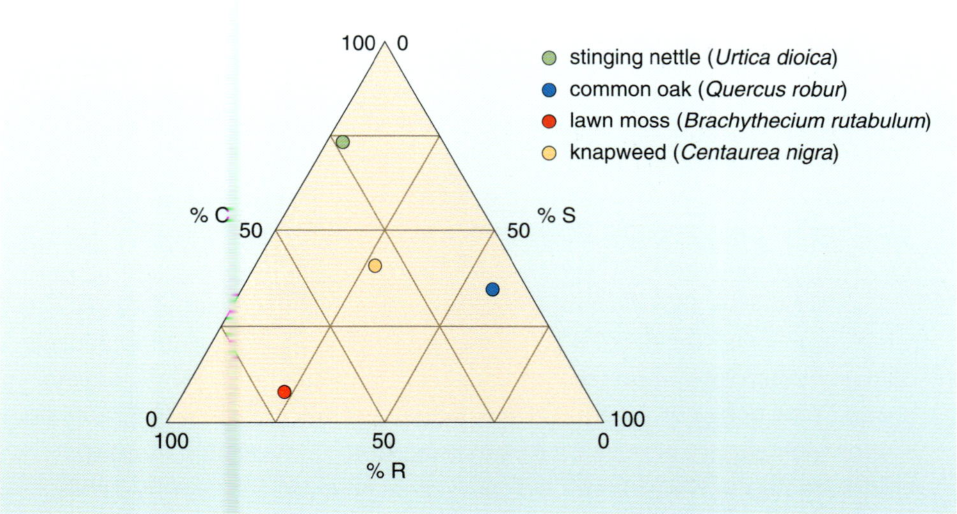

Figure 2.78 Grime's triangular diagram with positions of four named species marked. The three axes represent the proportion of competitive (C), stress-tolerant (S) or ruderal (R) traits exhibited by the species.

- ❍ Which traits would you expect to be exhibited by the species shown in red on Figure 2.78?
- ● The lawn moss shows primarily ruderal features, such as early reproduction, but also has some stress-tolerant traits, such as short stature.

Question 2.7

With reference to Activity 2.4 involving competition between heather and bracken, how would you categorize these two species with respect to their life strategy, and why?

Physiology of vegetation types

A second approach to the use of plant functional types is one proposed by Ian Woodward, also of Sheffield University. This approach contrasts with the C–S–R scheme, by covering a much larger scale and being rather coarser in its definition. It tends to consider the physiology of the dominant species within a broad type of vegetation. It is based on a global-scale analysis of data sets relating to the performance of vegetation under a range of abiotic conditions. By contrast, Grime's approach was based on detailed data collected by sampling vegetation from a range of habitats in a single region, northern England.

The functional types proposed by Woodward can be applied more readily at the level of the vegetation type or biome than to individual species. The aim of Woodward's scheme is to make predictions about the response of the world's vegetation to changes in climate, rather than to local changes in management for which Grime's approach is better suited.

The major environmental variable considered in Woodward's scheme is temperature. He interprets the temperature requirements of plants on a number of fronts. Two important ones are described below.

- *Tolerance of cold*. There are several levels of cold tolerance. These are often determined by the composition of the plant's cellular membranes. Many tropical species are unadapted to cold and their membranes are damaged by exposure to temperatures below 10 °C, even for a few hours. Sub-tropical species have more resilient membranes, which can function at temperatures as low as 0 °C, but which are disrupted by ice crystal formation if the temperature falls lower. Species adapted to grow at higher latitudes can withstand freezing and can survive even when the soil freezes, preventing water uptake. It is possible to categorize vegetation types on this basis.
- *Temperature sensitivity of leaf growth*. Biological reactions controlling rates of leaf expansion, like most other reactions, proceed faster at higher temperatures (see Section 2.4). The precise response to temperature depends on the structure of the enzyme proteins that enable the growth reactions to occur. Some enzymes function at low temperatures but their activity rate does not increase greatly at higher temperatures. Others are capable of much higher maximum rates, but do not function at all at the lower temperatures (see Figure 2.79). This is called a '**trade-off**.' The enzymes involved are either constructed to have a low temperature threshold for activity, or a high maximum rate of activity. It seems they cannot have both. The response of leaf growth rate to temperature is another physiological trait that can be used to categorize vegetation types.

Using a combination of these two variables with others, such as the temperature dependence of water and nutrient uptake rates, species can be placed into a number of categories. These groupings broadly correspond to the vegetation types of the major biomes (see Table 2.13).

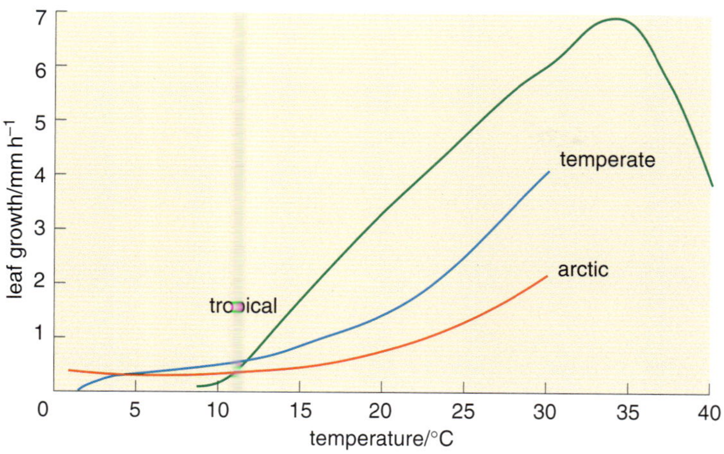

Figure 2.79 Rate of leaf growth as a function of temperature. The plots represent typical responses of leaves from three distinct climatic zones. Note that of the three, leaves from the Arctic grow faster than the other two at temperatures < 5 °C, whilst leaves from temperate regions outperform the others in the range 5–10 °C.

Table 2.13 The temperature tolerances of a range of vegetation types.

Biome	Minimum temperature tolerated/°C	Temperature threshold for leaf growth/°C
tundra	< −40	< 2.5
boreal coniferous forest	< −40	2.5 to 5
deciduous broadleaf woodland	−15 to −40	5 to 7
evergreen broadleaf woodland	−15 to +10	10 to 15
tropical rainforest	> +10	10 to 15

Individual biomes are summarized below and some of them will be addressed in more detail in Block 7.

- *Tundra* is the low-growing vegetation of the Arctic persisting on soils that are frozen for much of the year (Figure 2.80).
- *Boreal coniferous forest* is dominated by tree species such as spruce (*Picea*) and fir (*Abies*) (Figure 2.81).
- *Deciduous broadleaf woodland* is dominated by species of genera such as oak (*Quercus*), maple (*Acer*), beech (*Fagus*) and lime (*Tilia*) (Figure 2.82).
- *Evergreen broadleaf woodland* also has oaks, but these are the evergreen species such as holm oak (*Quercus ilex*), and in the Southern Hemisphere, eucalypts (*Eucalyptus* spp.) (Figure 2.83).
- *Tropical rainforest* is also composed of evergreen, broadleaf species but belonging to groups such as the dipterocarps or giant trees of the legume family (Figure 2.84).

Question 2.8

The lowest temperature recorded in England is −26 °C, whilst the mean temperature in the growing season is 12 °C. Using Table 2.13, to which biome would you expect the natural vegetation of England to belong?

Figure 2.80 A tundra landscape in the Arctic, where lichens dominate the vegetation.

Figure 2.81 A 'fish-eye' view of boreal coniferous forest in North America.

Figure 2.82 Deciduous broadleaf woodland in Europe.

Figure 2.83 Evergreen broadleaf woodland in southeastern Australia.

Figure 2.84 Tropical rainforest in South America.

In this way, species can be divided into broad categories based on their physiological tolerance of environmental variables and the categories then used in modelling studies to predict the response of vegetation to changes in the environment at a global scale. For example, using the estimates of meteorologists studying global climate change, Woodward has made a prediction of how the pattern of broad vegetation types may change in response to climate (Figure 2.85).

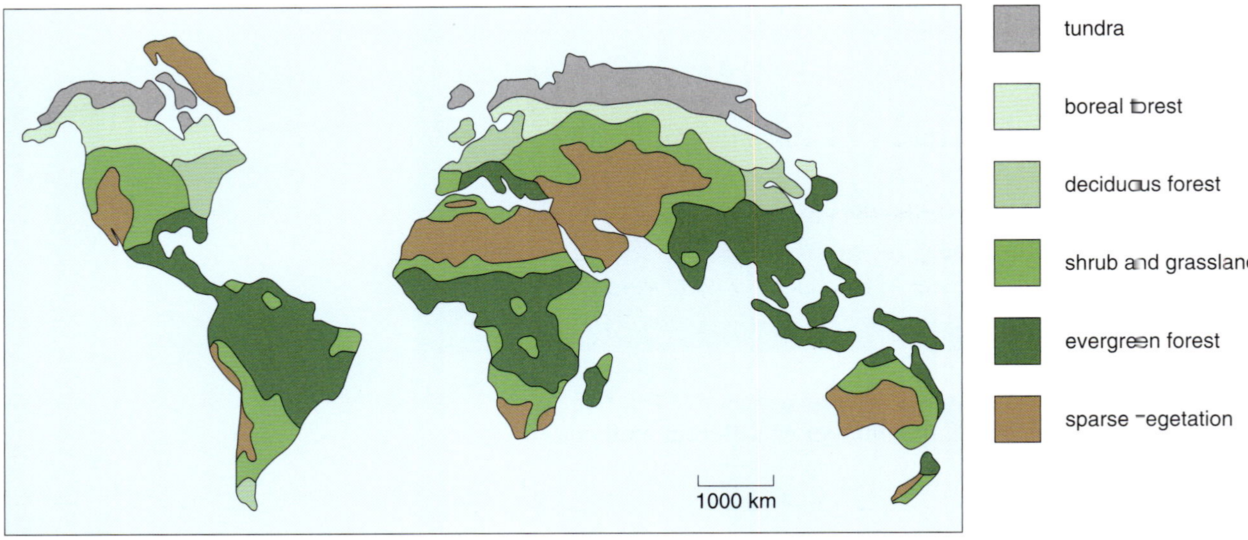

(a) vegetation now

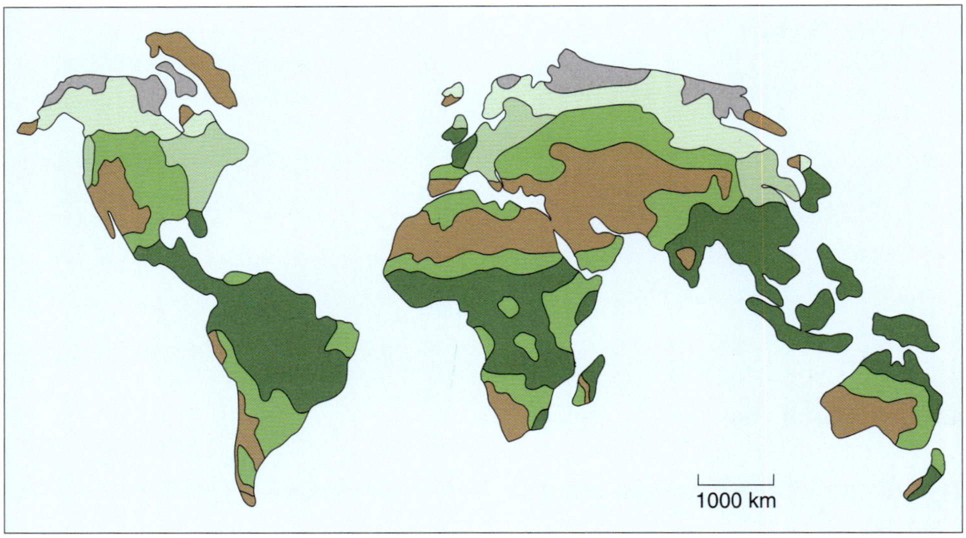

(b) 'greenhouse' world

Figure 2.85 Maps showing the distribution of the major vegetation types (a) now and (b) in the future, according to modelled predictions of climate change.

- ❍ What will be the changes in the distribution of the boreal forest according to Figure 2.85?
- ● It will move northward, displacing tundra over much of its range.

2.6.4 Summary of Sections 2.5 and 2.6

1. A species has a fundamental niche, which describes the range of environments within which it is able to survive. Its realized niche is a more restricted set of environments, ones in which it actually occurs. Many potential environments within its fundamental niche are barred to it by the presence of stronger competitors.
2. Although according to ecological theory, the species best suited to a habitat should come to dominate it and exclude other species trying to compete with it, in reality plants often occur in very species-rich assemblages. A number of mechanisms are proposed to explain this, but current evidence does not allow their relative importance to be assigned. In any single situation, it is probable that several of the mechanisms are acting together to allow coexistence.
3. The use of indicator values to describe the environmental tolerances of individual species is a useful concept in environmental science. The best-known ecological rankings are those of Ellenberg, which were developed for continental Europe but have been shown to be applicable in Britain too.
4. An alternative to community classification is the categorization of species according to functional groups in which species with similar traits are considered together. There are a large number of different methods for assigning species to groups, of which two were considered in more detail: the C–S–R strategy scheme of Grime and the physiological tolerance scheme of Woodward.

Learning outcomes for Section 2

After working through this section you should be able to:

2.1 Explain the central role of vegetation type in determining the distribution of organisms. (*Question 2.2*)

2.2 Describe and explain the relationship between the processes of photosynthesis and respiration.

2.3 Describe and give examples to illustrate the taxonomic hierarchy of classification.

2.4 Describe the hierarchy of the British NVC system. (*Question 2.1*)

2.5 Interpret a floristic frequency table in terms of community classification. (*Question 2.1*)

2.6 Comment on the usefulness of community classification compared to monitoring species populations. (*Activity 2.1*)

2.7 Define the term 'habitat' in two different ways, giving relevant examples of each. (*Question 2.2*)

2.8 Recall and explain, giving examples, the various ways in which an organism's environment can be subdivided and classified, with particular reference to an organism's requirements, the living and non-living components of the environment, and how elements of it act to control the organism's distribution and abundance. (*Activity 2.2*)

2.9 Explain the rationale behind the BMWP biotic index and be able to calculate one from sample data, using an appropriate scoring sheet. (*Question 2.3*)

2.10 Calculate a Spearman rank correlation coefficient and use it to assess whether there is a significant association between the abundance of an organism and an environmental factor. (*Activities 2.2 and 2.3*)

2.11 Describe, using examples, how biotic factors can affect an organism's distribution. (*Activities 2.3 and 2.4; Question 2.4*)

2.12 Interpret data presented in the form of graphs, distribution maps and tables and use the information to draw conclusions about reasons for changes in distribution and abundance of organisms. (*Activities 2.2 and 2.3*)

2.13 Define interspecific competition and explain the competitive exclusion principle. (*Activity 2.4*; *Question 2.5*)

2.14 List at least three mechanisms by which species are able to coexist and cite a relevant example of each. (*Question 2.6*)

2.15 Describe the use of indicator values for ranking the environmental tolerances of plant species.

2.16 Recall the C–S–R strategy approach to assigning species to functional groups, describe the main characteristics of each of the three primary strategies and cite relevant examples. (*Question 2.7*)

2.17 Comment on the broad changes one may expect in global vegetation in response to global climate change and the basis on which such predictions are made. (*Question 2.8*)

Resources to support life

3

In the previous section we saw how the distribution of vegetation within the environment is dictated by the availability of the basic resources required for plant growth and how vegetation patterns then underpin those of all other forms of terrestrial life. This section looks at the mechanisms by which plants capture resources and how they respond to variations in their availability. First we consider how vegetation captures energy. We then look at how this energy is distributed amongst different forms of life and at the various interrelationships that are essential for supporting the diversity of life. All forms of life have a requirement for energy to repair their structures and to maintain their viability. In addition to an energy source, vegetation requires supplies of water, carbon dioxide, oxygen and mineral nutrients to sustain itself. The availability and acquisition of these environmental resources are discussed.

3.1 Energy and materials for life

It is easy to see that organisms need supplies of energy and carbon when they are growing. The fundamental biological molecules that make up the bulk of the dry matter of all living things, such as protein, carbohydrates and fats, are all based on carbon skeletons (i.e. carbon atoms linked together into a huge variety chains and rings), and energy is needed to power biochemical synthesis. But why do all living things need a constant input of energy and carbon in order to maintain themselves, even when they have ceased to grow? The reason is that in nature, there is a constant tendency for order to be reduced to chaos, with a consequent release of thermal energy. Therefore, in maintaining a complex entity like a living organism, there is a need for a constant input of energy and materials; otherwise the body breaks down into its component parts and the energy stored within it is dispersed, i.e. the organism is degraded.

3.1.1 Primary production: energy capture

The concept of autotrophy was briefly mentioned in Section 2.2. Strictly speaking, green plants are **photoautotrophs**, because they use *light* energy to synthesize organic molecules from simple inorganic molecules (carbon dioxide and water), a process known as photosynthesis. (**Chemoautotrophs** obtain energy by oxidizing simple inorganic molecules, e.g. *Nitrobacter* in the soil; see Section 3.7.)

- ❍ Recall the chemical equation that summarizes photosynthesis.
- ● light energy + $6CO_2(g) + 6H_2O(l) = C_6H_{12}O_6(s) + 6O_2(g)$ (2.1)

There are two main sets of reactions involved in photosynthesis. In the so-called **light reactions**, energy from the Sun is converted to chemical energy in the form of complex energy-rich molecules such as **ATP** (**adenosine triphosphate**). These molecules can be thought of as an 'energy currency'. As the name suggests, these reactions take place in the light, and involve the light-harvesting pigments such as chlorophyll. During these reactions, water is split, separating H from O:

$$2H_2O(l) = 4H + O_2(g)$$

The oxygen is lost as a waste product and the hydrogen atoms are held by carrier molecules. The chemical energy is then used to drive the **dark reactions**. These reactions bring about the reduction of carbon dioxide by the hydrogen atoms, and its incorporation into sugars, such as glucose. All of these reactions take place in the **chloroplast** (Figure 3.1), a specialized organelle, several of which are found in all photosynthetic cells of plants.

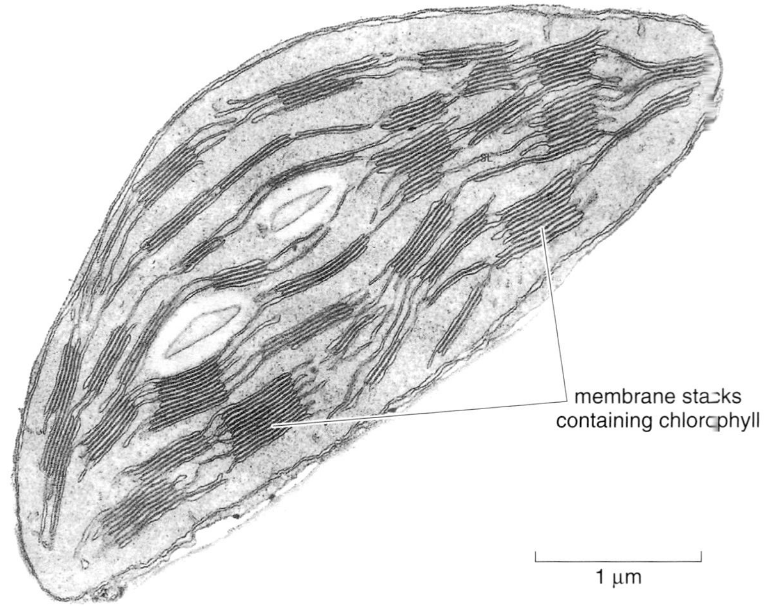

Figure 3.1 Transmission electron micrograph of a chloroplast, showing its internal structure.

3.1.2 The efficiency of energy capture

The carbohydrates produced as a result of photosynthesis and the huge range of substances derived from them are known as primary production. The **net primary production (*NPP*)** is the rate at which plants accumulate dry mass, or **biomass**, as it is called, and *NPP* is measured in units of $kg\,m^{-2}\,yr^{-1}$ (dry mass per unit area per unit time). The net primary production is the difference between the rate at which the plants photosynthesize (gross primary production, *GPP*) and the rate at which they respire (*R*): i.e. $NPP = GPP - R$. The glucose molecules produced in photosynthesis have two main fates: some are used to provide the energy for maintenance, growth and reproduction, and their energy is lost from the plant as heat during processes such as respiration. The rest are deposited in various forms in and around cells and represent stored dry mass.

The biomass produced on Earth each year is about 300 times the world's annual production of steel. But only 0.1% of the total sunlight energy entering the Earth's atmosphere is stored in plant mass in this way. As you know from Block 2, Part 1, as much as 90% of solar energy is absorbed or reflected by parts of the atmosphere, oceans or Earth, leaving just 10% reaching the land surface.

Furthermore, on a clear day, about 10% of this energy reaching ground level is in the ultraviolet (uv), and 45% is in the infrared (ir). This leaves just 45% in the visible wavelengths (400–700 nm), which can be used by plants in photosynthesis. However, much of this is wasted; a substantial proportion is not intercepted by leaves, but falls onto the ground and buildings, etc., and, of the light that actually does hit a leaf, 20% of it is either reflected or transmitted, and much of the remainder is not absorbed by the photosynthetic pigments. Under certain conditions, e.g. when the plant is under water stress or the light intensity is high, some of this absorbed energy may be wasted because the plant is already light-saturated. Consequently, an average of only 10% of the light reaching the surface of the Earth is used in photosynthesis, and less than a quarter of this ends up in the bond energy of molecules like glucose. The rest is lost in various stages of the photosynthetic reactions. Much of the energy stored in glucose is released in cellular respiration and other reactions necessary for the plant to maintain itself and grow. The rest, about 14% of the light used in photosynthesis, represents the net primary production (Figure 3.2).

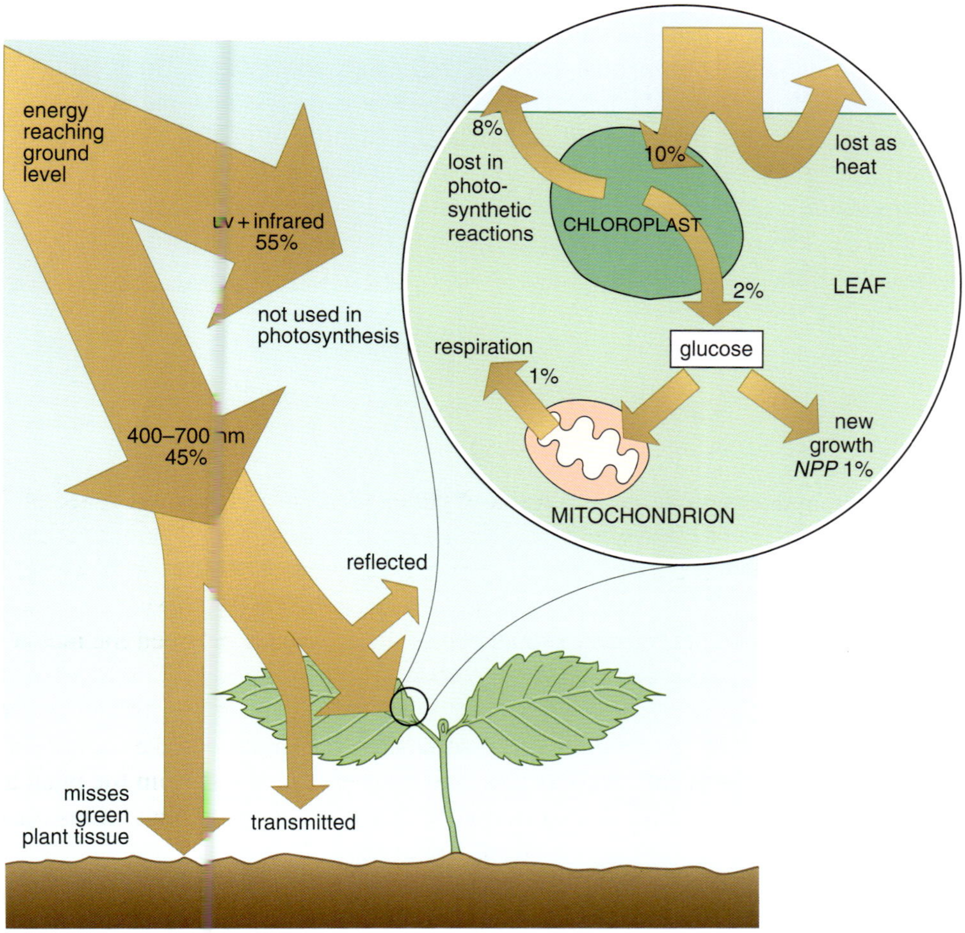

Figure 3.2 The average efficiency with which radiation that reaches ground level is trapped in net primary production (*NPP*).

The maximum conversion efficiency achieved by a crop over a short period of time is 5.5% of the light reaching the ground. However, most fields are uncultivated for part of the year, and when the crop plants are small they do not intercept much of the incoming light. Over the whole year, the proportion of the light reaching the surface of the Earth that is converted into energy in stored carbon compounds, is around 1% in temperate crops and 1.5% in tropical crops such as sugarcane.

Even after the light lost through absorption, reflection and just plain missing the plant are taken into account, the proportion of light actually captured by photosynthesis is very low. Some of the reasons why plants seem so inefficient at turning solar energy into sugars are addressed below.

3.1.3 Solar radiation: friend and foe

Plants, and thereby all animals and fungi, depend on light as a source of energy for life. However, light is also very dangerous. As you are undoubtedly aware, uv radiation can be hazardous. It is able to cleave chemical bonds directly and thereby disrupt long-chain molecules. The most common cause of light-induced damage, however, is the formation of highly dangerous forms of oxygen known as free radicals, which are especially harmful to membranes. It is perhaps surprising that the plants most at risk of such damage are those that habitually grow in the shade of the forest floor, as we shall see later.

The light reaching a plant growing on the shaded forest floor differs in three respects from the light hitting the leaves of the canopy above them. The most obvious difference is that the light at the forest floor is far *less intense* than that at the top of the canopy: in a tropical rainforest, the quantity of photosynthetically active light at the canopy is typically 160 times greater than at the forest floor. However, the *quality of light* at the two locations is also very different, because the canopy does not absorb light equally across the spectrum. Figure 3.3 shows how light beneath a canopy differs from light above it.

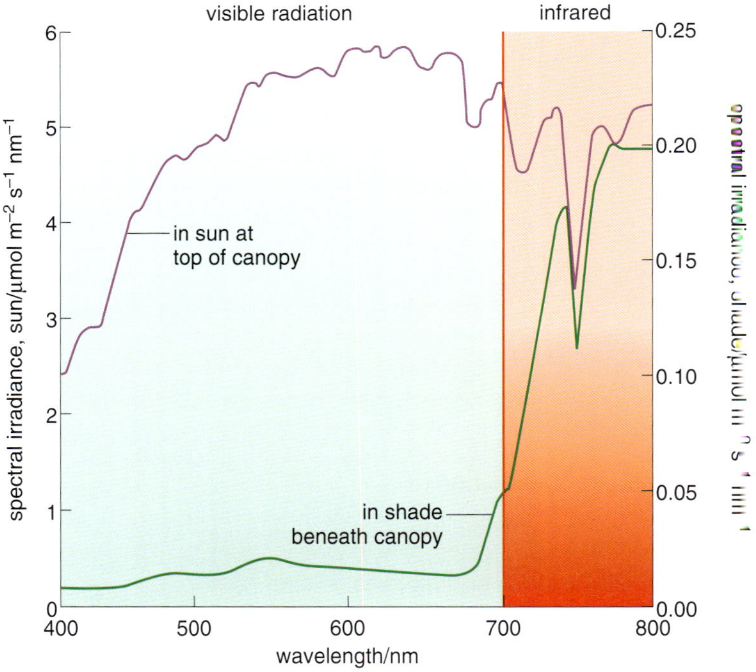

Figure 3.3 Spectral distribution of sunlight at the top of a canopy and under the canopy. For unfiltered sunlight, the total irradiance is 1900 μmol m^{-2} s^{-1}; for shade, 17.7 μmol m^{-2} s^{-1}. Notice that the units used to measure photosynthetically active radiation are moles per unit area per unit time. This quantifies the number of photons reaching the surface of the plant. Most of the photosynthetically active radiation is absorbed by leaves in the canopy, whilst most of the infrared radiation (heat) passes through them.

- ❍ From Figure 3.3, which wavelengths of light are absorbed most by the canopy, and which least?
- ● The visible wavelengths (400–700 nm) are absorbed proportionately more than the infrared wavelengths (> 700 nm).

The third way in which light at the forest floor differs from that above the canopy is that light *conditions may change abruptly* during the day. Light penetrates gaps in the canopy to produce **sunflecks**, the position of which alters over the course of the day. Additionally, wind can cause the sunflecks to move about very rapidly. Consequently, leaves may, for periods ranging from fractions of a second to several minutes, receive a 10-fold increase in light flux from sunflecks. Furthermore, because the light from sunflecks is unfiltered, it has the spectral composition of full sunlight.

In *all plants*, the photosynthetic systems used in the light reactions have a finite ability to absorb light. Sudden changes in the intensity and spectral composition of light can lead to over-excitation of the photosynthetic system, which results in **photoinhibition**, a reversible inhibition of photosynthesis. More extensive exposure to intense sunlight can lead to more serious and irreversible damage, in the form of **photo-oxidation** by free radicals, the most visible symptom of which is bleaching of the leaves. Free radicals can be produced whenever the photosynthetic apparatus is capturing more energy than it can use to fix carbon. They will react with almost anything they come in contact with and so have the potential to be very destructive within a cell. The problem is particularly acute for plants growing in shade because, in order to make the most of the small amount of light they receive, they tend to be very efficient light-harvesters. Sunflecks often represent a high proportion of the light available to forest-floor plants, yet sunflecks also induce the sudden changes in intensity that cause most damage. Nevertheless, plants flourish over the full range of light conditions. They have evolved a variety of mechanisms that prevent or minimize the harm that excess light can cause, while maximizing photosynthesis. These mechanisms are described below.

1 *Photorespiration*. **Photorespiration** is a process that acts as a safety valve, protecting against photoinhibition and photo-oxidation, when light energy is readily available but the supply of CO_2 is limiting the rate of photosynthesis. Photorespiration got its name because of its apparent similarity to true respiration. Around 1970 it was observed that the rate of respiration, as measured by oxygen consumption and carbon dioxide liberation, could be twice as great in light than in darkness. However, the similarity ends there; photorespiration does not liberate useful energy and when discovered did not appear to provide any beneficial function. Plant breeders tried hard to develop strains of crop that did not photorespire, thinking their yield would improve in the absence of such an apparently wasteful process.

When light intensity is high, the production of energy-rich molecules by the light reactions of photosynthesis may exceed the rate at which they can be utilized by the dark reactions. By releasing CO_2 from sugars, photorespiration creates an internal supply of CO_2 for the plant, which 'mops up' high-energy molecules as it is converted back to sugars in the dark reactions of photosynthesis (Figure 3.4).

Photorespiration appears to be destroying what photosynthesis is creating, but this is a price worth paying to protect the delicate photosynthetic machinery (see Box 3.1).

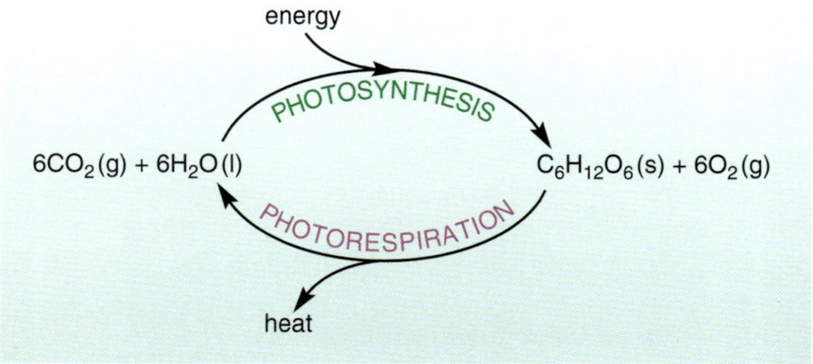

Figure 3.4 The opposing processes of photosynthesis and photorespiration create a so-called 'futile' cycle, whose only net effect is to convert potentially damaging light energy into thermal energy, but nevertheless has an important protective role.

Box 3.1 Explaining photorespiration using an analogy

Let us use an analogy to help us understand why plants photorespire. Imagine the photosynthetic apparatus in a cell is a delinquent paper-boy. Whilst he is delivering papers he is performing a useful function. When his bag is empty, he gets bored and turns to smashing up telephone boxes as an outlet for his energy (Figure 3.5a).

To avoid this criminal behaviour, the boy's mother follows him on his paper round, surreptitiously removing every second or third paper from its letter box and slipping it back into his bag. In this way, the boy never runs out of papers before the newsagent's van arrives with the next consignment (Figure 3.5b).

Figure 3.5 A paper-round analogy for photorespiration. (a) Scenario 1: the paper boy exhausts his supply of papers, then uses his energy less constructively! (b) Scenario 2: the boy's mother slips the papers back into his bag, thereby keeping him at work.

In this analogy, the papers represent CO_2, a raw material that is used to provide a useful function. The mother represents photorespiration. She is apparently doing something entirely counterproductive, but in fact is protecting the neighbourhood and her own son. The phone box represents the complex internal structure of the cell, which is susceptible to damage, and finally, the newsagent is the CO_2 delivery mechanism.

2 *CO_2 concentration*. Whilst almost all temperate plants employ photorespiration as a protective mechanism, some tropical species, which are at even higher risk from light-induced damage, have developed an alternative to it. They concentrate CO_2 in those cells that are carrying out photosynthesis, to ensure a constant supply. The process of concentration is itself very energy-demanding and therefore is only a viable option when there is excess light energy available, i.e. reliable bright sunshine.

To use the analogy from Box 3.1, the concentrating mechanism could be thought of as the newsagent in his van, following the boy to ensure the bag was always kept resupplied. In this way, the papers can be delivered faster and there is no need for the mother to worry about her son becoming bored, but it is an inefficient use of the newsagent's time.

Plants that employ the photorespiration safeguard are labelled C3 species, whilst those using the concentrating mechanism are labelled C4 species. Almost all of the British flora are C3 species, whilst tropical grasses in particular tend to belong to the C4 group (see Figures 3.6 and 3.7). The labels C3 and C4 refer to the biochemistry of the photosynthetic pathway, the details of which need not concern us here.

3 *Production of protective molecules*. Plants can produce special pigments, such as **carotenoids**, and other molecules such as ascorbic acid (commonly known as vitamin C), which react very readily with free radicals. Such molecules can be viewed as 'sacrificial.' They are made with the purpose of being destroyed in order to protect the more valuable photosynthetic structures.

4 *Reduction of light interception*. Plants that are habitually exposed to high light intensities are often protected by a shiny layer of thick wax, called a **cuticle**, on the leaf surface. This has the dual benefit of reflecting a proportion of the incident light and reducing water loss. (It can also protect the plant from insect or fungal attack.) Shade plants tend not to employ this adaptation.

❍ Suggest a reason why shade plants tend not to protect themselves with layers of shiny wax on the leaf surface.

● Wax reflects light, and shade plants need to be able to absorb as much light as possible when intensities are low.

❍ Which of the following strategies appears to have been promoted during evolution: (a) using all available light for photosynthesis but risking damage to sensitive molecules, (b) wasting light but protecting sensitive molecules from damage?

● Strategy (b): plants seem to have evolved strategies that give priority to protecting sensitive molecules, even if light available for photosynthesis is wasted.

Figure 3.6 Wheat (*Triticum aestivum*) is an important C3 crop species, which is usually grown in temperate areas.

Figure 3.7 Maize (*Zea mays*) is an important C4 crop species, which is more typical of the tropics. Maize has a higher potential *NPP* than wheat, but requires warmth and strong sunshine to achieve it.

3.1.4 Summary of Section 3.1

1. Unless there is a constant input of energy and materials, organisms break down into their component parts and the energy stored within them is dispersed.
2. Plants use the light and dark reactions of photosynthesis to convert light energy and atmospheric carbon dioxide into sugars (gross primary production), which act as a source of carbon and energy to all living tissues.
3. Some sugars are used to provide the energy for maintenance, growth and reproduction, and their energy is lost from the plant as heat during processes such as respiration. The rest are deposited in cells as stored dry mass (net primary production).
4. Only a tiny proportion of solar radiation ends up as primary production.
5. Light can cause serious damage to plants, which have evolved a number of mechanisms of protection, e.g. photorespiration, CO_2 concentration, shiny leaves and 'sacrificial molecules'.

3.2 Heterotrophy: energy transfer and release

Heterotrophs (all animals and fungi, most microbes and even some plants) obtain energy and carbon in the form of complex organic molecules (food) which are usually synthesized by other organisms. In respiration, the chemical energy in food molecules is used to synthesize ATP: it powers a wide range of energy-requiring processes and reactions, and when ATP is used in these ways, energy is released as heat.

- ❍ Recall the chemical equation summarizing aerobic respiration.
- ● $6O_2(g) + C_6H_{12}O_6(s) = 6CO_2(g) + 6H_2O(g) + \text{energy}$ (2.2)

Energy is stored within organisms predominantly in the form of carbohydrates and fats. You have already met some plant structures that are largely composed of storage tissues.

- ❍ Recall from Section 2.2 some plant structures that are rich in storage tissues.
- ● Structures such as bulbs, corms, tubers, rhizomes and also seeds contain large reserves of energy-rich molecules.

Plants tend to store energy in the form of complex carbohydrates such as starch. In contrast, animals often store energy in the form of fats just under the skin. When tissues are eaten by animals, the large masses of complex carbohydrates and fats have to be broken down into smaller molecules, such as glucose, before the energy can be released by the chemical process of respiration. This initial phase of digestion is brought about through a combination of mechanical (i.e. through the grinding action of teeth, or the crop in birds) and enzymatic digestion.

Complex biological molecules, such as starch and fat, store lots of chemical energy in a very stable form, which can be difficult to access. They can be likened to monetary deposit accounts which require some period of notice before withdrawals can be made. Simple carbohydrates such as glucose also contain lots of energy, and this molecule is stable enough to be transported easily inside an organism, but can be readily metabolized when required, rather like a current account with instant access from a wide variety of outlets. Small amounts of energy are carried by molecules like ATP, which move energy over very short distances, such as those involved in biochemical pathways. These 'energy currency' molecules are a bit like the small change we keep in our pockets for instant small transactions. (However, none of the energy 'accounts' have any interest accruing!)

3.2.1 Secondary production

Heterotrophs are divided into **consumers** (animals), which ingest organic matter, and **decomposers** (primarily fungi and bacteria), which break it down externally, so that they can absorb the soluble products through their membranes. Consumers are further subdivided into those that eat relatively fresh, organic matter that has normally been 'harvested' in some way, i.e. herbivores, which eat plants, and **carnivores**, which eat animals, and those called **detritivores** ('de-try-tee-vores'), which consume dead organic matter.

❍ From general knowledge, classify the following organisms as herbivores, carnivores or detritivores: earthworm, rabbit, cow, domestic cat, dung beetle (e.g. *Aphodius rufipes*) and greater horseshoe bat.

● The earthworm and dung beetle are detritivores; the rabbit and cow are herbivores; the domestic cat and greater horseshoe bat are carnivores.

Secondary production refers to the rate at which heterotrophs accumulate dry mass (biomass) or energy. Animals do not eat all the biomass available to them. Cows eat only the tender shoots of grass, not the bases of the leaves or the roots. Birds may eat the berries from trees but not the shoots or leaves. Therefore, just a fraction of *NPP* is actually consumed. Furthermore, only a small proportion of the energy that an animal eats becomes deposited in the mass of its tissues. Of the food consumed (C), some will pass through an animal's gut and out at the anus without having been absorbed through the gut wall (faeces, F). The proportion of the food that is absorbed across the wall of the gut is said to be **assimilated** (A). This relationship can be expressed in terms of an equation written in units of dry mass or energy:

$$C = F + A \qquad (3.1)$$

Some of the assimilated food is used in cellular respiration (R) to provide energy for movement and the manufacture of new chemical compounds. Some is removed in the nitrogenous waste which is excreted, known as urine in many animals (U). The rest is stored in the dry mass of new tissues as secondary production (P). This relationship is summarized in the equation:

$$A = U + R + P \qquad (3.2)$$

P is secondary production, i.e. the mass or energy available to other consumers.

3.2.2 Efficiency of energy transfer

The efficiency with which energy is transferred from food source to consumer varies from situation to situation. Some figures for a bullock grazing in a meadow have been worked out (Figure 3.8).

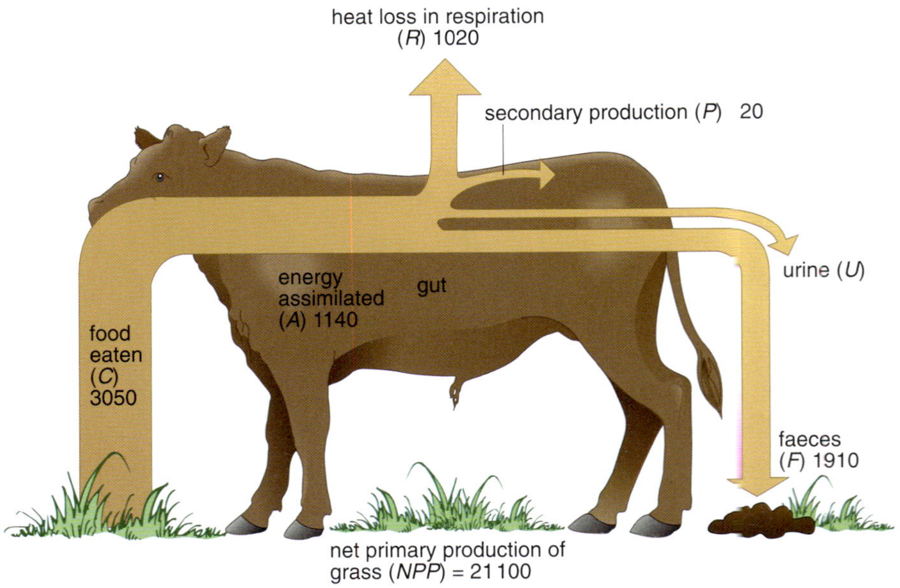

Figure 3.8 The energy budget for a bullock grazing in a meadow. The figures are energy flows in kJ m^{-2} yr^{-1}. Note the figure given for energy lost as heat also includes energy lost in urine.

The *NPP* of the vegetation is about 21 100 kJ m^{-2} yr^{-1}. The bullock eats only the leaves of the most palatable plants, and much of the energy trapped in photosynthesis must enter dead leaves or roots, which bullocks do not eat. Thus, only about 3050 kJ m^{-2} yr^{-1} are consumed (*C*). Well over half of this (63%) passes straight through the gut and is lost as faeces or methane gas (*F*). In common with many herbivores, bullocks rely on other organisms to assist with the digestion of cellulose. **Cellulose** is a large, complex carbohydrate molecule: it is the major component of plant cell walls and is very difficult to digest.

Parts of the bullock's gut are inhabited by specific protoctists that are able to break down the cellulose into smaller, more readily digested subunits. Methane is produced as a by-product of the process. The methane produced in the animal's gut by these mutualistic organisms, about 60 litres h^{-1}, is enough to light a small house continuously (if burnt in an electricity generator). The other 37%, amounting to 1140 kJ m^{-2} yr^{-1}, is assimilated, which in this case means taken through the gut wall into the bloodstream.

Most of the assimilated energy (89%, 1020 kJ m^{-2} yr^{-1}) is expended in cellular respiration (*R*) to keep the animal alive, or lost as urine (*U*). A bullock has a particularly high respiration rate because it is an endotherm with a constant body temperature of about 39 °C. The other 11%, about 120 kJ m^{-2} yr^{-1}, is incorporated in new cells as secondary production (*P*), of which much will be available to humans as food when the animal is slaughtered.

❍ Calculate the percentage of the original net primary production that ends up in the bullock's tissues, i.e. the efficiency of conversion of net primary production into secondary production.

● Approximately $\frac{120}{21100} \times 100\% = 0.6\%$ of the net primary production of the grass is incorporated into the secondary production of the bullock.

If this level of efficiency is representative of most examples of conversion of primary production into secondary production, then it seems to suggest that many more humans could be supported on a vegetarian diet than one containing a high proportion of meat. However, humans, like most omnivores, do not have mutualistic gut protoctists that enable them to digest cellulose. Therefore, humans assimilate a far smaller proportion of the primary production than true herbivores like cattle. Furthermore, not all herbivores are such poor energy converters as bullocks. Table 3.1 shows the influence that factors such as whether the consumer is an ectotherm or an endotherm, or how much of its diet it can assimilate, have on the transfer efficiency.

Table 3.1 The efficiency of assimilation and secondary production in selected animal species (values are percentage of energy consumed). The top four organisms are all ectotherms, while the bottom three are endotherms.

Animal	Feeding preference	Faeces (F)	Assimilated (A)	Respired (R)	Secondary production (P)	R/P ratio
grasshopper	herbivore	63	37	24	13	1.8
caterpillar	herbivore	59	41	17.5	23.5	0.7
wolf spider	carnivore	8.2	91.8	64.8	27	2.4
perch	carnivore	16.5	83.5	61	22.5	2.7
owl (adult)	carnivore	15	85	85	<1	very high
elephant	herbivore	66	33	32	1	32
cow	herbivore	60	40	39	1	39

❍ How do endotherms and ectotherms differ in terms of the proportion of assimilated energy (A) that is used in cellular respiration (R)?

● The proportion of assimilated energy used in cellular respiration is much higher in endotherms (>97%) than ectotherms (<74%).

Endotherms maintain a body temperature of 35–42 °C, and the respiratory costs of doing this are clearly very large. This is particularly true in cold climates or cold seasons of the year. The energy that ectotherms 'save' can be used to produce new cells.

❍ Using the data in Table 3.1, what effect does feeding preference have on the proportion of the food consumed that is assimilated?

● Carnivores assimilate a much higher proportion of the food they consume (over 80%) than herbivores do (<42%). (Recall $C = F + A$.)

Carnivores have high-protein diets, which are easily digested and absorbed. They often assimilate over 80% of the energy in their diets. Much of the energy in a herbivore's food is present in the glucose units that make up the cellulose of cell walls. Even cellulose-digesting specialists, such as termites and our old friend the cow, are unable to assimilate more than 40% of the energy in their food.

Like primary production, much of the energy available to herbivores and carnivores is wasted. Only a small proportion of it ends up as part of an animal's tissues. So what happens to all of the 'wasted' biomass, i.e. the uneaten remains, urine and faeces?

- ❍ What is likely to be the fate of the energy in the cowpats produced by the cow?
- ● Faeces serve as a food source (and breeding site) for many other organisms (including the dung beetle, *Aphodius rufipes*), and are often colonized by fungi such as *Stropharia semiglobata* (Figure 3.9).

Figure 3.9 The fungus *Stropharia semiglobata*, a decomposer that can exploit animal dung as an energy source.

Organisms such as the dung beetles, that eat the dead remains, are detritivores, and the microbes such as *S. semiglobata*, that digest the remains, are decomposers. Together, these two groups of organisms play a vital role in releasing the energy, carbon (and mineral nutrients) trapped in the remaining **detritus** (dead organic matter).

Question 3.1

In an area of African savanna, the *NPP* has been estimated as 12 700 kJ m^{-2} yr^{-1}, of which 8% is consumed by large herbivores, such as wildebeast and zebras, whose secondary production was estimated as 3% of their intake. Local lions were monitored over a period of a year and thought to have consumed flesh equivalent to 0.280 kJ m^{-2} yr^{-1}. Their growth over the year accounted for 0.004 kJ m^{-2} yr^{-1}, and no offspring were born. Assuming there were no other predators eating the herbivores:

(a) Calculate the percentage of *NPP* that ends up as herbivore flesh, and the percentage of this flesh that ends up as lion.

(b) If the lions are assumed to have assimilated 80% of their food intake, what percentage of the assimilated energy is stored in their own tissue at the end of the year? Comment on this value, and explain what has happened to the rest.

3.2.3 Detritivores and decomposition

Gravity ensures that detritus tends to move downwards, towards the soil or the bed in rivers and oceans. It is in these regions that the detritivores and decomposers are most abundant. The pattern of decomposition for the leaf litter in a forest is shown in Figure 3.10. The dead leaves are first attacked by detritivores such as worms, springtails and mites. They use some of the energy and materials as food, but their activities also break up the detritus into smaller particles, which increases the surface area available for chemical decomposition. The decomposer fungi and bacteria secrete enzymes, which break down the organic compounds in the detritus. Released from the remains of the cells by the action of these enzymes, the soluble organic subunits and nutrient ions are absorbed by the decomposers. The nutrient ions, essential for plant growth, are eventually released to the soil, although some particularly intractable organic compounds remain in the soil as humus. In contrast, the chemical energy in organic matter is converted to heat when this material is decomposed by microbes, and because organisms cannot recapture this energy, it is lost to the environment.

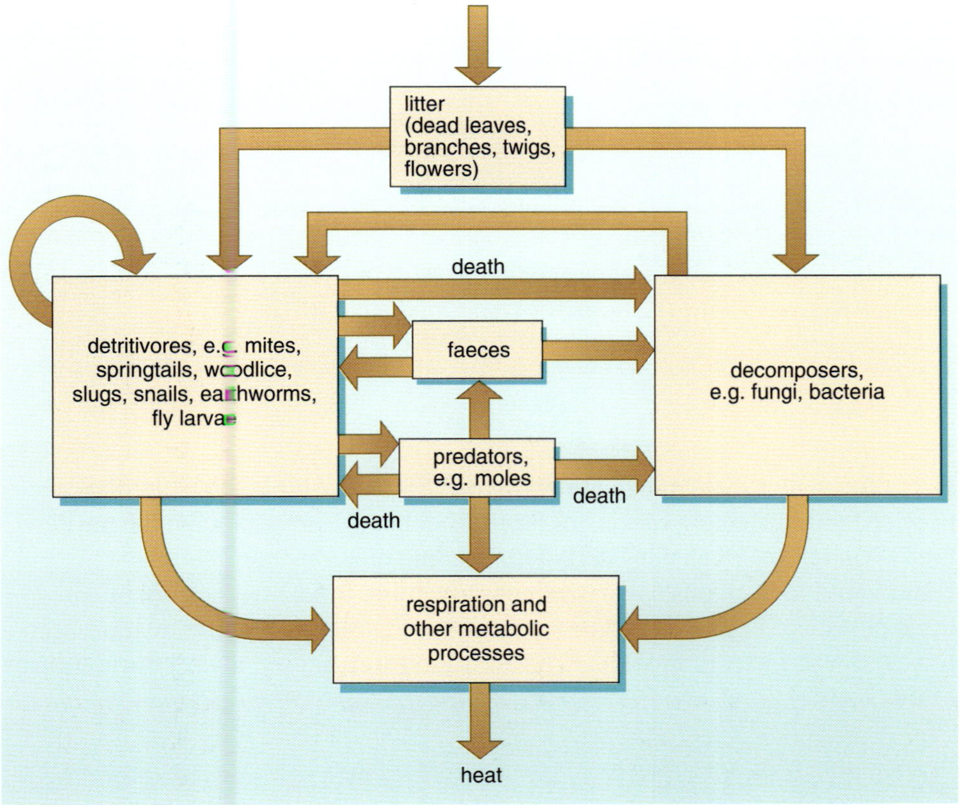

Figure 3.10 Energy flow in the decomposition of leaf litter in a forest. Arrows point in the direction of energy flow. Organic compounds released to the soil from each box form humus. Nutrient ions are released to the soil from each box, but mostly from the decomposers.

3.2.4 Food chains

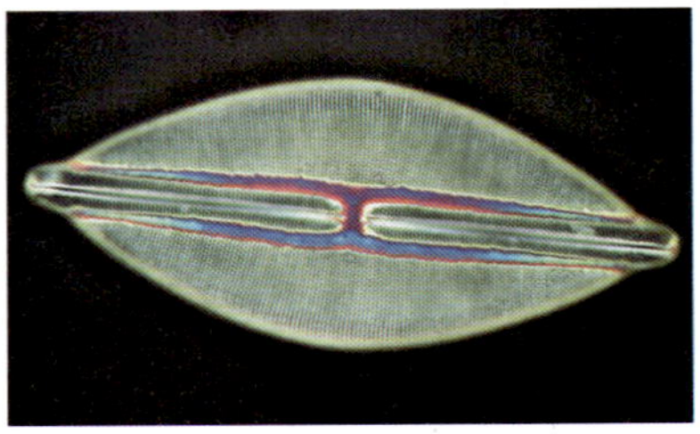

Figure 3.11 A diatom (*Navicula* sp.), which is a unicellular alga of freshwater streams. It is a member of the kingdom Protoctista.

Food chains can be used to trace the fate of particular units of primary production. In freshwater habitats, such as the River Teign, diatoms (a group of unicellular algae) form the basis of some food chains. Diatoms such as *Navicula* spp. (Figure 3.11) form an algal slime on rocks and boulders in brooks and streams.

The diatoms are grazed by the larva of the mayfly *Baetis rhodani* (Figure 2.54b), which may be eaten, in its turn, by the larva of the caddis-fly *Rhyacophila* sp. (caseless caddis). The food chain does not necessarily stop there.

- Recall from the Teign Valley virtual field trip the name of a bird that feeds on freshwater invertebrates.
- The dipper, *Cinclus cinclus* (see Figure 2.3b).

Figure 3.12 The sparrowhawk (*Accipiter nisus*), a top predator of the Teign catchment.

The dipper may itself be taken by a bird of prey such as the sparrowhawk (*Accipiter nisus*, Figure 3.12). We can represent this food chain as:

Navicula sp. (diatom) ⟶ *B. rhodani* (mayfly) ⟶ *Rhyacophila* sp. (caseless caddis) ⟶ *C. cinclus* (dipper) ⟶ *A. nisus* (sparrowhawk)

The sheep's fescue grass (*Festuca ovina*) is a good example of a terrestrial 'solar trap' (photosynthetic primary producer). The cockchafer (*Melolontha melolontha*) uses such grasses as a food source, and in its turn becomes a meal for the greater horseshoe bat (*Rhinolophus ferrumequinum*).

Food chains follow a set pattern: each consists of a linear relationship in which energy fixed by a plant is consumed by a herbivore, which is then followed by one or more carnivores:

$$\text{plant} \longrightarrow \text{herbivore} \longrightarrow \text{carnivore}_1 \longrightarrow \text{carnivore}_2 \longrightarrow \longrightarrow \text{carnivore}_n$$

In nature, the food and feeding relationships of plants and animals are rarely as simple as these examples suggest. You already know that a significant proportion of each step in the chain is either not consumed, or if it is, is not assimilated. The resultant detritus forms the basis for some food chains, e.g.:

faeces ⟶ *Coprinus niveus* (fungus) ⟶ *Lumbricus* sp. (earthworm) ⟶ *Turdus merula* (blackbird) ⟶ *Felix domesticus* (domestic cat)

Furthermore, the greater horseshoe bat eats a wide variety of insect foods, and its diet varies with its age and the time of year. Similarly, the cockchafer does not restrict itself to a single species of plant food, but grazes widely on the meadow herbs. Therefore, when considering the feeding relationships of a whole community, food chains are an oversimplification. In reality, they are often interconnected. Figure 3.13 shows part of the complex of feeding relationships of a freshwater stream in Wales. The term **food web** is used to describe diagrams of this sort.

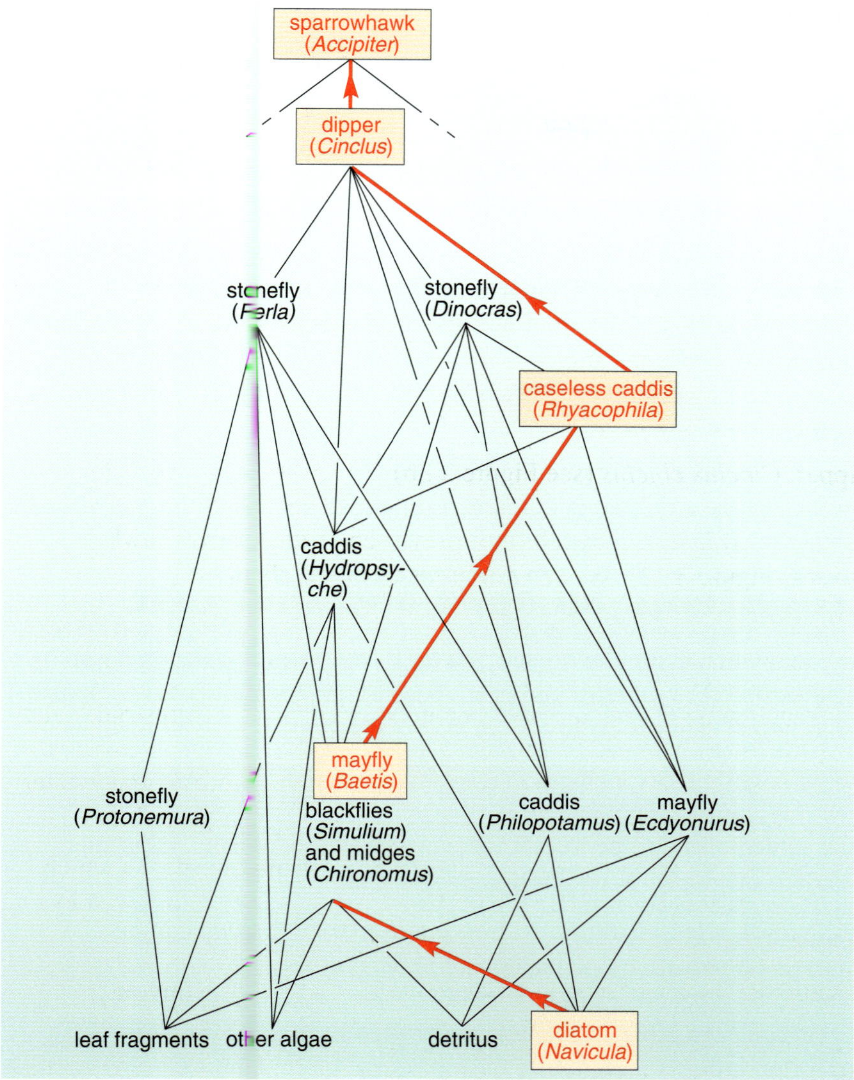

Figure 3.13 A simplified model for part of a food web in a freshwater stream.

- ❍ According to Figure 3.13, what foods do *Baetis* eat in addition to diatoms?
- ● *Baetis* eat other algae, leaf fragments and detritus.

In fact, the animal communities in freshwater habitats are good examples of communities supported by large inputs of dead organic matter, rather than primary production within the water body. This is especially true of those animals that exist in upland streams and pools that are often low in nutrients (**oligotrophic**). Primary production tends to be higher in lowland streams that are enriched by sewage effluent and nutrient-rich runoff from the surrounding agricultural fields, etc. Such waters are described as **eutrophic**.

Diagrammatic representation of food webs can grow so complicated that it becomes very difficult to unravel individual links and yet retain an appreciation of the whole system. In Figure 3.13, animals with similar feeding habits and similar predators have been grouped together. Even so, the model is still complex, and comparisons of one food web with another are difficult because different species are involved in each case. New techniques and concepts are required for comparing energy flow and productivity in diverse communities.

3.2.5 Summary of Section 3.2

1 Heterotrophs obtain energy and carbon in the form of complex organic molecules that have been synthesized by other organisms. Respiration is the process that releases energy and carbon from the organic molecules.

2 Heterotrophs can be classified as consumers and decomposers. Consumers include: carnivores, herbivores and detritivores, according to whether they eat animals, plants or dead organic remains. Decomposers, such as fungi and bacteria, digest detritus externally using enzymes, which they secrete.

3 The efficiency with which primary production is converted into secondary production (i.e. consumer biomass) depends on a number of factors, including feeding preference and whether the consumer is an endotherm or an ectotherm.

4 Feeding relationships within a community can be expressed as food chains or food webs, but these normally represent a simplified version of the true situation.

3.3 Ecosystems: an open-and-shut case

You already know that energy in organic matter is lost when it is decomposed by microbes and respired by animals and plant, because organisms cannot recapture energy lost as heat. Therefore energy flows in one direction through living systems:

radiant energy (sunlight) ⟶ chemical energy (organisms) ⟶ heat (respiration)

❍ Can the same be said to be true of the carbon released during respiration, i.e. does carbon inevitably flow through ecosystems?

● No, the carbon released during respiration as carbon dioxide can be recaptured and reused, by plants during photosynthesis.

Therefore, carbon (and other nutrients) *cycle* round living systems. An **ecosystem** is defined as a community of organisms, interacting with one another and their abiotic environment (e.g. atmosphere, soil), with its own characteristic pattern of energy flow and nutrient cycling.

Therefore an ecosystem is more than the sum total of the organisms present; it also includes the abiotic environment in which the organisms live. Ecosystems can also be seen as units that use energy and process matter — a definition that emphasizes function rather than components. Ecosystems require a constant energy input. Much of the focus of ecosystem research is the quantification of energy input and the relative amount of energy transferred from autotrophic to heterotrophic components. However, before considering this subject in more detail, it is necessary to consider some of the properties of ecosystems further.

The original concept of ecosystems envisaged that they would be self-supporting, i.e. that they would be **closed**; the green plants would take up from the environment the energy, ions and compounds that the other organisms need. However, research has shown that in practice most ecosystems do exchange energy and materials with their wider environment, and as such are **open** to some degree. Figure 3.14 illustrates diagrammatically how an ecosystem is thought to work.

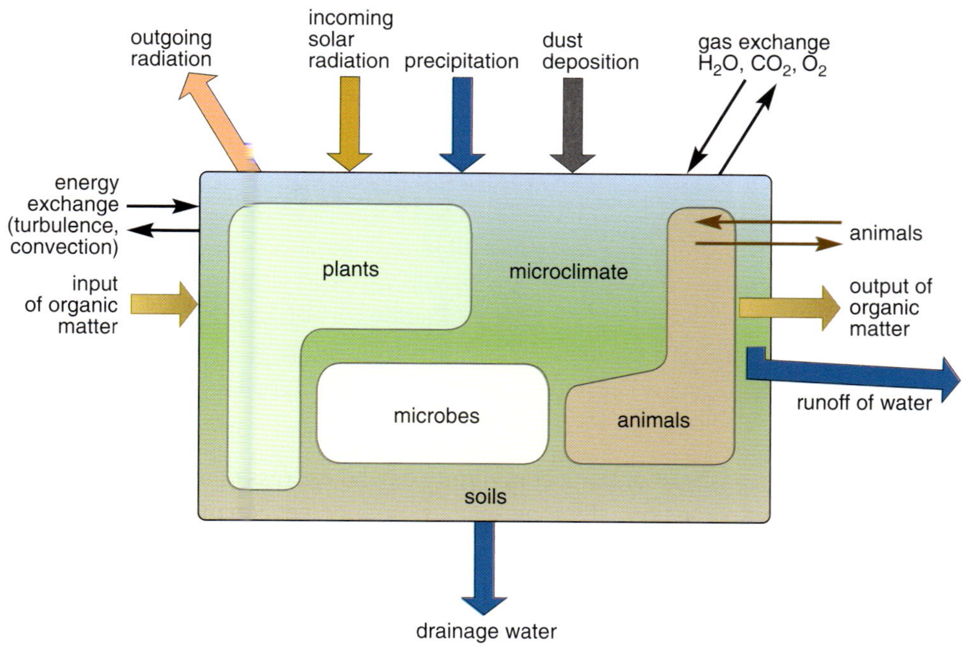

Figure 3.14 Diagrammatic representation of an ecosystem (framed area) exchanging energy and materials with its environment.

The smaller the extent of the exchanges with the environment outside the ecosystem, the more 'closed' the ecosystem is.

- ❍ Which of the following ecosystems do you think is the more 'closed': (a) a small isolated island in the Pacific Ocean, or (b) a stream flowing through lowland Britain?

- ● The small isolated island is more closed, as it receives few inputs from the surrounding air and ocean and makes few outputs. A lowland stream in Britain receives many inputs in the form of nutrient-rich runoff, sewage and industrial effluents, and organic matter such as dead leaves from neighbouring trees; outputs include the removal of fish and invertebrates by terrestrial organisms.

The essential components of an ecosystem are producers, decomposers, and inorganic compounds such as water, carbon dioxide and nutrient ions. The primary producers, usually green plants, extract the inorganic compounds from their environment. The decomposers return these materials to the environment. Animals, as consumers, are not essential components of ecosystems. In most cases, there is some exchange of carbon compounds, nutrient ions, gases and water between ecosystems.

3.3.1 A question of scale

An ecosystem can be of any size. It can range over several biomes. It might cover only a tiny area, like a small pond or a puddle. In many ways, ecosystems are defined by the environmental scientist, in order to answer specific questions relating to a particular scale in time and space. If the questions relate to nutrient cycling in woodlands, like those you encountered on the virtual field trip to the Teign Valley, then a study of Yarner Wood alone would not be large enough to give you representative data. You would need to study a greater range of woodland types and conditions. An entire forested catchment would be a more appropriate spatial scale, and a five-year study period would enable you to obtain some understanding of how the forest retains and cycles soluble nutrients. However, to understand how rock weathering and changes in the soil influence these processes, you would need to extend the study period to many decades or even centuries. Ecosystems change with time and very slow processes cannot be assessed over short time-scales.

Choice of appropriate scales of space and time is especially important when ecosystem disturbances are large. A period of a few months would be inappropriate for the study of a small stream, because of the effects of the seasons on flow rates and the invertebrate fauna. A period of a year would give a better indication of how the ecosystem as a whole functions. However, the climatic variations during the last part of the 20th century (Figure 3.15) illustrate how a year of study may still be inadequate. A year studying a stream during the dry 1980s would give a very different picture of ecosystem function compared with the year 2000, the wettest year on record! When disturbances are even rarer, e.g. major forest fires or volcanic eruptions, it is even more difficult to get a true picture of ecosystem function.

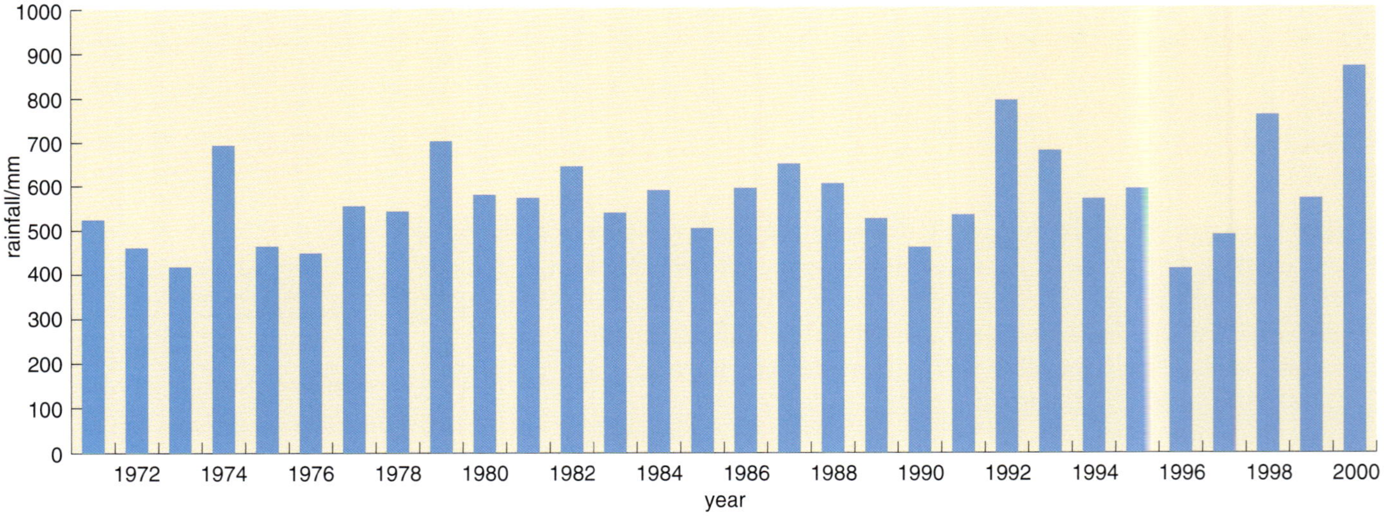

Figure 3.15 Annual rainfall totals over a 30-year period for a location in Bedfordshire, England. Note the variation between years; for example, in the year 2000 this site received more than twice the rainfall it had in 1996.

3.3.2 Studying ecosystems

An Oxford ecologist, Charles Elton, noted in the 1920s that '...the animals near the base of a food chain are relatively abundant while those at the end are relatively few in number, and there is a progressive decrease in between the two extremes'. This **pyramid of numbers**, as it is called, is a common feature of ecosystems the world over. General comparisons of communities can be made if organisms with similar food habits are grouped together.

When constructing a pyramid such as those in Figure 3.16a, the total number of autotrophs occupies the primary producers box at the base of the pyramid (P). In the case of this river at Silver Springs, the primary producers are tiny algae as well as larger water plants. The total number of herbivores occupies the primary consumers box (C_1), one step up the pyramid, and the carnivores that eat the herbivores occupy the secondary consumers in box three (C_2). Boxes for any additional consumers (C_3, etc.), as well as for the detritivores and decomposers (D) are included as necessary. The term **trophic level** may also be used to describe each successive layer in the pyramid. Primary producers usually represent the first trophic level, with herbivores and then the secondary consumers occupying the second and third trophic levels etc. In Figure 3.16b, the primary producers are large, in this case trees, and an individual plant may support many herbivores. However, the pyramid tapers upwards from the primary consumer level.

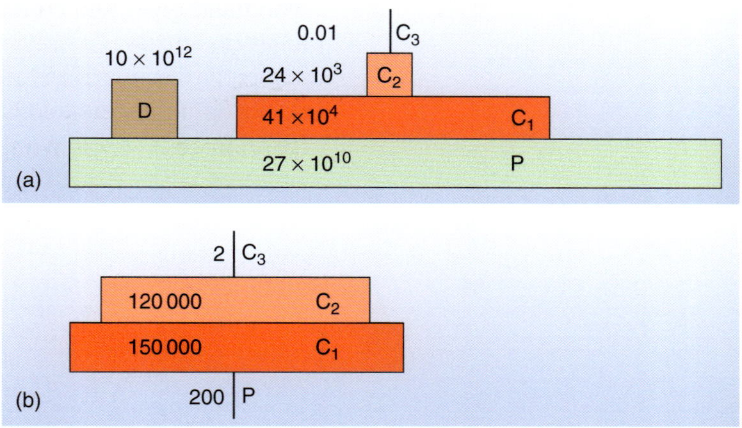

Figure 3.16 Pyramids of numbers: (a) temperate river at Silver Springs, Florida, USA, where primary producers are small; (b) temperate oak forest at Wytham Woods, England, where primary producers are large. Numbers refer to 1 m^2 (a) and 1000 m^2 (b).

There are problems inherent in the comparison of pyramids of numbers: consider the difficulties there are in comparing ecosystems based on diatoms with those based on trees, or cases where the primary consumers are elephants or mayfly larvae.

- ❍ Suggest an alternative parameter that would overcome this problem inherent in comparing differing sizes of organisms.
- ● Using the mass of organisms rather than their numbers might permit more sensible comparisons to be made between differing ecosystems.

Pyramids of biomass of organisms are constructed by taking samples of the flora and fauna in an ecosystem, and then drying and weighing them. The results are then expressed as mass per unit area, e.g. grams per square metre or tonnes per hectare. Figure 3.17 illustrates three pyramids of biomass: one for the river at Silver Springs, Florida, one for the oakwood at Wytham and one for the English Channel.

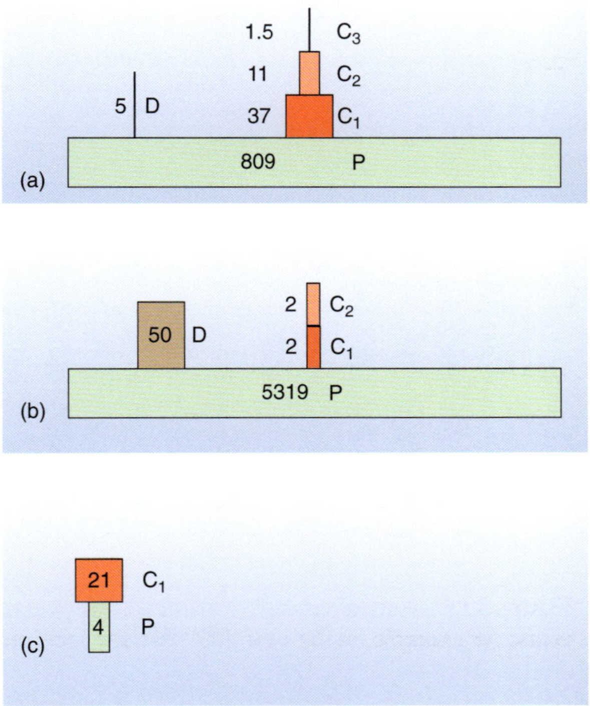

Figure 3.17 Pyramids of biomass for: (a) a river at Silver Springs, Florida, USA; (b) oakwood (Wytham Woods, England); (c) the English Channel. Units are g m^{-2}.

Notice how the pyramid for the oakwood looks like a more conventional pyramid, now that the effect of the huge differences in size of the individual organisms has been eliminated.

- ❍ How do the shapes of the pyramids for the river and the English Channel compare (Figure 3.17a and c respectively)?
- ● The pyramid of biomass for the river is upright (i.e. tapers upwards), while that of the English Channel is inverted.

The pyramid of biomass for the English Channel suggests that a lower mass of producers supports a higher mass of consumers! The explanation for this apparently illogical state of affairs is that the pyramids of biomass fail to take into account the amount of time over which the measured biomass has accumulated. Trees in a forest, for example, build up their store of biomass over a hundred years or so. The primary producers in the English Channel, the phytoplankton, may take less than a day to reproduce themselves. Furthermore, the biomass of plankton fluctuates seasonally, and phytoplankton have a much more rapid turnover rate. If one added up the total biomass produced by the phytoplankton over a year, it would exceed the total biomass gained by the primary consumers over the year. The amount of material present at any one instant is know as the **standing crop**. It gives no indication of the rate at which that material is being produced (productivity). Therefore, both pyramids of numbers and of biomass are limited in their usefulness.

A third type of ecological pyramid overcomes many of the problems associated with the other two types, this is a **pyramid of energy**. Figure 3.18 shows a pyramid of energy for three different ecosystems: a river at Silver Springs in Florida, tropical rainforest in Puerto Rico and arctic tundra in Canada. Each bar represents the total amount of energy utilized by the different feeding types in a square metre over a year.

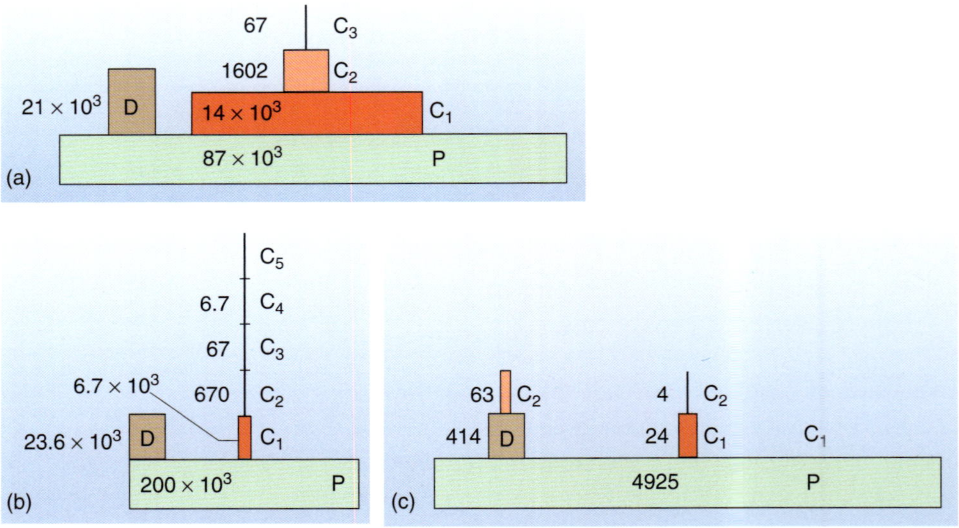

Figure 3.18 Pyramids of energy for: (a) a large flowing stream at Silver springs, Florida, USA; (b) tropical rainforest in Puerto Rico; (c) arctic tundra in Canada. Units are kJ m^{-2} yr^{-1}.

The values quoted for the primary producers give the energy captured by photosynthesis over the whole year. The figure for each of the consumer levels is the energy entering the mouths of the organisms each year.

❍ How do the pyramids compare for tropical rainforest and tundra in terms of the number of consumer levels?

● The food chains in the tropics are relatively long, with five successive consumer levels. In arctic tundra only two consumer levels are shown.

The short growing season, and the low intensity of solar radiation in the Arctic probably limit primary production and thereby the energy available for consumers. There may also be problems with maintaining viable populations of higher carnivores. It is likely to be very difficult for the higher carnivores to find either mates or sufficient numbers of the sparse prey to support themselves.

- ❍ Describe how the proportions of the pyramid of energy for Silver Springs (Figure 3.18a) differ from the pyramid of biomass for the same site (Figure 3.17a). What does this difference tell you about the energy value of the tissues of the primary producers and the secondary producers?
- ● When compared to the boxes for the secondary producers, the box for the primary producers is relatively narrower on the pyramid of energy than it is on the pyramid of mass. This difference implies that the tissues of secondary producers (basically animals) are more energy-rich, gram for gram, than the tissues of primary producers (plants and algae).

3.3.3 Ecosystems and energy budgets

Energy budgets for ecosystems are an alternative and more informative way of looking at ecosystem energetics. When the energy budgets for ecosystems are drawn in the same format (as shown in Figure 3.19) they can be readily compared.

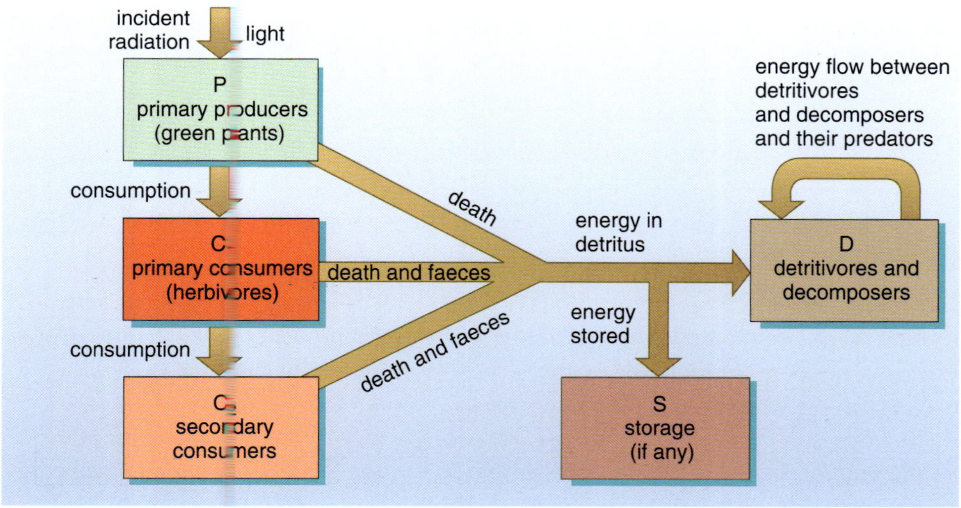

Figure 3.19 Format for the energy budget of an ecosystem. Losses of energy through 'respiration' are not shown but can usually be calculated if the data are fairly complete.

Energy captured by plants after entering an ecosystem may be lost as thermal energy through respiration, be consumed, pass directly to decomposers, or become stored (e.g. as wood or peat). There are three main types of ecosystem depending on which of these fates is dominant.

Grazing ecosystems

In grasslands, freshwater and the sea, much of the net primary production is eaten by herbivores (Figure 3.20). The rest enters the detritus food chain when the plants die. Over the whole year the ecosystem is in balance. The energy entering it equals the energy leaving it.

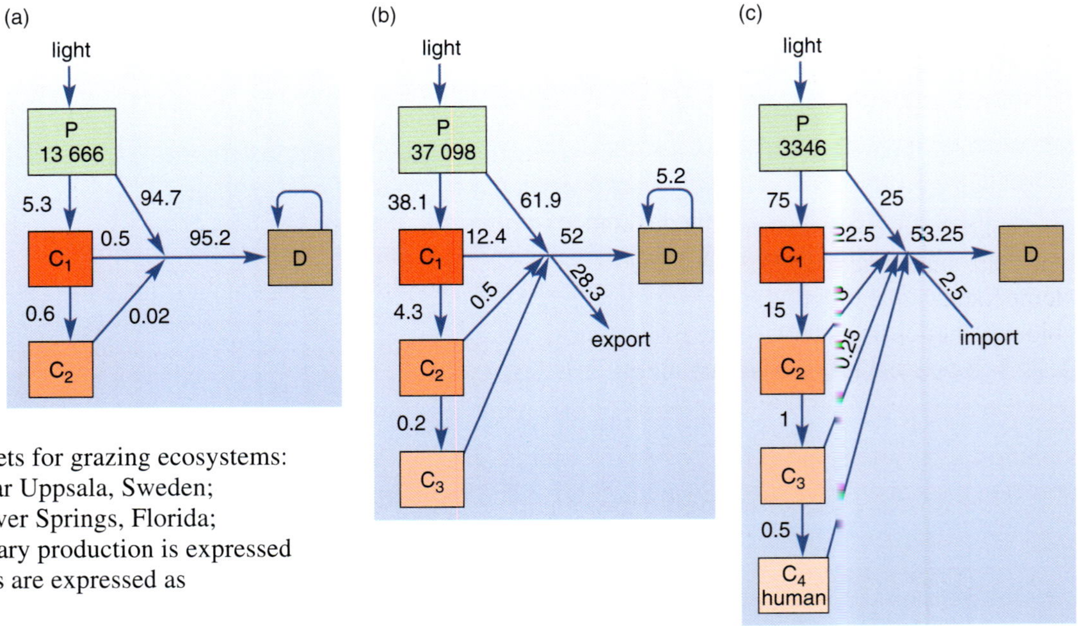

Figure 3.20 Energy budgets for grazing ecosystems: (a) temperate grassland, near Uppsala, Sweden; (b) a freshwater stream, Silver Springs, Florida; (c) the North Sea. Net primary production is expressed in kJ m^{-2} yr^{-1}. Other figures are expressed as percentages of *NPP*.

The proportion of biomass being consumed by herbivores on the grassland (Figure 3.20a) is only 5%, which is usual where large grazers are absent. Typically, large grazers like cattle (see Figure 3.8) can consume around 15% of the biomass available.

But this figure of 15% is still probably an underestimate of the total proportion of *NPP* being consumed where large herbivores are present, because it does not take into account the potentially numerous invertebrate herbivores in the ecosystem, in which case, 20% is probably a more realistic figure.

- ❍ Compare the three ecosystems shown in Figure 3.20. What appears to be the relationship between the percentage of biomass being consumed by herbivores and the number of trophic levels?
- ● As the percentage of biomass being consumed increases (from 5% in the grassland, through 38% in the stream to 75% in the North Sea), so the number of trophic levels increases (from three, through four to five respectively).

The greater the amount of energy available for consumption, the longer the grazing food chain that can be supported.

Storage ecosystems

In the early stages of the development of peat bogs, heathlands and forest, the energy entering the system each year is greater than the energy leaving it (Figure 3.21). The balance of the energy is stored. Little of the biomass is eaten, because there are few herbivores. Much of the rest is stored either as dead organic matter or as biomass, e.g. as soil organic matter, peat, or in the trunks, roots and branches of shrubs or trees in heathland or woodland.

In peat bogs, decomposition is inhibited by the acid, anaerobic conditions and low temperatures that often exist there. Much of the biomass builds up as peat. In the British Isles, as much as 9 m of peat has accumulated in many raised bogs since the last glaciation. The stored energy is released when peat is burnt.

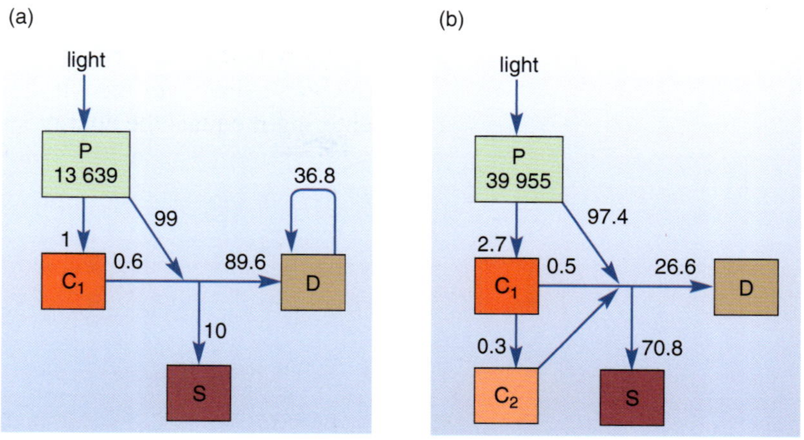

Figure 3.21 Energy budgets for storage ecosystems: (a) peat bog, Moor house, England; (b) immature forest (*Picea omorika*) more than 21 years old, Ross-on-Wye, England. Units as for previous figure.

- ❍ How does the proportion of biomass entering storage compare in the bog and the young forest?
- ● Over seven times more biomass enters storage in the immature forest than in the bog (71% as opposed to 10% respectively).

Detritus ecosystems

Once plants in heathlands and tropical and temperate forests have matured, a high proportion of the net primary production enters the detritus food chain, sometimes referred to as the decomposer subsystem (Figure 3.22).

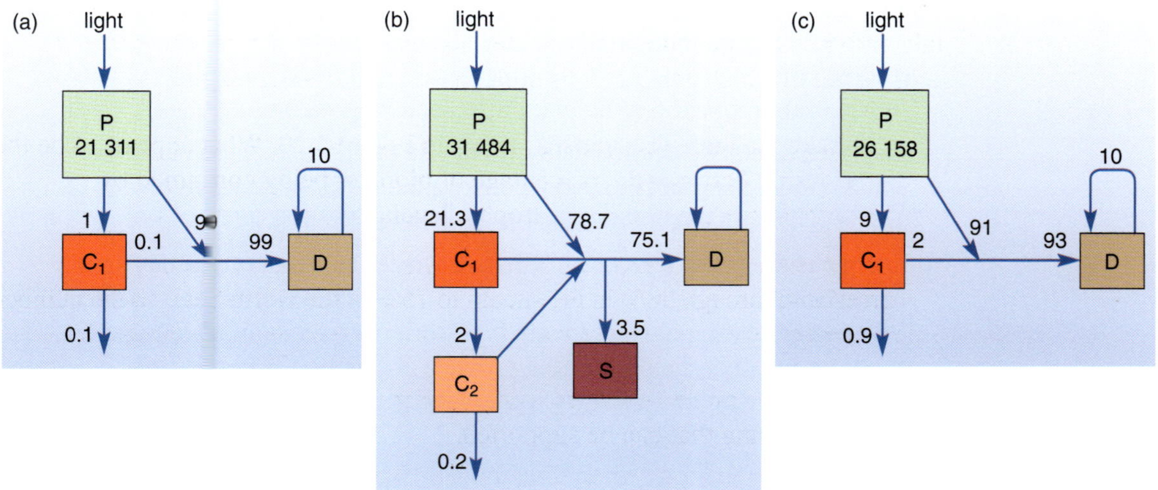

Figure 3.22 Energy budgets for detritus ecosystems: (a) mature heathland (25 years old), Studland, England; (b) tropical rainforest, El Verde, Puerto Rico; (c) temperate forest, Cove Forest, Smoky Mountains, Tennessee. The arrows below the bottom consumer boxes indicate small amounts of energy presumed to be eaten by further consumers. Units as for previous figure.

The grazing food chain is weak. Again, over the whole year, the energy captured by green plants in photosynthesis equals the energy lost in the respiration of all the organisms in the community.

- ❍ Based on the three energy budgets in Figure 3.22, what is the mean percentage of the biomass that ends up as detritus?
- ● 89% of biomass ends up as detritus in these ecosystems.

In fact, it is thought that decomposers account for 90% of the energy flow through ecosystems. In a temperate forest, most of the detritus is in the form of dead leaves.

- ❍ What would happen if there was no decomposition?
- ● Leaves would rapidly accumulate to smother the wood.

Around 354–290 million years ago, a similar situation did occur: huge masses of forest died and were not decomposed. They were buried, and gradually, over the aeons, anaerobic processes converted the dead vegetation to coal.

In their early stages of development, trees are growing rapidly; therefore much of the biomass is stored as wood. However, as the forest matures, the rate of photosynthesis does not increase much, so much more of the biomass is respired by the increased mass of plant tissues, and therefore less is available for growth. Eventually, the production of new wood is balanced by the decomposition of old wood. The ecosystem as a whole stores no energy — a **stable equilibrium** is attained.

3.3.4 Summary of Section 3.3

1. An ecosystem is a community of organisms interacting with one another and their abiotic environment to form a unit with its own characteristic pattern of energy flow and nutrient cycling. Most ecosystems are open in that they exchange energy and materials with their wider environment to some extent.
2. What constitutes an ecosystem is largely determined by a scientist who wishes to answer a specific question relating to energy flow and nutrient dynamics. Ecosystems can be of any size, but should be big enough, and studied over a sufficiently long time period to provide data that reflect a substantial proportion of the naturally occurring variation.
3. Ecosystems can be studied using pyramids of numbers, biomass or energy, which model the relative sizes of the trophic levels present according to different criteria, each of which has advantages and disadvantages.
4. Energy budgets are a more informative way of representing ecosystem energetics that enable comparisons of different ecosystems to be made. Using energy budgets, three main types of ecosystem can be recognized, based on whether the bulk of the primary production is consumed by herbivores, goes into storage or enters the decomposer food chains.

Question 3.2

Examine the three hypothetical energy budget diagrams (a), (b) and (c) in Figure 3.23 and identify what type of ecosystem each one represents, giving reasons for your decisions.

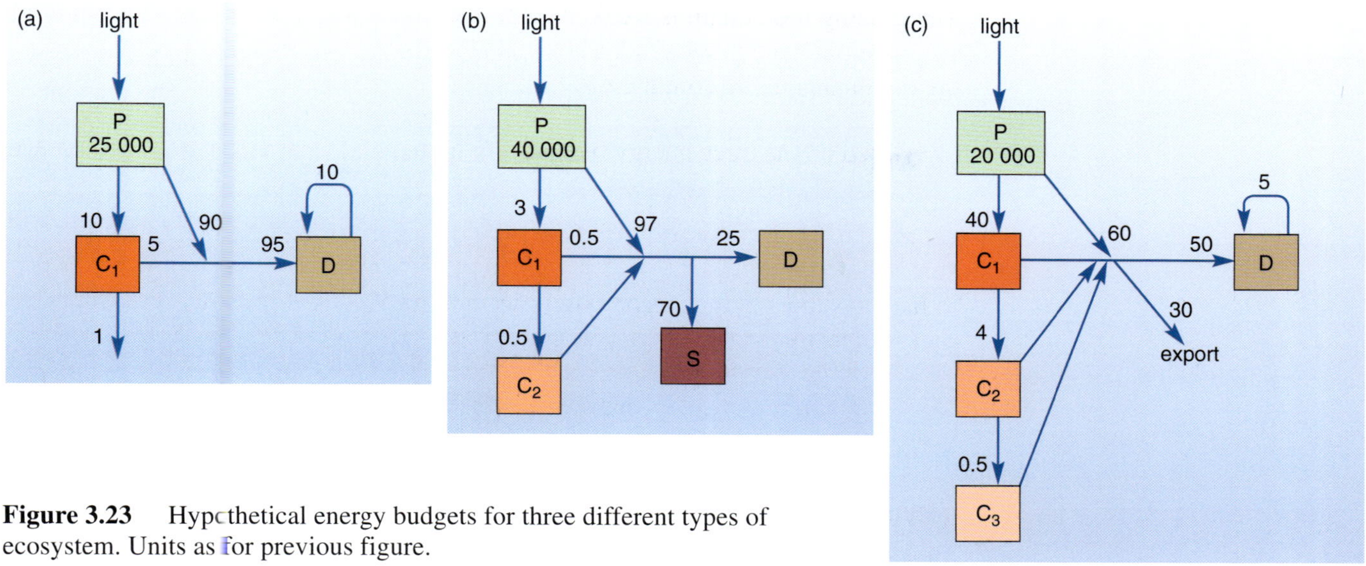

Figure 3.23 Hypothetical energy budgets for three different types of ecosystem. Units as for previous figure.

3.4 Water as a resource for plants

All life depends on water. For terrestrial organisms, the acquisition and retention of water is a major problem. They are basically packages of water surrounded by an environment that is, more often than not, 'demanding' water from them.

- ❍ Recall the main reasons why living organisms require water.
- ● Water is a major structural component of all organisms; it is the medium for all their chemical reactions; it is a raw material for many reactions including photosynthesis; many organisms require external water to reproduce (Section 2.4).

The Earth's atmosphere is occasionally saturated with water vapour, in which case there is no drying effect, but this is usually only true for brief periods or for relatively small parts of the Earth's surface. In most places and for the majority of the time the atmosphere 'demands' water from any wet surface. This means that the air has the capacity to hold additional water vapour. It is described as being unsaturated. (Refer back to Block 2, Part 1 for further detail on humidity and vapour pressure.)

The process by which liquid water enters the gas phase and moves off into the atmosphere is called evaporation. On land, much of the evaporation is via the leaves of plants. Plants tend to act as a conduit for water stored in the soil to move up through the plant's tissues to be released into the atmosphere by the leaves. This particular type of evaporation process is called transpiration. For many purposes, it is difficult to distinguish whether water vapour entering the atmosphere from a land surface has come via a plant or directly from a wet surface such as the soil. To recognize the importance of the route via plants, but to avoid separating the two processes, the term evapotranspiration (introduced in Part 1, Section 2.4.2) is often used.

3.4.1 Water movement in the soil–plant–atmosphere system

Water flows from a reservoir in the soil through the roots, stem and leaves of a plant and is released into the atmosphere as vapour (Figure 3.24). The amount of water lost to the atmosphere is of great importance, particularly when studying the water cycle (in Part 1 of this Block) or energy balances (Block 2, Part 1). In terrestrial systems, typically 75% of the water entering the atmosphere does so via plants. How can we describe the flow of water through the soil–plant–atmosphere system?

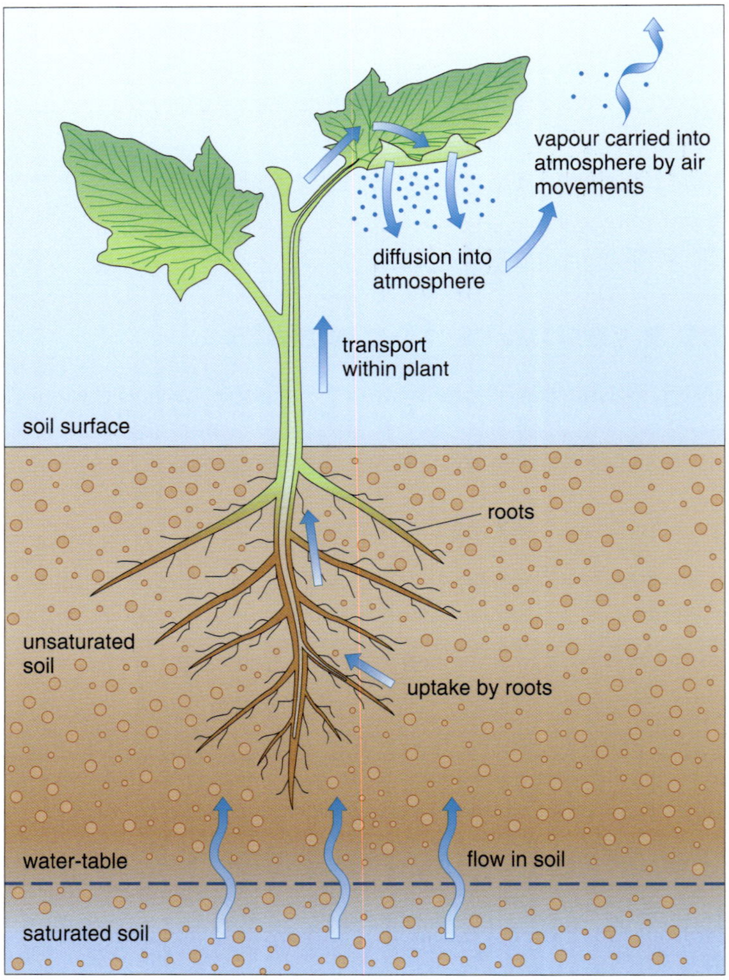

Figure 3.24 Pathway of water movement from soil to atmosphere.

There are three variables to consider. You have already encountered Darcy's law (both earlier in this Block and in Block 2, Part 2), where it was used to describe water movement through soil.

- ❍ Recall the three terms that were used in the equation describing this law.
- ● (1) the force driving the flow (hydraulic gradient, h/l); (2) the conductivity of the soil, K; (3) the flow rate (or flux), v.

Darcy's law is in fact just a form of the general flux equation:

flux = driving force × conductivity

This relationship applies not only in the soil, but can be extended to the plant itself and its interface with the atmosphere too. We explore each of these variables in the following sections, in order to appreciate one of the most fundamental processes in the biosphere. First we concentrate on the driving force and its measurement in different parts of the soil–plant–atmosphere system.

3.4.2 The driving force for water movement

The driving force is created by a change in pressure through the system. In this context, pressure is not as easy to observe as, for example, air pressure is in a car tyre. It is often present as potential pressure. Water flowing over a waterfall is a useful example (see Figure 3.25).

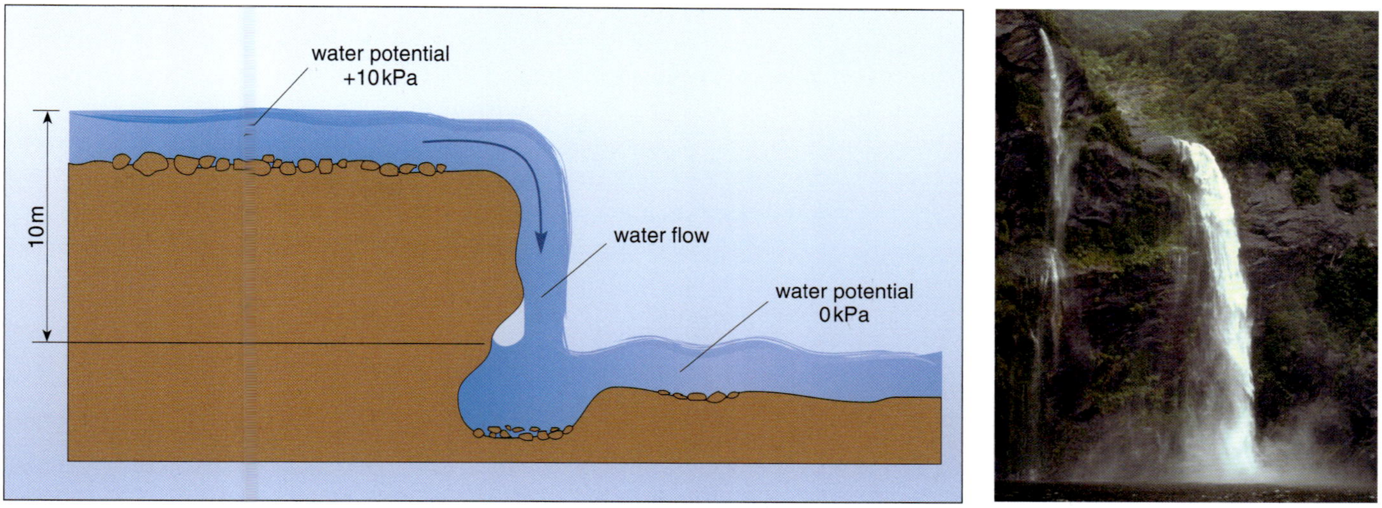

Figure 3.25 Forces involved in water passing over a waterfall.

The water above the fall has the 'potential' to fall and to release energy in doing so. It is therefore said to have potential energy. Let's take the water downstream of the fall as a reference point and describe it as having zero energy. Each cubic metre of water above the fall has the potential to drop 10 m. A cubic metre of water has a mass of 1000 kg. If 1000 kg falls 10 m under gravity, the amount of energy released can be calculated. (Refer back to the introduction to energy in Block 2, Part 1.)

$$\text{energy } (E) = \text{mass } (m) \times \text{gravitational constant } (g) \times \text{vertical distance } (h)$$

Recall that $g = 9.81\ \text{m s}^{-2}$, but as this is a back-of-the-envelope calculation, lets call it $10\ \text{m s}^{-2}$.

So, for this example

$$E = 1000\ \text{kg} \times 10\ \text{m s}^{-2} \times 10\ \text{m} = 100\,000\ (\text{kg m s}^{-2})\ \text{m}$$

But we have seen earlier that a mass (kg) multiplied by an acceleration (m s^{-2}) is a force (recall $F = ma$) and that this force is measured in newtons (N), such that $1\ \text{N} = 1\ \text{kg m s}^{-2}$, so

$$E = 100\,000\ \text{N m}$$

From Block 2, remember that $1\ \text{N m} = 1\ \text{kg m s}^{-2}\ \text{m} = 1\ \text{kg m}^2\ \text{s}^{-2} = 1\ \text{J}$, so

$$E = 100\,000\ \text{J or } 100\ \text{kJ}$$

Each cubic metre of water therefore has 100 kJ of potential energy relative to the reference state. A quantity of energy such as this, within a defined volume can be expressed as a pressure. (In Block 2, Part 1, you came across a definition of pressure as force per unit area, i.e. $p = F/A$, and as energy per unit volume or energy density, i.e. $p = E/V$.) If you would like to review the units in which it is measured and convince yourself that pressures can express not only force per unit area, but also energy density, then read Box 3.2.

Box 3.2 Using pressure to describe water potential

Pressure is usually defined as force per unit area and it is expressed in the unit of pascals (Pa).

If an elephant stands on your hand, then pressure is applied! Let's assume the elephant has a mass of 1000 kg — quite a small elephant! It is balancing on one foot on top of your hand, as in Figure 3.26. The area of contact between hand and foot is 10 cm × 10 cm. We can now calculate the pressure being applied:

$$\text{force} = \text{mass} \times \text{acceleration}$$

$$= 1000\,\text{kg} \times 10\,\text{m}\,\text{s}^{-2} = 10\,000\,\text{N}$$

$$\text{area} = 0.1\,\text{m} \times 0.1\,\text{m} = 0.01\,\text{m}^2$$

$$\text{pressure} = \text{force/area}$$

$$= 10\,000\,\text{N}/(0.01\,\text{m}^2) = 1\,000\,000\,\text{N}\,\text{m}^{-2} \text{ or } 1\,000\,000\,\text{Pa}$$

(Remember 1 Pa = 1 N m^{-2}.)

So the pressure being applied to your hand is 1 000 000 Pa or 1000 kPa.

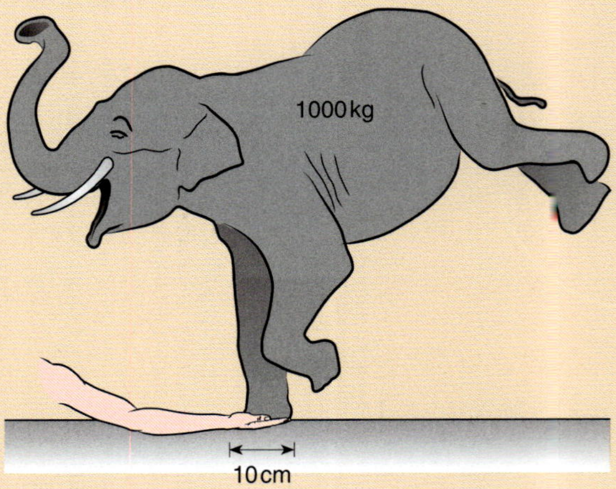

Figure 3.26 A graphic illustration of pressure as force per unit area!

Most of us can easily relate to this type of pressure — it would hurt! There is another way to express pressure, however. It can be thought of as energy per unit volume rather than force per unit area. To convince ourselves, let us consider the units it is measured in. The unit of pressure is a pascal (Pa), which is defined as one newton (N) per one square metre (m^2; metre × metre) of area. This is quite a small amount. It is equivalent to the weight of a small apple spread over the area of a dining table.

We have learnt that energy was measured in joules (J) and that a joule is defined as one newton multiplied by one metre. In other words, it is the energy you need to lift our small apple off the floor and put it on the table. Again, this is quite a small quantity. This is why the kilojoule (1 kJ = 1 000 J) is the more commonly seen unit. Volumes, meanwhile, are measured in cubic metres (m^3; metre × metre × metre).

Let's start by expressing one unit of pressure as a force per unit area. Mathematically, we get

$$1\,\text{Pa} = \frac{1\,\text{N}}{1\,\text{m}^2}$$

Now for the cunning part: we could multiply both top and bottom of the fraction on the right-hand side of the equation by 1 metre, whilst leaving the left-hand side unchanged, because multiplying both parts of a fraction by the same quantity has no net effect (they cancel each other out). So, we can write

$$1\,\text{Pa} = \frac{1\,\text{N}}{1\,\text{m}^2} \times \frac{\text{m}}{\text{m}}$$

Remember that 1 N × m (1 N m) is 1 joule (J) and 1 m^2 × m is equivalent to (metre × metre × metre) or m^3. So we can write

$$1\,\text{Pa} = 1\,\text{J}\,\text{m}^{-3}$$

One pascal is equal to one joule per cubic metre; in other words, pressure can be expressed as an amount of energy per unit volume, i.e. $p = E/V$.

Pressure = energy/volume and is measured in units of pascals (1 Pa = 1 J m^{-3}). Therefore the water above the waterfall can be said to have 100 kJ m^{-3} or 100 kPa of potential pressure relative to the reference state, which has 0 kPa. This pressure is called the **water potential**. It is a measure of the water's capacity to move through a hydrological system. Water at any point in the soil–plant–atmosphere system can be ascribed a water potential, the standard notation for which is Ψ (the Greek letter 'psi', pronounced 'sigh'). Water potential is always expressed relative to a reference state, which is defined as pure liquid water at a reference height (which should be defined for the particular system you are working with). One cubic metre of water upstream of the waterfall has a positive potential energy of 100 kPa, so its water potential can be defined as

$$\Psi = +100\,\text{kPa}$$

- ❍ If 100 m^3 of water were being held in a tank at the top of a water tower 20 m high, what would its water potential be relative to ground level? (You can do this calculation back-of-envelope style!)

- ● First calculate its potential energy ($E = mgh$):

$$\text{energy} = 100\,000\,\text{kg} \times 10\,\text{m}\,\text{s}^{-2} \times 20\,\text{m}$$

$$= 20\,000\,000\,\text{J}$$

Now this amount of energy is spread over a volume of 100 m^3, so to calculate the water potential as a pressure, we divide energy by volume ($p = E/V$):

$$\text{pressure} = 20\,000\,000\ \text{J}/(100\ \text{m}^3)$$

$$= 200\,000\ \text{Pa or } 200\ \text{kPa}$$

Therefore the water potential of the water at the top of the tower is 200 kPa.

So far, we have considered positive water potentials. However, in soils, plants and the atmosphere, water is more often held under **tension**, which equates to a negative water pressure (sometimes called a suction). To visualize this, think of a damp sponge placed on a table. The water does not run out of it; it is being held under tension. If you press on the sponge lightly, it is compressed, but water still does not come out. If you increase the pressure, a pool of water will eventually form on the table. The pressure you need to apply to expel water from the sponge is equal and opposite to the tension under which the water was being held.

Question 3.3

In the experiment illustrated in Figure 3.27, a weight of 10 N was needed to start expelling water from a sponge 0.20 m long and 0.10 m wide. What was the water potential of the water in the sponge at the start of the experiment?

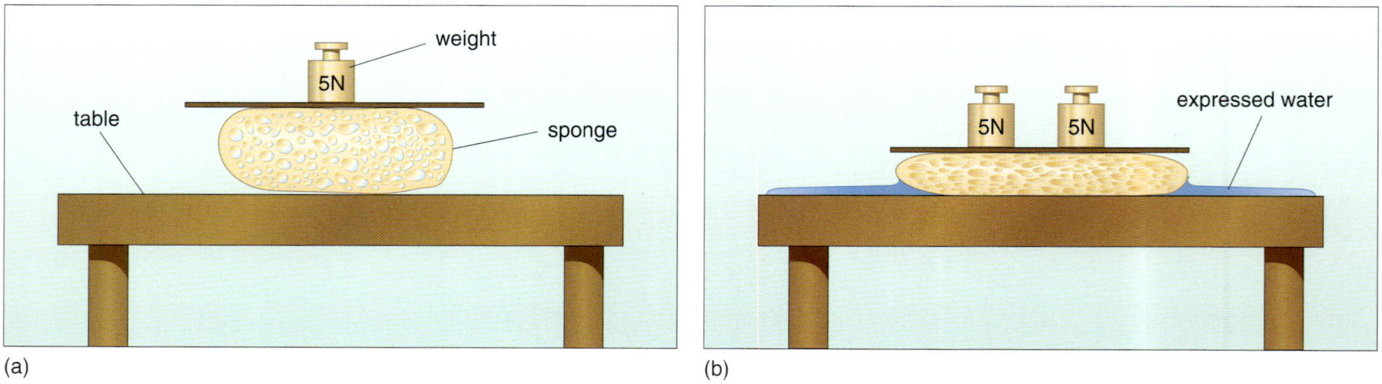

Figure 3.27 A damp sponge on a table with various weights stacked on to it. In (a) the weight (5 N) is insufficient to expel water from the sponge. In (b) the weight of 10 N squeezes water out of the sponge and a pool of water forms.

3.4.3 Water potential in plant tissue

The water potential within a piece of plant tissue may be measured in a similar way to the sponge example above, by using a pressure chamber known as a **pressure bomb**. Compressed gas is fed into a chamber containing a piece of plant tissue until the pressure is sufficient to push water out of the cut stalk (Figure 3.28). The pressure as read from a gauge then gives you a direct measure of how strongly the water was being held within the plant. A gauge reading of 100 kPa would indicate that the plant tissue's water potential was −100 kPa. This is a typical value for water held in a plant. The water is being held 200 times more tightly than the water in a sponge. If you squeeze a wet sponge, water will come out. A leaf also contains water-filled tubes, but if you cut a leaf off a plant and squeeze it between your hands, it is not easy to force water back out of the cut end. You would need to apply 100 kPa of pressure.

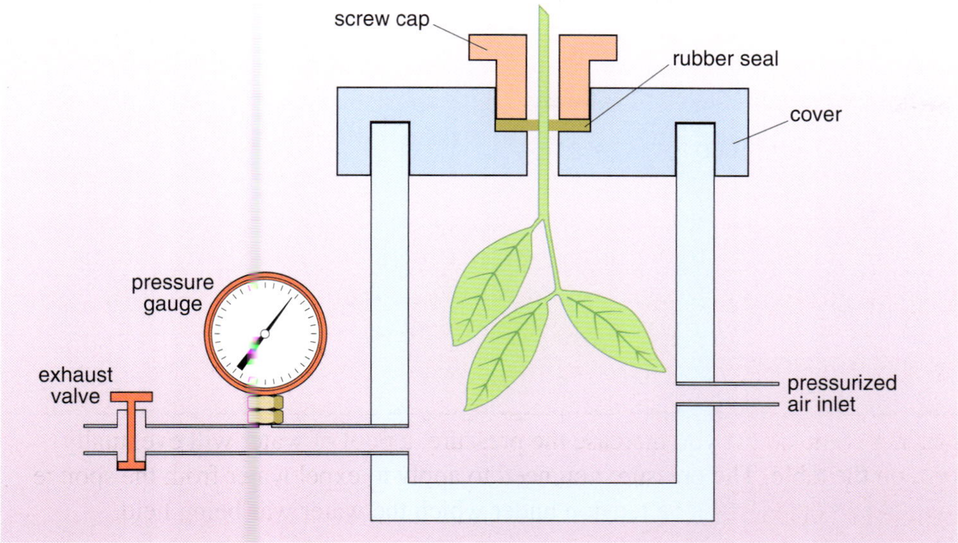

Figure 3.28 A pressure bomb used for measuring water potentials in plants. The vessel is pressurized until water emerges from the cut stalk.

Water is held tightly by the leaf in part because of its adhesion to the fine-bore tubes containing it, but also because it is retained within cells by the presence of **solutes**. Solute is the term used to describe a substance such as a salt or a sugar, when it is dissolved in water to form a solution. Solutes have the effect of lowering the water potential of the solution in comparison to pure water. It is not necessary within this course to consider the physics of this statement, but in layman's terms the solutes can be thought of as helping to bind water molecules more tightly together, therefore rendering them less available for movement or evaporation. A solution is said to have an **osmotic pressure** (π), which is the component of water potential that accounts for the presence of solutes. Expressed simply, the more concentrated the solution, the higher the osmotic pressure (see Box 3.3). The effects of osmotic pressure are best observed when two solutions with different osmotic pressures are separated by a **semipermeable membrane**, which allows water molecules (relatively small) to pass through it, but not solute molecules, which are generally much larger.

There are two other components of water potential. **Hydrostatic pressure** (***P***) corresponds to pressure we think of in the everyday sense. Water in a bottle of sparkling mineral water is under positive hydrostatic pressure, due to the pressure of gas in the bottle acting on it. If you made a hole in the bottom of the bottle, water would spurt out much faster than by gravity alone. It is also possible to have negative hydrostatic pressure (sometimes referred to as tension). When you suck water up a straw, the water inside the straw is under negative hydrostatic pressure. If you made a hole in the side of the straw, air would rush in and the water would fall back.

The third component we will consider is **matric pressure** (***m***), which accounts for the tendency of water to 'cling' to solid surfaces by adhesion; it was matric pressures we were considering in the example with the sponge. Mathematically water potential is defined as

$$\Psi = P - \pi - m$$

P is a positive component of the equation because water with high hydrostatic pressure tends to move to zones of lower pressure. Osmotic pressure is a negative component because solutions with high osmotic pressure tend to attract water toward them (see Box 3.3) and matric pressure is also negative because zones with high matric pressure such as an almost dry sponge or a nearly dry soil tend to attract water into them.

All the components are expressed relative to pure water, which has an osmotic pressure of zero ($\pi = 0$), a matric pressure of zero ($m = 0$) and a hydrostatic pressure of zero ($P = 0$), if it is at the reference height, which is normally taken as ground level. So for pure water at ground level, the water potential is found by an easy calculation:

$$\Psi = 0\,\text{Pa} - 0\,\text{Pa} - 0\,\text{Pa}$$
$$= 0\,\text{Pa}$$

Box 3.3 A brief introduction to osmosis

Osmosis is the process of water movement between solutions of different concentrations. If we have a compartment (compartment 1) containing a litre of water in which we have dissolved 1 g of table salt, then it will have an osmotic pressure (π). A separate compartment (compartment 2) also contains a litre of water, but this time has 2 g of salt dissolved in it. Its osmotic pressure is higher. Let us assume that the two compartments are at the same elevation and we set that level as our reference height, then the hydrostatic pressure term (P) for both compartments is set to zero.

Remember $\Psi = P - \pi$, so the water potential of both solutions is negative.

Water will potentially flow down the gradient of water potential from compartment 1 (less negative) to compartment 2 (more negative.) There is no pathway for water actually to move between them, so there is only a potential for movement. However, if we replace the barrier with a semipermeable membrane, similar to a biological cell membrane, then water movement can occur. The semipermeable membrane allows water to pass through it, but not salt. Water now flows down its water potential gradient from compartment 1 (dilute solution) to compartment 2 (concentrated solution), but the salt is retained. This process is **osmosis** (Figure 3.29). (For practical proof that this process really occurs, try Activity 3.1 below.)

The osmotic pressure can be quantified by measuring the piston pressure required to stop water moving. In this example, the difference in osmotic pressure between the two solutions was found to be 82 kPa.

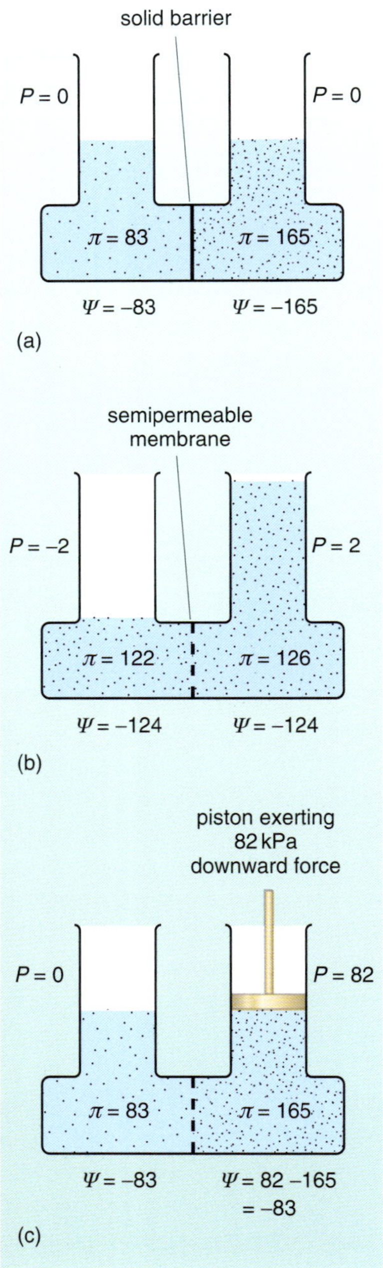

Figure 3.29 Illustration of the relationship between osmotic pressure (π), hydrostatic pressure (P) and water potential (Ψ). (a) Two solutions with different solute concentrations (higher on the right than on the left). (b) The same solutions, but with the barrier between them replaced by a semipermeable membrane. Water has moved down a water potential gradient, from the dilute to the concentrated solution, via the membrane, until the pressures equilibrate. (c) If we now press down on the piston to force water back through the membrane, the pressure that needs to be applied to restore the situation in (a) is equal to the original difference in osmotic pressures between the solutions.

Activity 3.1 Potato osmometer

In this activity, you make a system with a potato acting as the semipermeable membrane separating the two aqueous compartments of different osmotic pressure (water, and salt solution), and actually watch the osmosis happening.

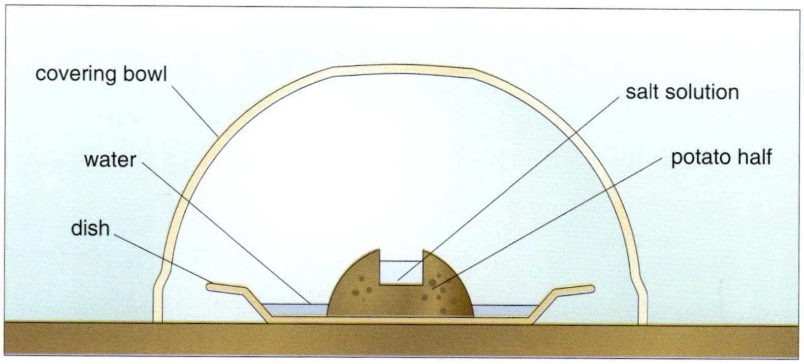

Figure 3.30 The experimental set-up needed to observe osmosis in action.

Take a large, fresh potato and slice it in half. With the flat side down (Figure 3.30), bore out a hole in the top of one half as shown in the diagram, using a small knife or an apple corer. Place the potato with its flat side down in a shallow dish of water. Now completely dissolve a heaped teaspoon of table salt into a quarter of a cup of water and carefully transfer a teaspoon of the resultant solution into the hole. Note the position of the solution's surface in the hole. Mark it if you can. Now cover the whole thing with a big plastic bowl or cling-film to stop water evaporating, then wait for a couple of hours and return to see if the solution level has changed.

It is expected that the water level will have risen in the hole. Water has moved upward against gravity but down a water potential gradient.

The potato is simply replacing the semipermeable membrane in Figure 3.29. The cell membranes in the potato allow water to pass through them, but not salt.

❍ Would a baked potato work just as well?

● No. Cooking disrupts membranes. (If you want proof, repeat the exercise with a cooked potato.)

Think about the following questions:

1 Which solution has the greater osmotic pressure: the tapwater in the dish or the salt solution in the potato well?

2 Which solution has the higher water potential?

3 In which direction has water moved?

Figure 3.29 showed one way of measuring osmotic pressure in a solution. Plant physiologists use a similar technique to measure the water potential within plant tissues. It employs the instrument we met earlier called a pressure bomb (Figure 3.28). Instead of a piston it uses compressed gas to apply a pressure to the tissue. Water is therefore forced out of the bomb under hydrostatic pressure. Both the osmotic pressure within the leaf cells and the matric pressure within the fine xylem tubes resist this movement. When free water first appears at the cut end of the tissue outside the bomb, the applied pressure is said to be equal and opposite to the original water potential of the leaf.

3.4.4 Water potential in soil

The pressure at which water is held in a moist but freely draining soil is approximately −5 kPa, within plant tissue it is normally in the range of −50 to −200 kPa (though it can fall to −1500 kPa or occasionally even lower) and the atmosphere may typically have a water potential anywhere between 0 and −100 000 kPa. The direction of water movement can be predicted using one simple rule. Water will move, either as a liquid or as a vapour, from a zone of high water potential to a zone of lower water potential. In the case of the waterfall in Figure 3.25, water moves from the zone of +100 kPa to the zone of 0 kPa. That is to say, it goes over the waterfall — which is not a surprise!

In the case of the soil–plant–atmosphere system, it may move from a potential energy status of −5 down to −200 and finally to −20 000 kPa (Figure 3.31). This gradient of water potential is the driving force behind evapotranspiration. Note the extremely low water potential of a dry atmosphere. This is a typical example of a warm, dry day in the UK. When the atmosphere is saturated with water vapour, such as when it is raining, its water potential is zero and evapotranspiration will cease due to the lack of any driving force.

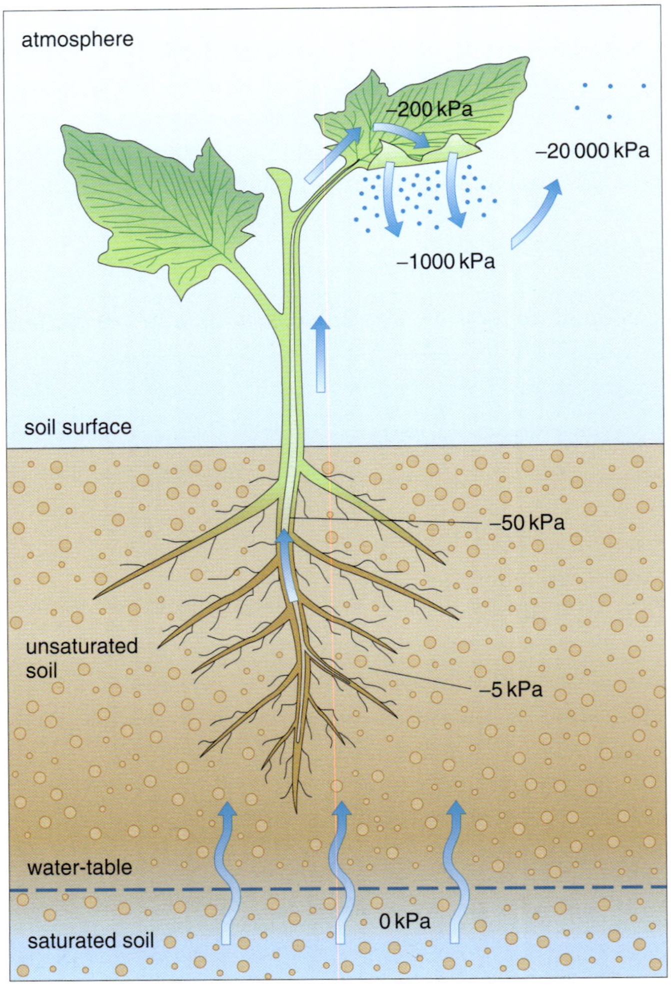

Figure 3.31 Diagram of water fluxes through the soil–plant–atmosphere system showing the gradient of water potentials along which the water moves.

As a soil dries, the residual water within it is held increasingly tightly in the small spaces between the soil particles. The water potential of the soil is said to become more negative due to the increasing matric pressure. The soil is acting like the sponge we encountered earlier; it is holding water by adhesion to its surfaces. Plants will continue to extract water from it until its water potential has fallen as low as −1500 kPa. They do this by accumulating salts, sugars and similar solutes in their tissues, thereby lowering their internal water potential to a value lower than that of the soil water. The water will then flow from soil to plant. If the soil becomes drier than this, most plants can no longer obtain water from it. Although they are able to lower their internal water potential below this value (sometimes to as low as −4000 kPa), water is almost unable to move in soil with matric pressures as low as −1500 kPa, as it is so tightly bound to soil surfaces. It is therefore unable to reach the root.

If the soil contains salt, the ability of plants to extract water from it is reduced. The water being held by matric pressure (e.g. 10 kPa) in the small pores of the soil would also have an osmotic pressure (e.g. 90 kPa) in the presence of salt, making the total water potential even more negative. ($\Psi = P - \pi - m$, therefore $\Psi = 0 - 10 - 90 = -100$ kPa.)

In order to extract water from a saline soil, plants have to lower their internal water potential by accumulating even more internal solutes before water will flow in. But if too many solutes are accumulated, the normal functioning of the plant cell becomes impaired. This is a major problem in areas of irrigated agriculture in some parts of the world. Salts from the irrigation water accumulate in the soil and lower the soil water potential, making it increasingly difficult for plants to draw water out. Eventually the land becomes barren as crops are no longer able to extract water from the soil.

- If the air is saturated, is there likely to be any movement of water in the soil–plant–atmosphere system? (Drawing a diagram labelled with water potentials would help you answer this.)

- Yes. Water could be absorbed from the atmosphere by plants and even conceivably be released back to the soil, if it were dry — transpiration in reverse! This is because the gradient in water potential now runs from the air (0 kPa) to the plant (perhaps −50 kPa) to the soil (e.g. −100 kPa).

In a few arid environments, such as the Namib Desert in southern Africa, some plants (notably the shrub *Trianthema hereroensis*, Figure 3.32) rely on absorbing moisture directly from the atmosphere in this way, especially when cold nights cause dew to form on leaves.

With the normal daytime water potential of the atmosphere in the range of −10 000 to −100 000 kPa, the atmosphere has tremendous potential to take water from other sources. It is from this drying atmosphere that all life needs to protect itself.

Figure 3.32 The Namib Desert shrub *Trianthema hereroensis*, which survives for most of the year by ‘drinking’ dew.

3.4.5 Resistances to the flow of water

The resistances to flow in the soil–plant–atmosphere system are not evenly spread. Water is able to move relatively freely within the soil, if it is damp. There are gradients of water potential within the soil, along which water flows. Likewise inside the plant, water can move relatively freely within a plumbing system of water-filled tubes (xylem vessels). It is the root–soil interface that usually creates the largest resistance to water movement in the liquid phase. Water can flow into the root only where there is physical contact between root and soil water (see Figure 3.33). This may be only a small proportion of the root, especially in a coarse-textured soil. A large flux of water entering the root system via a small area of root surface creates a resistance. You can visualize a bath-full of water trying to leave via a small plughole. There is a resistance, which means the bath takes some time to empty. A plant compensates for this factor by growing a hugely extensive network of fine roots (Figure 3.34). Even a small grass plant may have a root surface area of around 100 m^2 — the area of a badminton court!

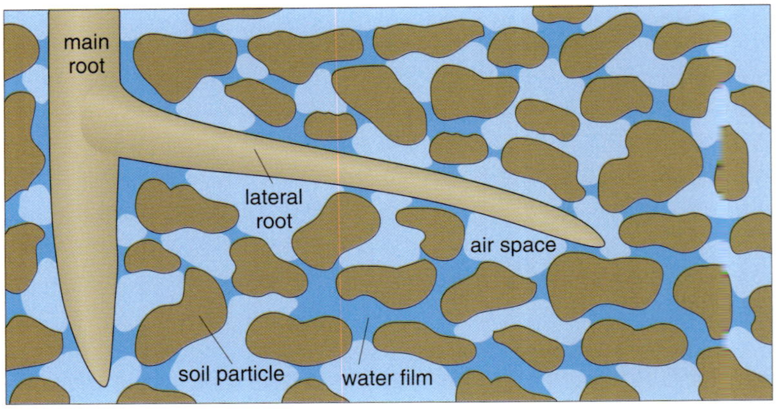

Figure 3.33 A cross-section through a root and its surrounding soil. Much of the soil is air-filled, so water can only move to the root as a liquid where there are points of contact. This is why the root–soil interface can represent a major resistance to water movement.

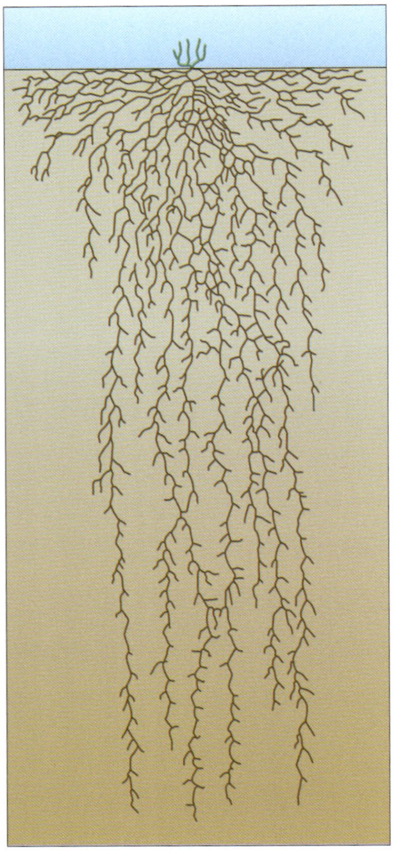

Figure 3.34 The greatly branched system of fine roots typical of grass plants.

Once inside the plant, the water has to cross at least one cell membrane, which creates a further resistance, and it travels from roots to shoots via the xylem vessels. Water movement in such narrow tubes causes resistance to flow as the water 'sticks' to the sides of the pipes. All the resistances to water movement in this liquid phase are inconsequential, however, in comparison to the resistance at the leaf–atmosphere boundary. We saw in Figure 3.31 that the biggest 'jump' in water potential was between the leaf and the atmosphere. The driving force for water movement is therefore often very high, but so is the resistance to flow at this step. The movement of water vapour away from the evaporating surface inside the leaf, to the air outside, is the most tortuous part of the whole pathway from soil to air, creating resistance to water movement. Water leaves the leaf predominantly through small pores in its surface called **stomata** (sing. stoma), because the rest of the leaf surface is covered by a waxy layer or cuticle, which forms a barrier to water loss (Figures 3.35 and 3.36). By opening or closing the stomata, the plant can exert some influence over its rate of water loss and thereby be able to protect itself from dehydration.

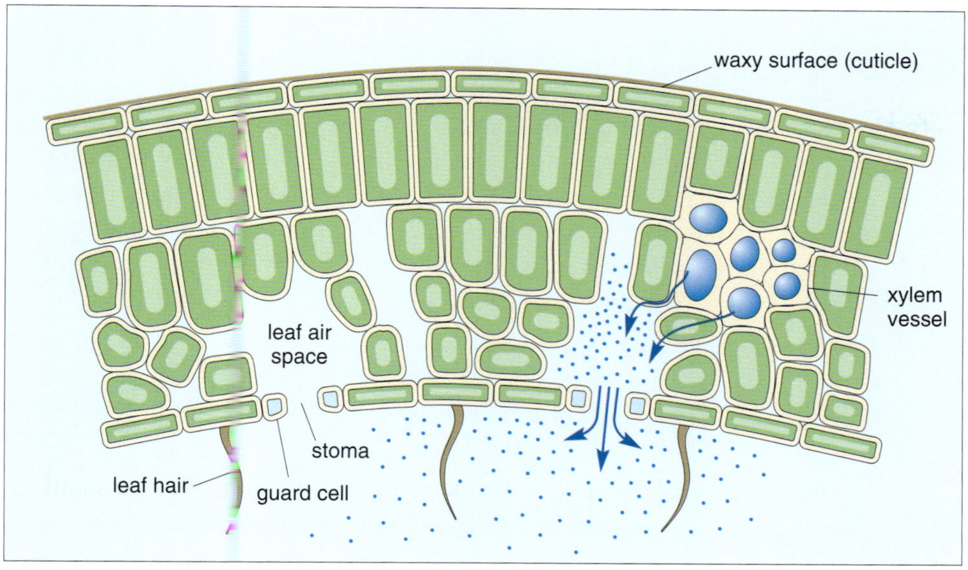

Figure 3.35 A transverse section through a typical leaf, showing the waxy surface layer (the cuticle) and the stomata.

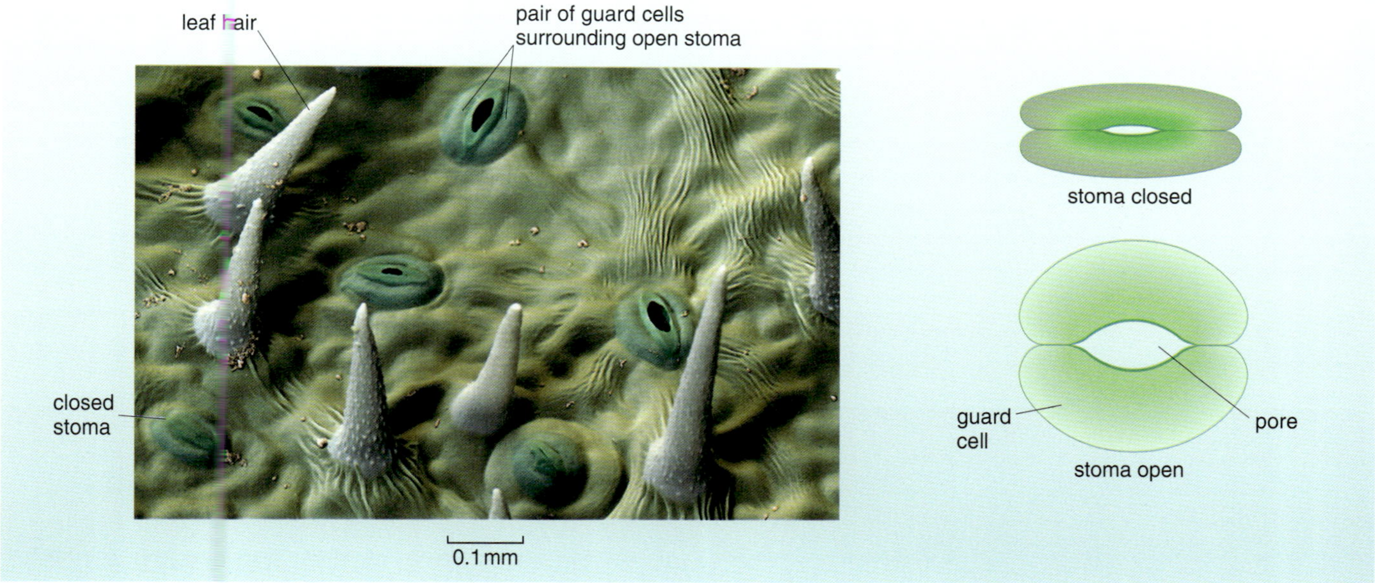

Figure 3.36 The underside of a leaf showing a stomatal pore. On the right are diagrammatic representations of an open and a closed stoma. The plant opens its stomata by pumping ions into the guard cells, which then attract water by osmosis and expand.

3.4.6 The flux of water through the system

The flux of water through the system is called the **transpiration stream**. The movement of water from the soil through the plant to the atmosphere involves energy. The process that is most demanding of energy is the phase change, from liquid water to gaseous water vapour, which occurs inside the leaf.

- ❍ Where does the energy for this phase change comes from?
- ● The source of energy to drive the process is radiation in the form of sunlight.

The rate of transpiration on a bright sunny day can be several times that on an overcast one, even though temperatures and humidities may be similar. Plants absorb a large proportion of the Sun's radiation and the energy is largely used in the evaporation of water within the leaf. In this way, plants can be regarded as cooling machines. A single large tree can consume and evaporate hundreds of litres of water per day (see Activity 3.2). The conversion of this quantity of water from liquid to gas requires energy (Figure 3.37). It can be calculated that the tree has absorbed over 400 000 000 J of radiation energy from sunlight. The energy and water relations of vegetation can have a profound effect on climate at a regional or even a global scale.

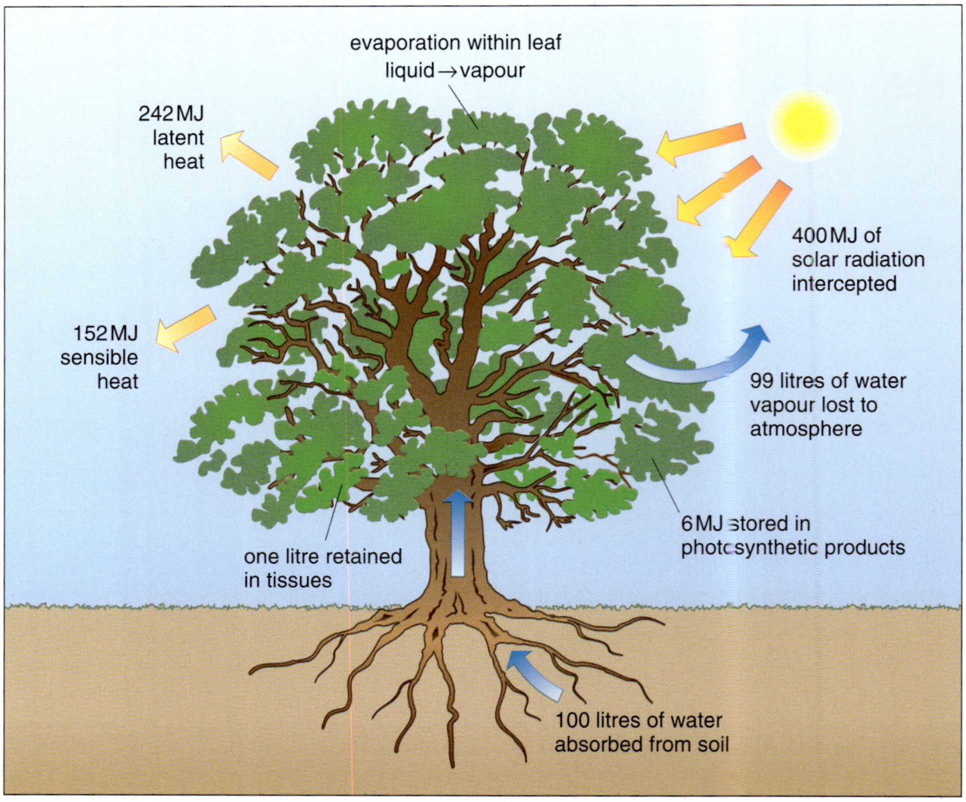

Figure 3.37 A single tree showing the fluxes of water and energy involved in transpiration. Note what a small proportion of each quantity is actually retained within the tree.

We have looked at the physical system by which water is released into the atmosphere by transpiration. We will now consider the biological context for this process.

❍ If plants were entirely covered in an impermeable cuticle that prevented any movement of gases into or out of the leaves, they would lose very little water and water availability would not be such a constraint on their growth. Therefore, why do plants have stomata, which provide holes in the cuticle through which water escapes? (*Hint*: refer back to Section 2.2.1.)

● The reason plants have stomata is to absorb carbon dioxide from the atmosphere. The simultaneous loss of water vapour from within the leaf is an unavoidable by-product of carbon acquisition.

The plant can be regarded as 'paying for' the carbon dioxide it gains from the atmosphere with the water vapour it loses. Carbon dioxide is 'expensive.' Typically, for each molecule of CO_2 entering the leaf, 50 molecules of water leave (Figure 3.38).

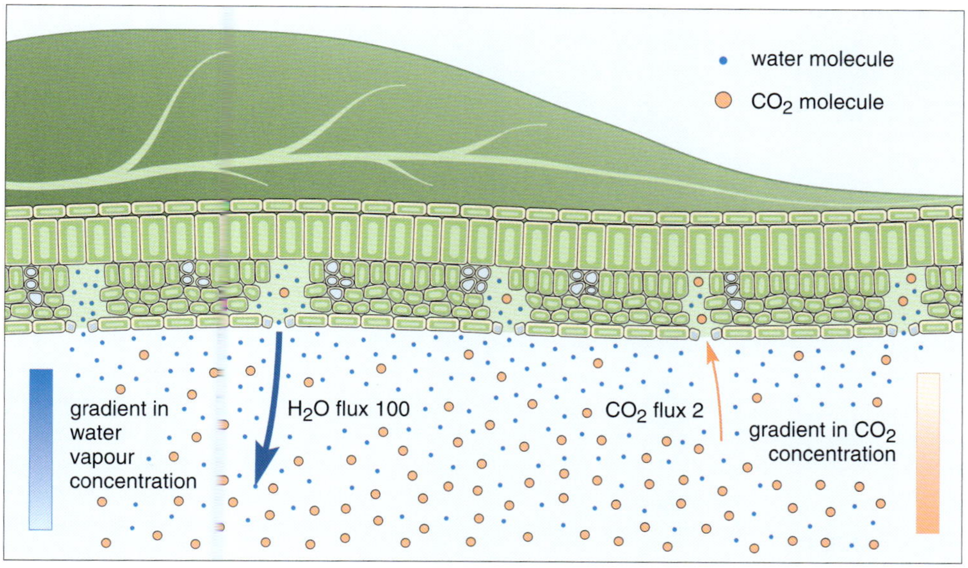

Figure 3.38 The relative magnitude of diffusion between carbon dioxide entering a leaf and water vapour leaving it.

- ❍ What is creating the gradient in CO_2 concentration between the leaf and the atmosphere?
- ● The consumption of CO_2 by photosynthesis within the leaf.

Activity 3.2 The transpiration stream

On the DVD there is a three-minute video clip about the transpiration stream in trees. As you watch, consider the following points:

- the structure of wood, which allows water to flow;
- the role of the leaf pores (stomata) in the regulation of transpiration;
- the nature of the force driving water movement — this is not explicitly described in the clip, but the preceding text has discussed it.

There are three other video clips of similar length within this part of Block 3. They can be usefully watched either as you encounter the relevant activities or all together at the end, as revision.

3.4.7 Conservation of water

In many environments, the availability of water limits the growth of plants. It is a scarce resource that plants have become adapted to conserve. Some plants are able to improve their efficiency by opening stomata at night instead of during the day.

- ❍ Why would this behaviour result in the loss of less water? (Recall the description of humidity that you read in Section 2 and think about what happens to relative humidity as the temperature drops at night.)
- ● Relative humidity tends to be higher at night, therefore the atmosphere's demand for water is lower. There is no solar radiation at night, so the energy driving transpiration is much reduced.

These plants can be more than ten times as efficient in their use of water than 'normal' plants, which open their stomata during the day (Figure 3.39). Cacti and other succulent species growing in deserts employ this mode of carbon dioxide uptake. CO_2 is taken up at night whilst the stomata are open (Figure 3.40). This method of photosynthesis is Crassulacean acid metabolism, usually referred to as **CAM**.

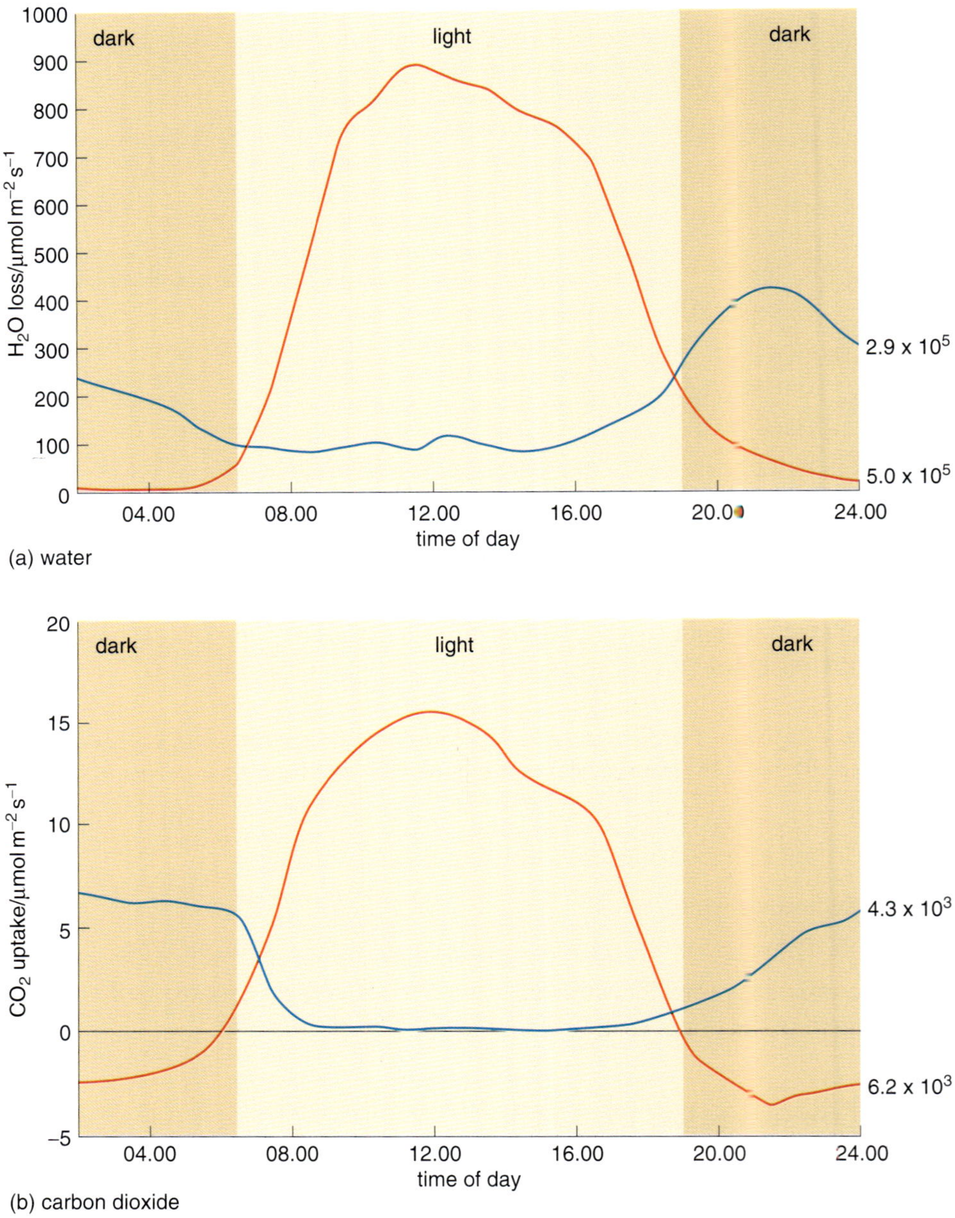

Figure 3.39 Fluxes of (a) water and (b) CO_2 versus time of day, for plants with the normal (C3, red lines) and the CAM (blue lines) photosynthetic pathways. The value at the end of each line represents the area under the curve and estimates the total flux over a 24-hour period. All values are expressed in units of $\mu mol\ m^{-2}\ s^{-1}$.

- From Figure 3.39, which pathway has accumulated the most CO_2 over a 24-hour period, and what are the implications for the plant's competitiveness?

- The normal (C3) plant accumulates 50% more CO_2, which would allow it to grow faster and outcompete the CAM plant in a non-stressed environment.

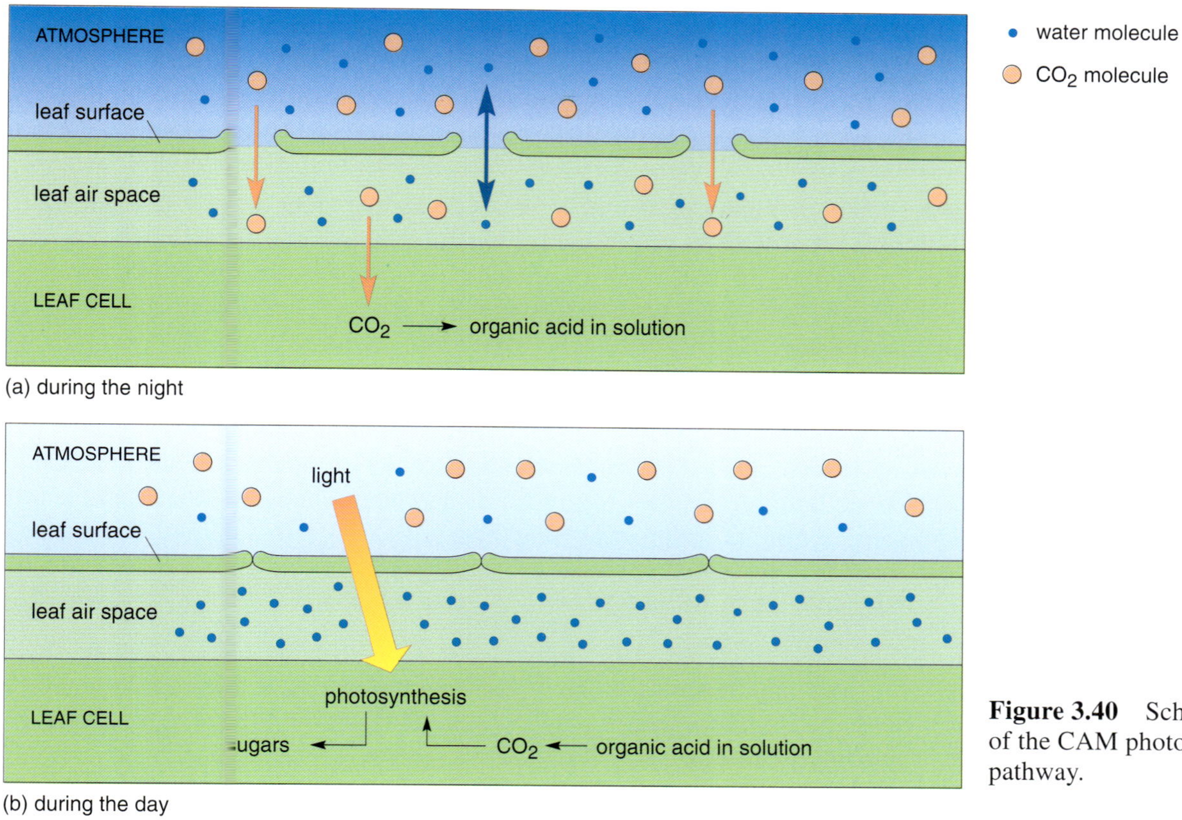

Figure 3.40 Schematic diagram of the CAM photosynthetic pathway.

The reason cacti contain so much water in their tissues per unit of photosynthetic area (Figure 3.41) is not purely to store water for use during droughts, but also to create a larger reservoir in which to temporarily store CO_2 over the night/day cycle. The CO_2 is stored as an acid and requires large volumes of water within the cells to dilute it. In deserts, plants are generally not limited in terms of light availability and so it is often the plant's capacity to store CO_2 that limits the amount of carbon it can fix. All CAM plants are succulent (they have a high water content).

Figure 3.41 A cholla (pronounced 'choya') cactus (*Opuntia imbricata*), showing its fleshy tissues with high water content. Leaves are reduced to spines and the stems take on the photosynthetic role to reduce surface area and water loss.

Question 3.4

Water use efficiency (WUE) is the ratio used to describe the amount of CO_2 absorbed per unit of water lost:

$$\text{WUE} = \frac{\text{moles of CO}_2\text{ absorbed}}{\text{moles of H}_2\text{O lost}}$$

Using the information in Figure 3.39, calculate the WUE over a 24-hour period for both the CAM and the C3 plant. Which uses its water more efficiently?

Where water resources are the overriding constraint on growth, plants that open their stomata only at night have an advantage over their day-opening counterparts. In other environments (such as a reedbed), it is more important to maximize growth rates (i.e. open stomata in the day when CO_2 can be fixed as it is absorbed) than it is to maximize water use efficiency, because competition from other plants is a greater immediate threat to survival than is dehydration.

- ❍ Which of the C–S–R growth strategies encountered earlier (Section 2.6.3) would describe the behaviour of the desert perennials?
- ● Stress tolerator. They are making the most conservative use of a limiting resource at the expense of a rapid growth rate.

Why do not all plants take in their carbon dioxide at night in order to conserve water? In order to make use of the carbon dioxide, plants need energy in the form of sunlight. If the CO_2 is absorbed at night, it has to be stored within the leaf until the Sun comes up, and then released so that it is available for photosynthesis.

Plants in temperate climates tend not to have such extreme adaptations as the cacti, because taking up CO_2 at night rather than during the day has several costs. The main one is the limited storage capacity of the plant tissues for CO_2. This sets a limit on the plant's growth rate and therefore makes it susceptible to competition from faster-growing neighbours.

- ❍ Why does nocturnal uptake of CO_2 create a problem for the photosynthetic reactions?
- ● CO_2 can only be converted to sugar using the energy of daylight. It therefore has to be stored within the plant for several hours until the photosynthetic machinery can start.

Plant species of less arid areas are nevertheless often adapted to cope with limited water availability. In an area such as the Teign catchment, the amount of rainfall on an annual basis exceeds the evaporative demand of the atmosphere on an annual basis. (The evaporative demand of the atmosphere is the 'potential evapotranspiration', a term you met in Part 1 of this Block.) During an average year in the Teign catchment, over 900 mm of rain falls, whilst only approximately 640 mm is potentially released back to the atmosphere as evapotranspiration. One may suppose that for plants growing in the region, water would not be a limiting resource. However, the vast majority of species, even in areas as wet as the Teign catchment, do have adaptations to prevent water loss, although their adaptations are less effective than those of cacti. For example, after a meadow has been cut for hay, the cut material from almost all of the 50 or more species present, has become severely wilted after just 24 hours (Figure 3.42). This demonstrates that once the supply of water is cut off, the tissues of most plants rapidly dehydrate, so not even the cuticle and closed stomata block water loss entirely. Cacti can survive for months without a water supply, but most British native plants would die the same day if their water supply were entirely stopped.

Figure 3.42 A recently cut hay meadow with rapidly dehydrating tissues.

- By studying Figures 3.43 and 3.44, during which period of the year are plants most at risk of dehydration in (a) Devon and (b) Cambridgeshire?

- (a) June–July; (b) April–August.

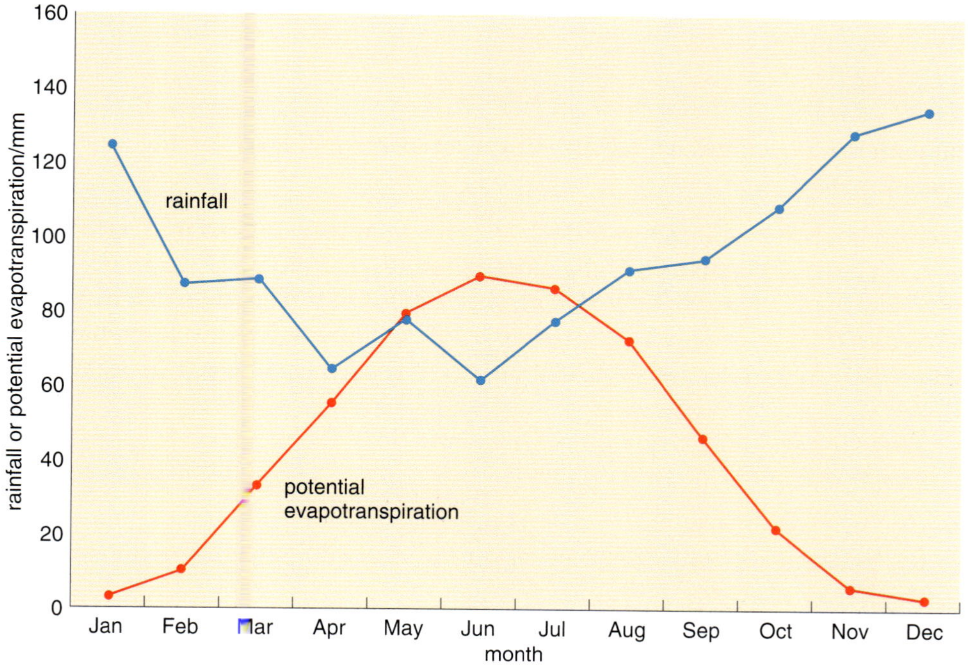

Figure 3.43 Plots of monthly rainfall and potential evapotranspiration for Devon, England.

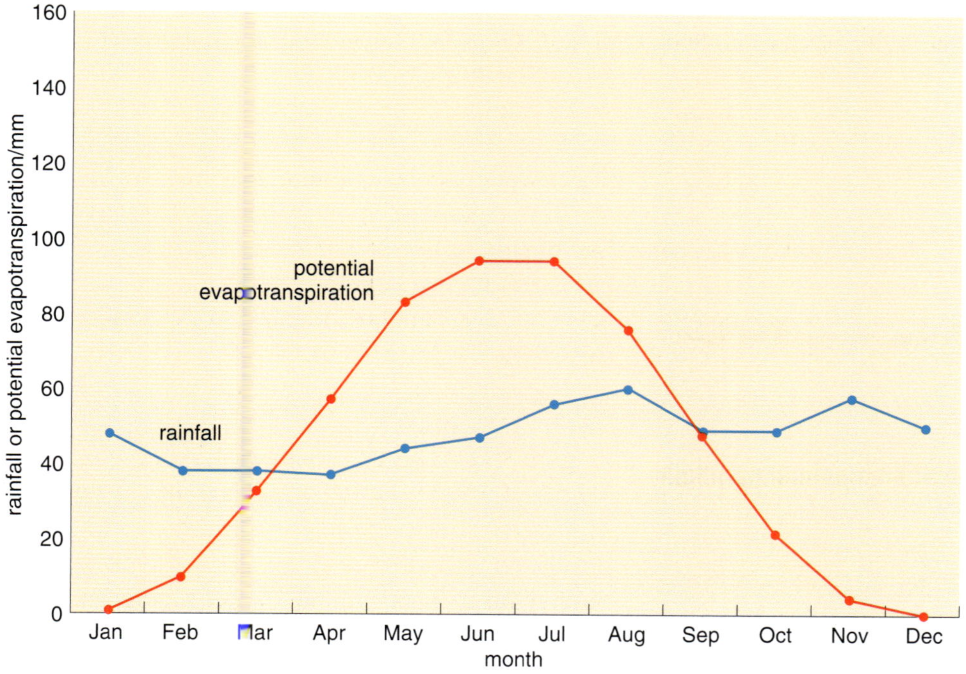

Figure 3.44 Plots of rainfall and potential evapotranspiration for Cambridgeshire, England.

Most plants need a continuous supply of water to the roots; only the most highly adapted can survive more than a few hours during the growing season without a source of water. Therefore, an excess of water on a whole-year basis is irrelevant to plants: they require a supply on a daily time-frame, not an annual one. Even in an area as wet as the Teign catchment, there will occasionally be a 10 or even a 20-day period in which no rain falls. It is then that plants rely upon the water storage capacity of the soil.

3.4.8 Water storage by the soil

Water retention and release are among the most important properties of a soil from a plant's perspective.

A freely draining soil, such as a sand, is unable to retain much of the water falling as rain during a wet period. At the other extreme, a very fine-textured soil such as a clay may retain a large volume of water, but holds it so tightly that plants are unable to extract it. To understand the usefulness of a soil type as a water store for vegetation, we have to consider its **moisture release curve** (Figure 3.45).

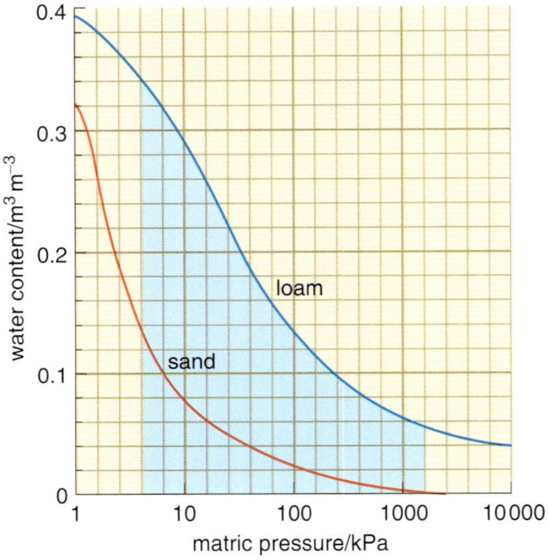

Figure 3.45 The moisture release curves for two soils, showing the available water capacity (see text for explanation).

- ❍ Using Figure 3.45, determine the proportion of the loam's volume that is occupied by water when the soil matric pressure is 100 kPa.

- ● The water content is approximately 0.14 or 14% by volume.

The soil water potential, Ψ_{soil}, describes how tightly water is held within the pores of the soil's matrix. (Remember $\Psi_{soil} = -m$.) If Ψ_{soil} is greater (more positive) than −5 kPa, then the water is being held so loosely that it will drain out under the force of gravity. (We will consider soils in which drainage is prevented later.) If Ψ_{soil} is less (more negative) than −1 500 kPa, then water is being held so tightly by the soil (by hydrogen bonding between the water molecules and the soil particles) that most plants are unable to abstract it. It is therefore the soil's ability to retain water that can be accessed by the plant, sometimes referred to as its **available water capacity** (AWC), which is of interest and it equates to the blue-shaded zone on Figure 3.45. Some typical water contents of soils of a range of textures and at the critical water potentials are shown in Table 3.2. The figures in the table refer to the proportion of the soil's total volume that is occupied by water.

Table 3.2 The moisture content of soils, expressed as the volume of water per volume of soil, for a range of textures at given water potentials, and the values of available water capacity (AWC) derived from these.

	Moisture content expressed as a proportion of the soil's volume		AWC
Texture	$\Psi_{soil} = -5\,kPa$	$\Psi_{soil} = -1500\,kPa$	
sand	0.10	0.05	0.05
silt	0.30	0.20	0.10
loam	0.35	0.15	0.20
clay	0.50	0.35	0.15

On soils with a low AWC, such as sands, plants are at risk of running short of water. If one can estimate the rooting depth of the vegetation (for example 1 metre), one can convert the AWC, given as a volume fraction (0.05 for a sand) to an actual volume of water available in a given area. So 1 m^2 of sandy soil, with vegetation rooting to a depth of 1 m, has a rooting volume of 1 m^3 (or 1000 litres). The AWC as a proportion of the volume is 0.05, so the available water per square metre of soil is $1\,m^3 \times 0.05 = 0.05\,m^3$ or 50 litres.

- Calculate the volume of water available to a plant in a square metre of the clay soil described in Table 3.2, assuming the rooting depth is 0.80 m.

- The rooting volume is $1\,m^2 \times 0.8\,m = 0.8\,m^3$. Using an AWC of 0.15, the volume of available water is $0.8\,m^3 \times 0.15 = 0.12\,m^3$, which is equivalent to 120 litres.

For ease of comparison with meteorological data, the volume of available water can be expressed as a depth of water instead of a proportion of the volume; using the sand example above, a volume of 0.05 m^3 of water standing on an area of 1 m^2 would give a depth of 0.05 m. Descriptions of soil types often give available water capacities in this format. What a water depth of 0.05 m means to a plant is that it can meet the demands of the atmosphere for a period in which potential evapotranspiration does not exceed 0.05 m. On a sunny summer's day in the Teign catchment, the evapotranspiration demand of the atmosphere would be approximately 0.004 m. Therefore if there were no rain for two weeks (14 days), the total demand would be 0.056 m, the soil moisture reserves would become exhausted and the vegetation might begin to dehydrate and wilt. Two weeks of dry weather is not an infrequent event in summer in Devon, so plants have become adapted to this constraint, by either reducing their transpiration such that the demand from the atmosphere is not met, or acquiring more water from deeper in the profile.

Question 3.5

Using the data in Table 3.2, estimate the number of days in the growing season each soil could fully supply the potential demand of the Teign catchment's vegetation for water. Assume that the soil is holding its maximum amount of water at the start of the period and that plants can extract water only from the top 0.80 m of the profile. Also assume that during the growing season, evapotranspiration exceeds rainfall by a constant rate of $0.002\,m\,d^{-1}$.

3.4.9 Plant responses to a shortage of water

There are three approaches by which a plant can ameliorate the problem of water shortage. They loosely correspond to the three C–S–R life strategies as described by Grime:

- The 'competitive' approach is to maximize the amount of water taken up. This involves sending roots deeper into the profile, growing a more extensive root system to gather water from a greater area and a denser network to explore the soil more thoroughly in order to extract all the available moisture from throughout the rooting volume.
- The 'stress-tolerant' approach, in contrast, is to minimize the water loss. The waterproof, waxy cuticle over the surface of the leaf is often thicker. Leaves may also grow a dense network of hairs both to reflect radiation, thereby keeping the leaf cool, and to trap a layer of moist air around the leaf, increasing the resistance of the pathway for water vapour to diffuse out of the leaf. These adaptations involve an investment by the plant and may compromise its growth rate, but the benefit they confer is allowing plants to grow on soils that become dry (Figure 3.46).
- The 'ruderal' approach is to grow rapidly whilst the soil is moist (usually in spring), set seed by mid-summer, then remain dormant in the soil seed bank during the late summer when soils are most prone to dry out. These ruderals are behaving like desert annuals. Their life cycles allow them specifically to avoid drought rather than to avoid disturbance, which is the more common driver for the ruderal strategy in UK, as exemplified by the arable weeds. An example of a species that completes its life cycle very early in the year, thereby allowing it to colonize drought-prone sandy soils, is the whitlow grass (*Erophila verna*, Figure 3.47), which is actually a member of the cabbage family, not a grass at all.

Figure 3.46 Wild thyme (*Thymus polytrichus*), a classic example of a species with a 'stress tolerator' strategy, The leaves have a dense covering of hairs, designed to minimize water loss. The scientific name '*poly-trichus*' is in fact Greek for 'many hairs.'

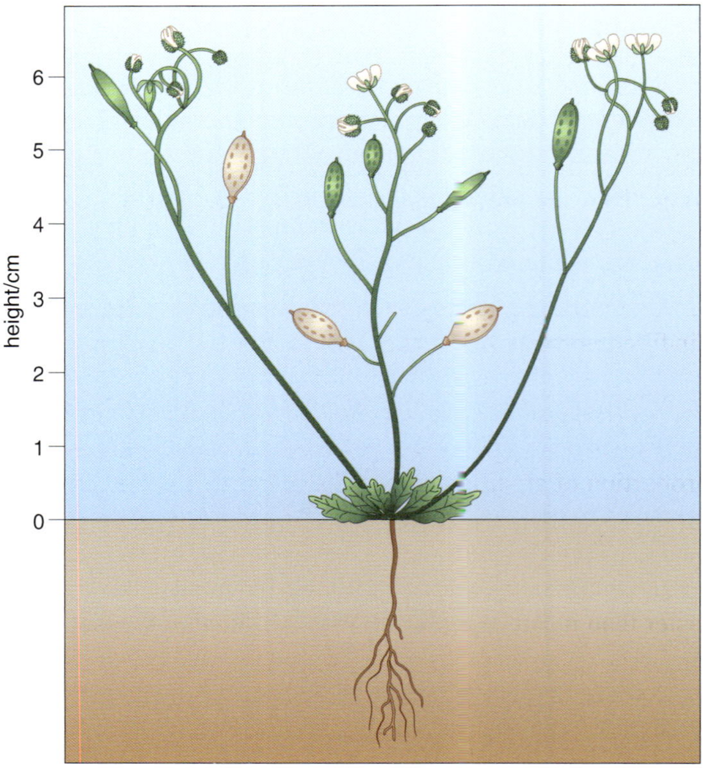

Figure 3.47 The whitlow grass (*Erophila verna*), which completes its short life cycle in early spring. Again the scientific name '*verna*', meaning spring in Latin, flags up this lifestyle. Note the relative sizes of the leaves and the reproductive structures.

3.5 Oxygen as a resource for plants

As you learnt earlier, all plant tissues respire (Section 2.2.1). This respiration is necessary to provide energy for the maintenance of all the vital processes at a cellular level. The normal respiration pathway in plants is much the same as in the other multicellular organisms; animals, fungi and protoctists. Depriving plant tissue of oxygen will cause suffocation, just as it does in animals. Plant shoots have no difficulty in obtaining oxygen from the surrounding atmosphere. The gas is able to diffuse freely into shoots via stomata or specialized breathing pores known as lenticels (Figure 3.48).

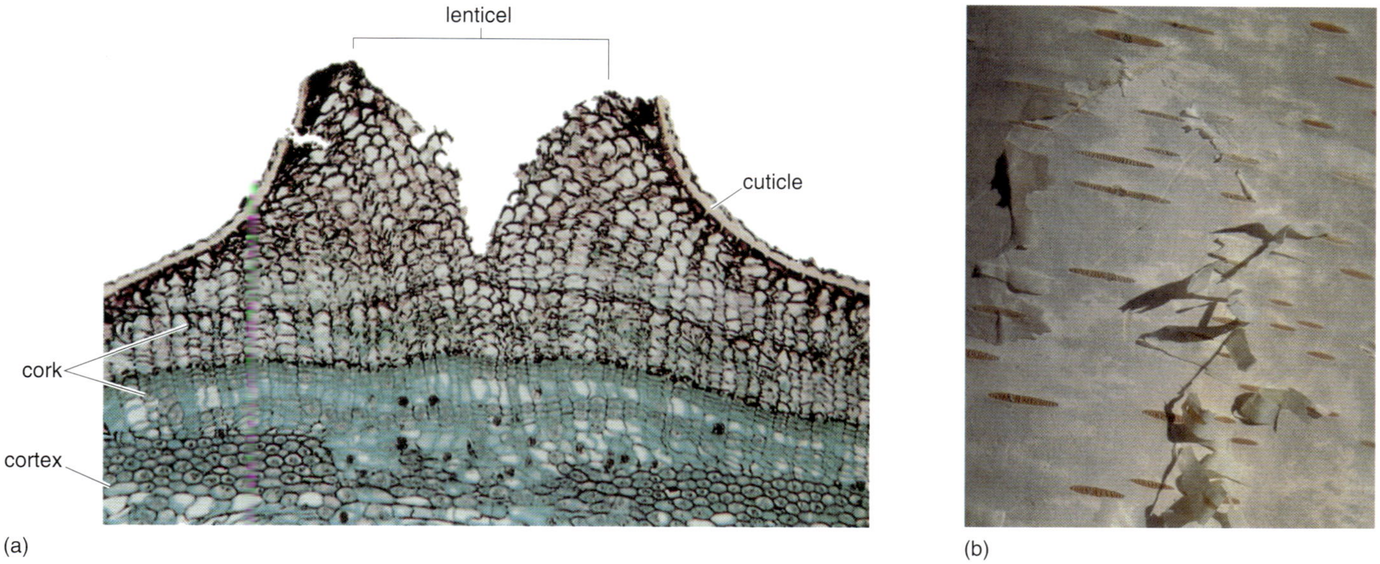

Figure 3.48 (a) A cross-section through a lenticel structure in the bark of a tree. The loosely packed cork cells allow air to enter, thereby supplying the underlying tissues, cortex, with atmospheric oxygen for aerobic respiration. (b) The external appearance of lenticels on the bark of a birch tree (*Betula* sp.).

Access to a sufficient supply of gaseous oxygen is more of a problem for the roots. They are buried under soil. However, soil is a porous medium and providing it is not waterlogged (a state in which all its pores are water-filled), then a proportion of its volume will be filled by air. This proportion is referred to as the **air-filled porosity**. Soil physicists have calculated that for most soils an air-filled porosity of 0.1 (i.e. 10% of the soil's volume occupied by air) is sufficient to allow oxygen from the atmosphere to diffuse relatively freely down into the root zone of the vegetation. As a result of this diffusion, the soil air can remain sufficiently oxygen-rich to allow roots to respire. However, if the proportion of air-filled pores falls below this value, diffusion is hampered and the soil may become depleted in oxygen, a condition described as **anoxic**. This occurs primarily as a result of the microbes, such as bacteria and fungi in the soil, that are living off decaying organic matter (decomposers) using oxygen faster than it can be replaced. In these situations, plant roots are at risk of suffocation. The root tissue of many plant species will die just a few days after the soil atmosphere exhausts its reserves of oxygen.

3.5.1 Life in waterlogged soils

The problems plants face in waterlogged soil do not end with a lack of oxygen supply. The decomposers (bacteria and fungi) in the soil rely on a supply of oxygen to break down organic matter efficiently. In anoxic soils, the recycling of nutrients is slowed down and often becomes limiting for plant growth. Some of the soil bacteria, when deprived of oxygen, are able to respire by using alternative pathways that do not involve oxygen. They are able to utilize compounds present in the soil in place of oxygen gas. The most accessible alternative for many types of bacteria is often the nitrate ion (NO_3^-). As can be seen from its chemical formula, it is composed of the elements nitrogen and oxygen. It is often the preferred form for the uptake of plants' most important mineral nutrient, nitrogen. When nitrate is used by respiring bacteria, the nitrogen it contains may be lost to the atmosphere as nitrogen gas. Plants growing in the waterlogged soil may experience nutrient limitation not only because increasing amounts of nutrient are locked up in undecomposed organic matter, but also because the supply of available nitrogen is being rendered unavailable as it is turned to gas.

Figure 3.49 A soil profile showing mottling of orange and blue-grey colours which are indicative of periodic anoxia within the soil.

Other bacteria are able to transform iron, one of the constituent elements of the clay minerals within the soil particles themselves. The iron atom (Fe) within compounds can exist in one of two forms. The first is the oxidized form, Fe^{3+} or Fe(III). This forms compounds that are often orange-brown in colour (e.g. rust) and rather insoluble in water. Bacteria make use of these oxidized compounds to enable them to respire in the absence of oxygen. The products that are formed by bacterial respiration have iron in its alternative, or reduced form, Fe^{2+} or Fe(II). This, in contrast to its Fe^{3+} counterpart, tends to form compounds that are blue-grey in colour and rather more soluble in water. The colour of a soil, which is often determined by compounds of iron, can tell the environmental scientist much about its oxygen status (Figure 3.49).

It is the differential solubility of the two forms of iron that is of relevance to plant growth. Although plants need to absorb some iron for use in their biochemical pathways, it is only a trace amount and the absorption of excess iron is toxic and often lethal for plants.

So, to recap, some soil bacteria make use of three important reactions in waterlogged soil:

1. $C_6H_{12}O_6 + O_2 \longrightarrow CO_2 + H_2O + \text{energy}$

   This has the effect of mopping up any residual oxygen left dissolved in the soil water, thereby preventing plant roots from using it for their own respiration.

2. $C_6H_{12}O_6 + NO_3^- \longrightarrow CO_2 + H_2O + N_2 + \text{energy}$

   This deprives plants of their principal mineral nutrient, nitrogen.

3. $C_6H_{12}O_6 + \text{Fe(III) compounds} \longrightarrow CO_2 + H_2O + \text{Fe(II) compounds} + \text{energy}$

   This releases iron into the soil solution at levels that may be toxic to plants.

(*Note*: the arrows indicate that the reactions are not shown as balanced equations.)

Waterlogged soils, therefore, are potentially very hostile environments for plant growth. Yet, consider Figure 3.50. It shows lush growth of reeds in a permanently waterlogged soil. These plants are fast-growing and productive. They adopt a competitive strategy, rather than the stress-tolerant one which you would expect in an environment hostile to plant growth. There is a paradox here. Our knowledge of soil processes suggests that plants should struggle to grow in a soil containing little air, yet our observations from many wetlands indicate the contrary. How is it that wetland vegetation can circumvent the problems of wetland soils?

Figure 3.50 Vigorous growth of the common reed (*Phragmites australis*) on waterlogged soils.

The answer is quite simple. Let us first consider a swimmer over a coral reef (Figure 3.51). Whilst the swimmer has her face in the water, she too, like our wetland plant roots is in an air-free environment.

- ❍ How does the swimmer obtain sufficient oxygen to sustain respiration?
- ● Via a hollow air-filled tube which is in gaseous connection with the atmosphere above (commonly known as a snorkel).

Figure 3.51 A swimmer snorkelling over a coral reef is entirely dependent on an air-filled tube for her oxygen supply.

Wetland plants use the same principle. They do indeed have air-filled tissue in their roots and stems, which allows oxygen to diffuse along a gaseous pathway from the atmosphere to the root surface (Figure 3.52). This tissue has the anatomical name of **aerenchyma** ('air-enn-kye-mah'). The advantage of having air-filled tubes is that oxygen diffuses in air 10 000 times faster than it does in water. Although oxygen molecules have to diffuse over a distance of 1 m or more through relatively narrow tubes, they provide a much more plentiful supply to the roots than could be obtained via diffusion through just a few millimetres of waterlogged soil. Much of the plant's root is dedicated to oxygen transport (Figure 3.53). A further advantage is that the oxygen is not 'intercepted' by soil bacteria on its way to the roots. The supply from the aerenchyma can be sufficient, not only to meet all the respiratory demands of the plant's roots, but also to allow oxygen to 'leak' out into the surrounding soil (the rhizosphere), where it favours bacteria that use oxygen for their respiration. These displace the anaerobic bacteria and halt the reactions described above, which make the soil rather hostile for plant growth. This can sometimes be seen in the soil profile by red-stained lines in an otherwise dark matrix. The red coloration is due to Fe(II) being reoxidized to its Fe(III) state.

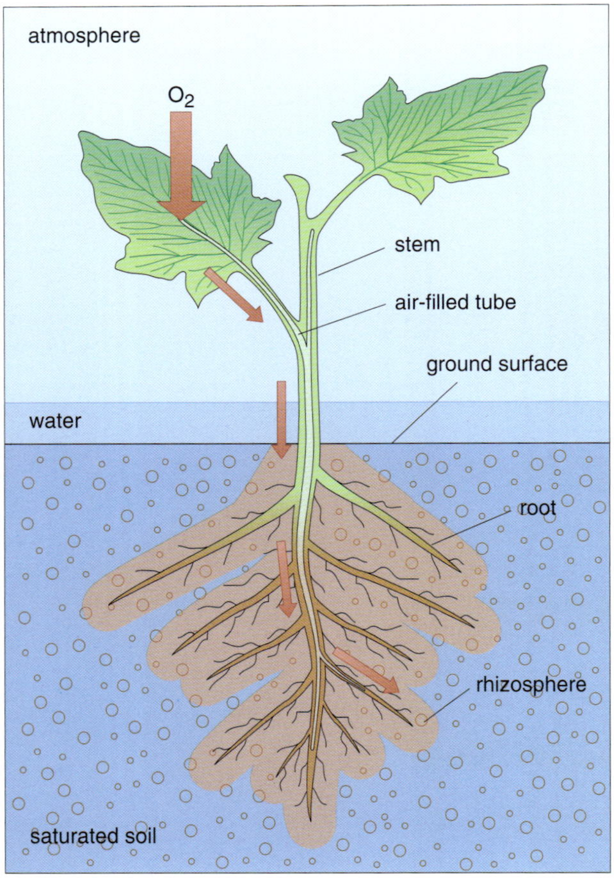

Figure 3.52 The pathway by which oxygen may diffuse from the atmosphere down to the rhizosphere via air-filled tubes (aerenchyma).

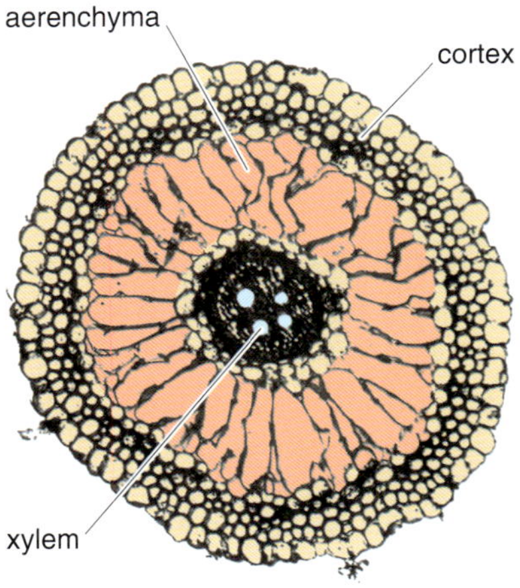

Figure 3.53 Transverse section through the root of the bog asphodel (*Narthecium ossifragum*). The species lives on almost permanently waterlogged soil and relies on aerenchyma (shown in pink) to supply its roots with oxygen. Note how much of the cross-sectional area is devoted to oxygen transport.

3.6 Vegetation and hydrology

We have seen in the previous sections that plants require rather different adaptations to survive in environments that are liable to be waterlogged compared to those that are liable to dry out.

- ❍ What structural adaptations have plants evolved to cope with (a) lack of soil water and (b) lack of soil air?
- ● (a) Deep roots, waxy leaf surface (cuticle) and dense hairs; (b) air-filled tubes (aerenchyma) in stems and roots.

It is possible for these two habitats to occur side-by-side, creating fine-scale variation within the environment (Figure 3.54). Species have tended to specialize their adaptations to cope with one or other of these potentially hostile environments. For that reason, there can be a significant change in the species composition of the plant community over short distances, in response to variation in the water regime. This enforced specialization is another example of trade-off (Section 2.6.3). Adaptations carry costs and so it is difficult to become specialized to more than one type of environment. The adaptation to drought, whether it is by deep and extensive rooting or by thick cuticles and reduced growth rate, carries a cost in terms of structural materials used and extra growth foregone. The main adaptation to waterlogging is the production of aerenchyma within stems and roots. This also creates a drain on structural materials. To attempt to adapt to both situations risks rendering the species uncompetitive in either.

Figure 3.54 A meadow showing distinct transitions between two plant communities. One is growing on ridges, the other in furrows. The elevation difference between the two zones is just 0.3 m, yet the communities differ markedly. This illustrates the influence of hydrology. Such ridge-and-furrow meadows are frequent in the English countryside. They characterize fields that were ploughed by oxen in medieval times, but which have subsequently returned to grassland.

Figure 3.55 Tufted hair-grass (*Deschampsia cespitosa*), which is often found on soils that experience waterlogging in spring and soil drying in summer.

There are some habitats that experience both waterlogging and drought. In a UK context, an example would be a soil having low AWC (see Section 3.4.8) that is liable to flood in winter and spring, but can dry out completely in summer. Some species do not conform to the trade-off concept and manage to adapt to these habitats because competition from other species is low. An example of such a species is the tufted hair-grass (*Deschampsia cespitosa*, Figure 3.55), which seems to have adapted to drought by an ability to develop a deep and extensive root system, whilst at the same time becoming adapted to avoid some of the problems connected with waterlogged soils, by excluding potentially toxic Fe^{2+} ions from entering its tissues.

3.7 The nutritional environment of plants

The distribution of plant species is strongly influenced by the availability of a number of mineral elements that they require for growth. The availability of these minerals is closely linked to the state of the soil environment.

3.7.1 Composition of plant tissue

Plants are not made from water and carbon dioxide alone. They are not purely a product of photosynthesis and water uptake (though admittedly this is often 98% true in terms of leaf fresh weight). The remaining 1–2% of the fresh weight comprises mineral nutrients.

- Recall from Section 2.4 the names of (a) two macronutrients and (b) two micronutrients.

- (a) Nitrogen and phosphorus; (b) manganese and iron.

A typical composition of a leaf in terms of its chemical elements is shown in Table 3.3.

The concentrations shown in Table 3.3 are by no means fixed. There is a good deal of variation between species and between tissues, and there is also some flexibility within a given tissue of a given species. If a mineral is readily available in the soil, plants will often take the opportunity to absorb it, even though it may not be utilized immediately in new growth. The internal concentration of that element therefore increases. This is sometimes referred to as **luxury consumption**. The portion not immediately required is stored within the plant's tissues for later mobilization.

Luxury consumption is an important strategy for many plants where important minerals such as nitrogen become available as a flush when soil conditions are appropriate for the breakdown of organic matter. Soluble nitrogen compounds are released from organic compounds and there is a finite time window in which a plant can capture the resource before it is lost from the soil by the processes of leaching and denitrification (both of which you met in Block 2, Part 2) or taken by competing plants.

In the other direction, plants can tolerate some dilution of their mineral content during periods of rapid growth, when rates of uptake from the soil cannot match

Table 3.3 A list of the elements that make up a typical leaf. There are 16 elements common to almost all plant tissue. Other elements do occur: some are essential for particular species (e.g. sodium and silicon) and some are simply contaminants from the environment (eg. lead or strontium).

Element	Symbol	Content/% leaf dry weight
carbon	C	40
oxygen	O	46
hydrogen	H	5.5
nitrogen	N	3.5
potassium	K	2.50
calcium	Ca	1.40
sulfur	S	0.50
phosphorus	P	0.35
magnesium	Mg	0.20
chlorine	Cl	0.02
zinc	Zn	0.005
boron	B	0.003
manganese	Mn	0.001
copper	Cu	0.0007
iron	Fe	0.0002
molybdenum	Mo	0.0001
others	Na, Si, etc.	0.02

demand from new tissue. Most nutrients can be remobilized and redistributed throughout the plant to supply the tissue in greatest need (usually the growing point). In these situations, old leaves may be plundered for their nutrient reserves and discarded in order to keep the new leaves at the top of the canopy well supplied. This would be typical behaviour in species following the competitive strategy (Figure 3.56). In contrast, species with a stress-tolerant strategy in the same situation would tend to conserve their older leaves and slow down or stop the production of new ones. Part of the reason for this is that some valuable minerals, such as calcium, cannot be remobilized and therefore are lost when old leaves are shed.

Figure 3.56 Yellowing lower leaves of a stinging nettle (*Urtica dioica*), showing the withdrawal of some essential nutrients from old tissues in order to keep growing points supplied.

3.7.2 The importance of nitrogen

As can be seen from Table 3.3, the most abundant element in plant tissue, after the three constituents of carbohydrate (carbon, hydrogen and oxygen), is nitrogen. This mineral nutrient is the one whose availability most often limits the growth of plants. It is also the most dynamic of the nutrients, existing in a number of compounds, which are in a constant state of cyclical flux.

Figure 3.57 A native British legume, bird's-foot trefoil (*Lotus corniculatus*).

Figure 3.58 A very different member of the pea family, a tropical tree of the African savannah, the camel thorn (*Acacia erioloba*).

Figure 3.59 A nodulated root. The nodules are composed of plant tissue but contain bacterial cells within them.

- ❍ Carbon, hydrogen and oxygen are all more abundant constituents of plant tissue than nitrogen. Why does their availability not limit growth more frequently?
- ● Hydrogen and oxygen are available from water; carbon and oxygen are obtained from gaseous CO_2. New supplies of both these compounds are regularly delivered to the plant by the atmosphere.

Almost 80% of the Earth's atmosphere is composed of nitrogen gas. Plants are therefore constantly bathed in this element, which is able to diffuse into their tissues via stomata and lenticels, just as are the other two atmospheric gases, oxygen and CO_2. Why then should this element be limiting for growth? The answer lies in the chemical stability of nitrogen gas.

- ❍ Recall from Block 2, Part 1, the reason why nitrogen is stable.
- ● Nitrogen is composed of molecules that each contain two nitrogen atoms joined by a triple bond (N_2). Plants, like animals, do not have any means of cleaving this very strong bond. The gaseous pool of nitrogen is therefore completely unavailable to them.

The availability of nitrogen compounds in the soil is variable in time and space and often limits plant growth. Soil nitrogen is taken up by roots, transported throughout the plant, then returned to the soil as plant litter. The litter is broken down by fungi and other decomposers, which has the effect of re-releasing the nitrogen in more soluble forms, making it available for plant uptake. (The detail of the nitrogen cycle will be addressed later in Block 5.) The important aspect to be aware of at this stage is that the plant–soil cycle is not a closed one. There is interchange with the wider environment.

Biological molecules capable of cleaving the strong triple bond of gaseous nitrogen do exist. They are enzymes produced by some species of bacteria. The bacteria that possess them are able to convert gaseous nitrogen from the atmosphere into organic compounds within their cells and therefore avoid limitation in growth due to nitrogen shortage. If a plant were able to mimic this capacity, it would be at a great advantage over its neighbours. Some plants have succeeded in this goal by harnessng the bacterial power of **nitrogen fixation** within a mutualistic relationship. A **mutualistic** association is one between organisms of two different species in which both partners benefit. The plants participating in this relationship are primarily from a single family, namely the pea family (*Fabaceae*), whose members are often referred to as **legumes** (Figures 3.57 and 3.58). They form partnerships with bacteria (e.g. of the genus *Rhizobium*), which possess the enzyme to cleave the strong triple bond present in gaseous nitrogen. In return for a nitrogen supply, the plant provides the bacterium with a home (knobbly nodules on the sides of the plant's roots, Figure 3.59), with sugars for energy and with an almost oxygen-free atmosphere. This third item is important as the essential nitrogen-cleaving enzyme is poisoned by oxygen. It may be for this reason that only bacteria can make use of this enzyme, as only they can survive prolonged periods without oxygen. The bacterial nodules are kept low in oxygen by a protein, produced by the plant, that is similar to the oxygen-carrying haemoglobin found in the red blood cells of animals. The plant protein is called leghaemoglobin and it acts by mopping up free oxygen in the nodule, thereby protecting the nitrogen-fixing apparatus.

Legumes do not hold a monopoly on nitrogen fixation within the plant kingdom. A relatively small number of other species, unrelated to the pea family, have formed similar relationships with other types of bacteria (e.g. *Frankia* spp.), which give them the same benefits. One of these others is the alder (*Alnus glutinosa,* Figure 3.60), a common riverside tree in Britain, which lives in waterlogged soils, which are prone to be deficient in nitrogen compounds. Another is the bog myrtle (*Myrica gale,* Figure 3.61), whose common name suggests its habitat — waterlogged peat soils. A particularly important group is the lichens, which although not within the plant kingdom, are regarded as a component of vegetation (and are discussed further in Section 4). Finally some strains of free-living bacteria, such as *Azotobacter* and *Azospirillum*, are able to fix nitrogen without being enclosed in a nodule. These bacteria tend to be closely associated with the root, living on sugars secreted by the plant, but tend not to be as efficient as the nodulated bacteria, and they may not make a major contribution to the nitrogen requirement of the plant with which they are associated. In aquatic systems, the free-living blue–green bacteria (cyanobacteria) are able to fix nitrogen and sometimes are so successful they create what are misnamed as 'algal' blooms.

Figure 3.60 A riverside alder tree (*Alnus glutinosa*) relies on mutualistic bacteria to supply it with nitrogen from the atmosphere.

- ❍ What aspect of waterlogged soils has caused several species of plant growing on them to develop mutualistic relationships with nitrogen-fixing bacteria?
- ● In waterlogged soils, other bacteria are using nitrate as part of their respiratory process. The nitrogen cycle between the plant and the soil is therefore broken and nitrogen is lost from the system to the atmosphere as nitrogen gas. By becoming a nitrogen fixer, a plant can replace this lost nitrogen.

3.7.3 Limiting nutrients

Figure 3.62 illustrates the differential response of a species to variation in phosphorus supply. The shape of the curve is typical of growth limitation due to nutrient supply. Above a threshold value of nutrient availability in the soil, the nutrient does not constrain the growth rate, but below this threshold the growth of the plant is restricted and ultimately stopped. Plants require a cocktail of nutrients from the soil to sustain active growth. The one whose availability is most restricting growth rate is referred to as the limiting nutrient. Small changes in its availability will have a direct effect on growth rates, whilst small changes in the availability of the others would be expected to have no effect on the plant's performance.

Figure 3.61 Bog myrtle (*Myrica gale*) lives in soils that are almost permanently waterlogged and nitrogen-deficient. It is another example of a plant that can 'fix' nitrogen via its association with bacteria, but it is not a legume.

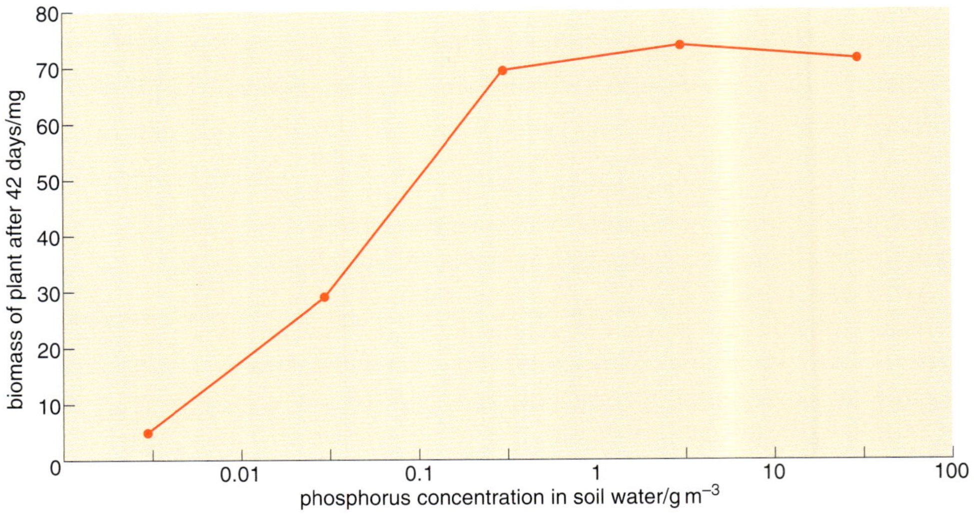

Figure 3.62 A plot of growth rate versus supply of the macronutrient, phosphorus, for the grassland herb, small scabious (*Scabiosa columbara*).

- Using Figure 3.62, at what concentration in the soil water does phosphorus cease to limit growth of small scabious?
- Between 0.1 and 1.0 g m^{-3}.

Question 3.6

Consider the three growth response curves for the stress-tolerant grass, *Anthoxanthum odoratum* (Figures 3.63 and 3.64). A colony of this species is growing on soil known to have the following mineral availabilities: 10 g m^{-3} Ca, 10 g m^{-3} Al, 5 g m^{-3} P. (i) Which element is likely to be growth-limiting?

If the soil were limed (the agricultural practice of increasing grassland productivity by spreading crushed limestone, $CaCO_3$, over the sward), the relevant availabilities would become: 30 g m^{-3} Ca, 1 g m^{-3} Al, 3 g m^{-3} P.
(ii) Which element would then be limiting?

(iii) Would liming increase or decrease the potential productivity of *Anthoxanthum*?

Figure 3.63 Sweet vernal grass (*Anthoxanthum odoratum*) is a common grass throughout Britain in non-intensive grasslands. It is the species that gives hay its characteristic smell.

3.7.4 Uptake of nutrients from soils

Plant roots are unable to explore the soil sufficiently intensively to extract all the available nutrients. Some very soluble minerals, such as the nitrate ion (NO_3^-), diffuse toward the root down a concentration gradient. These nutrients can be captured with relatively low rooting densities as each root is able to deplete a zone of up to 40 mm radius around the root. However, several nutrients (particularly phosphorus compounds) are not very mobile within soils, because they are either poorly soluble or they tend to adhere to the surfaces of soil particles and do not diffuse readily within the soil water. Therefore, unless a plant root comes into very close contact with the compound, it will not be absorbed. Roots exude a wide variety of chemicals from their surfaces, which act to solubilize nutrients in the soil and to assist in their uptake by the root. As you may recall from Block 2, Part 2, they acidify their immediate environment to help dissolve nutrient-containing compounds from the soil. This process is effective over a range of the order of a few millimetres, and the zone over which a root can deplete the phosphorus in the

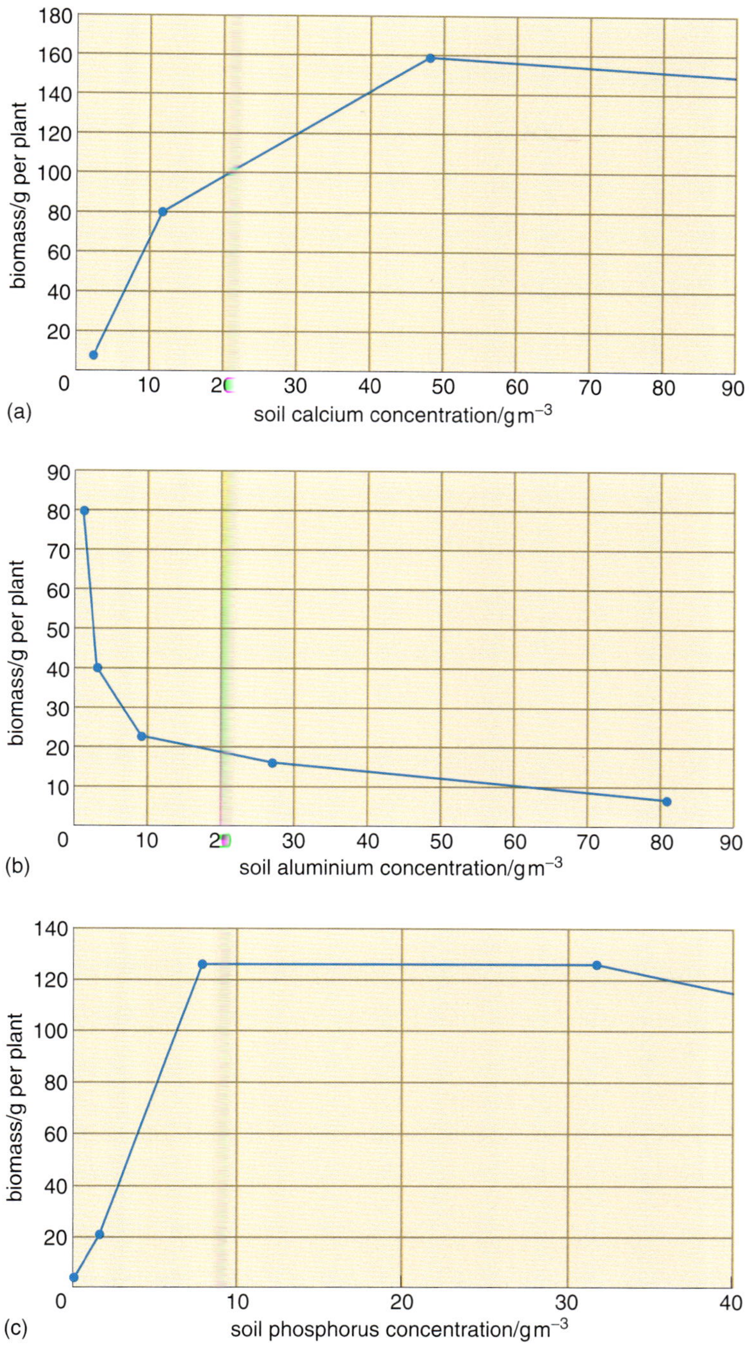

Figure 3.64 Growth curves for sweet vernal grass, showing the mean biomass accumulated by a single plant grown under a range of nutrient availabilities: (a) calcium; (b) aluminium; (c) phosphorus. In each case, one nutrient was varied whilst all others were maintained at optimal levels. Note that growth declines with increasing aluminium availability, indicating toxic levels.

soil may be as little as one millimetre in radius. A typical rooting density (length of root per volume of soil) is 1 km m^{-3}. A kilometre seems an astonishing length of root to pack into a single cubic metre of soil, yet even at this density, the average spacing between roots is still almost 30 mm. The root is perhaps 1.5 mm wide and its root exudates can mobilize phosphorus compounds for a further distance of 1.5 mm from its surface. Therefore the diameter of the cylinder of soil being exploited for phosphorus is 4.5 mm (Figure 3.65). So the proportion of the soil being exploited for phosphorus (assuming the roots are evenly spread throughout the volume) is just (4.5 mm/30 mm) × 100% = 15%.

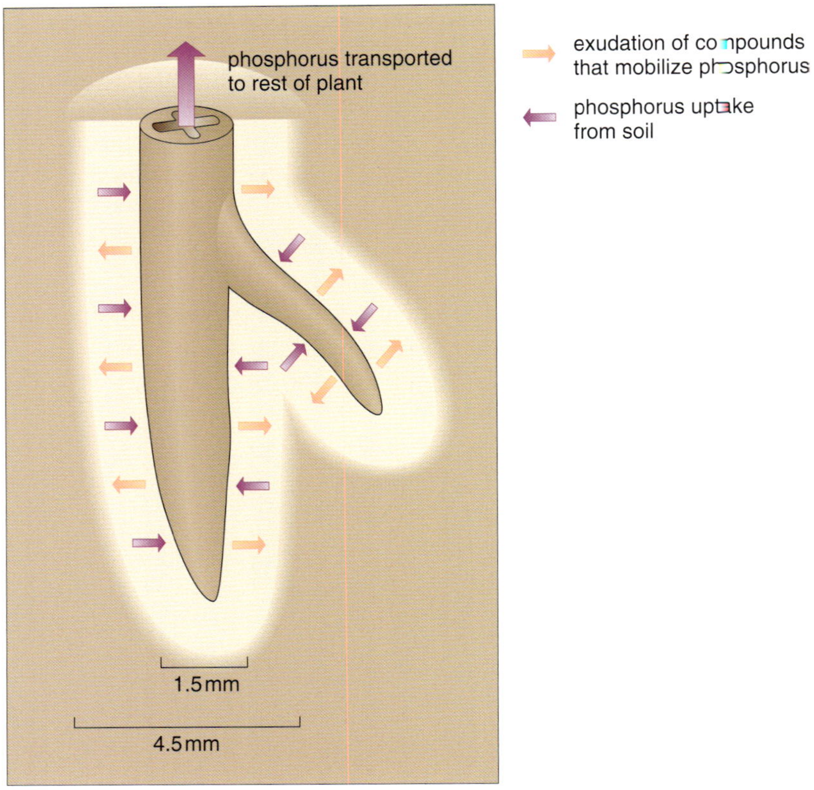

Figure 3.65 The cylindrical depletion zone around a root.

- Calculate the proportion of the soil being exploited for nitrate at the same rooting density, 1 km m^{-3}.

- The average spacing of the roots is 30 mm. The width of the depletion zone around a single root is twice the depletion zone radius plus the root width:

  $(2 \times 40\ \text{mm}) + 1.5\ \text{mm} = 81.5\ \text{mm}$

  Therefore the adjacent depletion zones will be overlapping to a great extent and potentially all the soil will be exploited for nitrate if the roots are uniformly distributed through it.

Figure 3.66 A cross-section through a grass sward, illustrating the high density of fibrous roots.

In the surface layer of a vigorous grassland community, the root density in the top 100 mm of soil may reach the equivalent of 40 km m^{-3}, or expressed another way, 4 km of root for each square metre of turf! (See Figure 3.66.) Yet this amazing density may still not be able to access all the potentially available phosphorus reserves, because the roots will not be arranged evenly throughout the soil and therefore will be competing with one another to deplete the same volume of soil in many cases, whilst other pockets will go unexplored. Sustaining such a huge length of living root creates a great drain on the plant. Roots, which are unable to photosynthesize themselves, rely on sugars being delivered from the shoot. Root respiration and exudation may account for up to half of the carbon assimilated by the shoot. Plants would benefit from being able to increase their nutrient uptake efficiency. The next section will reveal how they can achieve this.

3.7.5 The role of mycorrhizas in nutrient uptake

A **mycorrhiza** ('my-kor-rise-ah') is literally a fungus root (*mycor* = fungus, *rhiza* = root, in Greek). It is a mutualistic relationship between a species of fungus and a plant. The fungus receives sugars from the plant and in return facilitates the plant's uptake of mineral nutrients (Figure 3.67). It is a very successful partnership and the vast majority of plant species worldwide are mycorrhizal. There are several different forms of mycorhizas, depending largely on which species of fungus is involved in the partnership, but most have essentially the same functional relationship. The fungus is composed of very fine threads known as hyphae (about 4 μm in diameter), which are able to explore the soil much more extensively than plant roots (typically 400 μm in diameter).

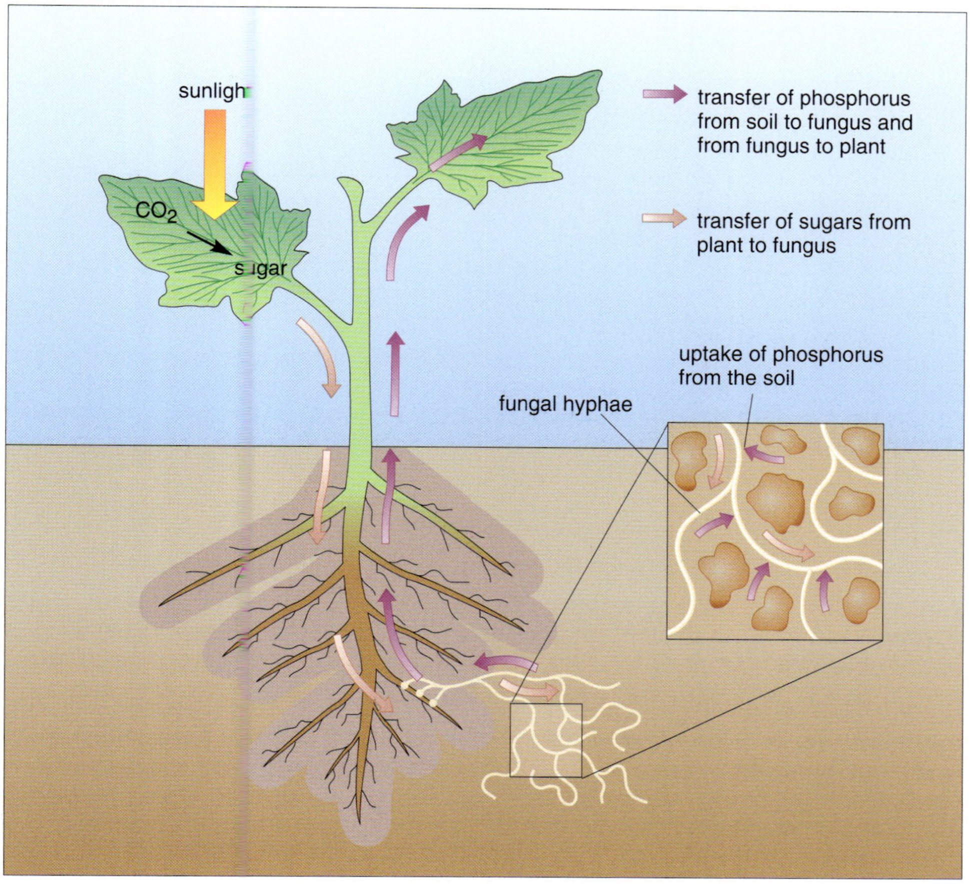

Figure 3.67 Mycorrhizal structure showing the flux of commodities between the two partners, plant and fungus.

Question 3.7

Per gram of tissue, how many times greater is the length of fungal hyphae than that of plant root? (Assume that 1 g corresponds to 1000 mm^3 of tissue, for both types, and that both tissues are cylindrical with a volume, $V = \pi r^2 h$, where π (pi) is a constant equal to 3.14, r is the radius (half the diameter) and h is the height of the cylinder — in this case the length of hypha or of root.

As can be seen from the above example, the fungus is able to come into close contact with a great deal more of the soil for a given amount of tissue. We have seen that for non-mobile minerals, this close contact is essential for uptake. The cost of maintaining tissue, in terms of the carbohydrate used in growth and respiration, is largely a function of its mass. Therefore, for a given amount of carbohydrate, the fungus may be up to 10 000 times as efficient at extracting minerals from the soil. In terms of their chemical exudates, fungi again appear to be better at releasing minerals (especially phosphorus) from tightly bound forms within the soil. By a combination of its greater length and better absorption properties, the fungus has access to a much larger supply of phosphorus than a corresponding mass of root. The fungus needs a certain amount of this mineral for its own needs, but it is able to absorb a substantial excess. Mycorrhizal plants are therefore often at a considerable advantage over their non-mycorrhizal neighbours (Figure 3.68).

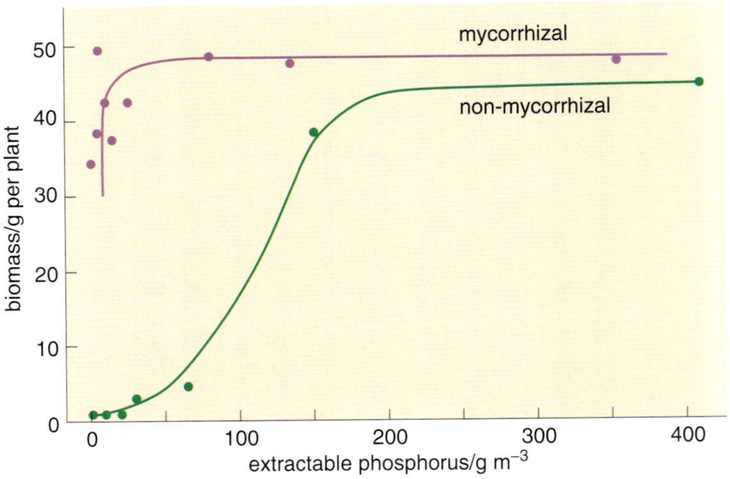

Figure 3.68 An illustration of the benefit derived by cassava (*Manihot esculenta*) from having a mycorrhizal association when growing in phosphorus-deficient soils. Cassava is the staple food for some 500 million people in the tropics, where it is generally grown without the aid of fertilizers, indicating the importance of mycorrhizas to human populations.

Mycorrhizas confer a number of other benefits to the plant. They may tap a larger volume of soil for water than the plant could do alone. They allow plants to survive in soils that are contaminated by heavy metals, by acting as a filter, immobilizing the toxin before it enters the plant. They can also protect the root system from infection by pathogenic strains of fungi.

The fungus is often not host-specific. It is able to form partnerships with many different species of plants. As a result, a single fungus may form a link between a number of plants of different species by which minerals and organic compounds can move from one plant to another (Figure 3.69 and Table 3.4). This phenomenon may have important ecological consequences in terms of a stand of vegetation truly behaving as a community.

- ❍ Based on the information in Table 3.4, what is the effect on the nitrogen content of the maize plant of applying nitrogen fertilizer to the soya bean?
- ● The nitrogen content increases, suggesting nitrogen was passed between the two species via fungal hyphae.

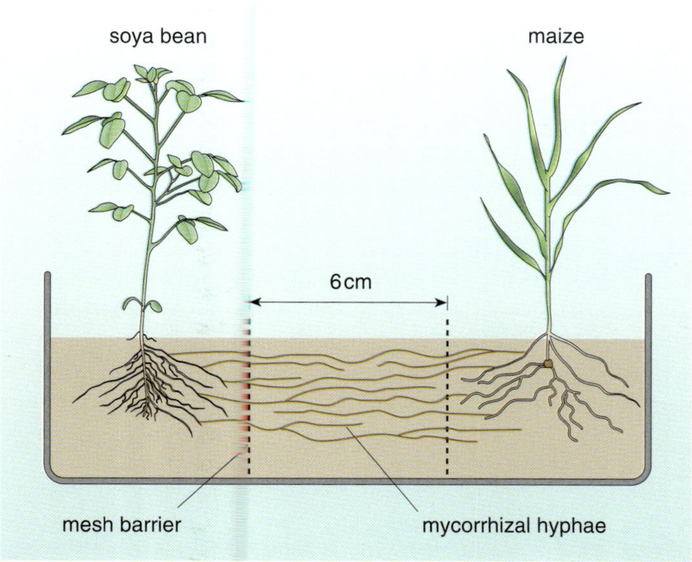

Figure 3.69 An experiment to demonstrate the transfer of nutrients between individual plants of different species (soya bean and maize) via their mycorrhizal connections. The dashed lines represent fine mesh barriers, which prevent plant roots from passing through but allow hyphal threads to pass.

Table 3.4 Results of the experiment shown in Figure 3.69.

N supply to soya bean	Dry weight/g per plant		N content/mg per plant	
	Soya bean	Maize	Soya bean	Maize
−N	3.9	7.2	30	33
+N	21.8	8.6	351	55

- If a plant community were able to share nutrients between not only individuals of the same species, but between species too, as this mycorrhizal evidence suggests, whose view of community does it support? Clements' or Gleason's? (Refer back to Box 2.1 if necessary.)
- Clements, who suggested a plant community could behave as an integrated whole.

The advantage to the plant of having a fungal partner is lost if nutrients are freely available, because the plant is able to absorb sufficient minerals through its own root system and there is no need to 'pay' a fungus to do it instead.

- Using Figure 3.70, identify the phosphorus concentration above which the mycorrhizal plant actually suffers from its association with a fungus.
- The plants with mycorrhizal associations grow less well when the phosphorus added is greater than about 10 g m^{-3}. This is presumably because the fungus is placing a drain on the plant's resources for no benefit in terms of phosphorus acquisition.

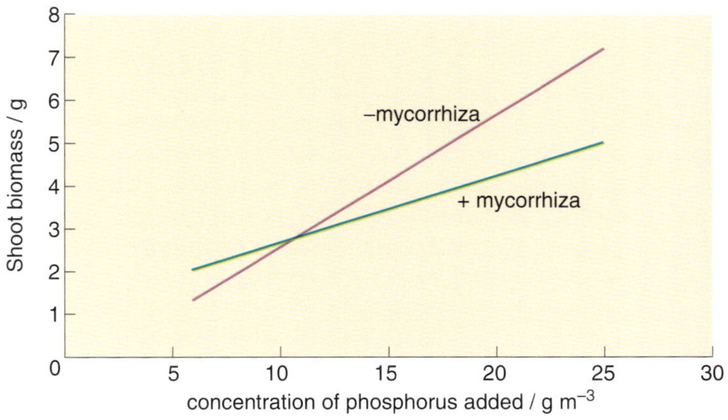

Figure 3.70 The yield of soya bean plants (*Glycine max*) grown with and wthout a mycorrhizal fungus. Note that the mycorrhizal association confers some advantage at low phosphorus levels but is a disadvantage to the plant when phosphorus is readily available from the soil.

Figure 3.71 The bee orchid (*Ophrys apifera*), a species of orchid growing wild in Britain. Along with most orchid species, this requires a close mutualistic relationship with a fungus throughout its life — many species fail to germinate if they cannot detect the presence of a fungus nearby.

In agricultural and horticultural soils with a long history of fertilizer addition, plants tend not to be reliant on mycorrhizas, whilst in nutrient-poor soils, they are completely reliant upon them. Some groups of plants, such as the heather family and the orchid family (e.g. Figure 3.71), have evolved in such close partnership with fungi that their root systems have become specialized for fungal infection and they cannot perform well without a fungal partner, even on nutrient-rich soils.

In terms of biomass, mycorrhizas are at their most abundant in deciduous temperate forests. The trees in this seasonal vegetation, which includes all the familiar British woodland trees such as oak, birch, and lime, have their root systems entirely sheathed by fungal tissue. The trees' young roots barely touch the soil at all. The fruiting bodies of the fungi associated with these types of mycorrhizas are the familiar toadstools, which appear on woodland floors in the autumn (Figure 3.72).

(a)

(b)

Figure 3.72 (a) Mycorrhizal fungi occasionally become visible, as in this case, where the fruiting body of the fly agaric fungus (*Amanita muscaria*), growing in association with birch trees, emerges from the soil. (b) A silver birch tree (*Betula pendula*), the host for *A. muscaria*, which is able to live on very nutrient-poor soils as a result of its association with the fungus.

Activity 3.3 Forest mycorrhizas

You could now view the three-minute video clip *Forest mycorrhizas*, which is about mycorrhizal associations within a spruce forest. Note the importance of fungi in the functioning of the forest, their diversity and why they are so rarely seen.

3.7.6 Plants of nutrient-deficient habitats

Some soils are so deficient in nutrients (particularly nitrogen and phosphorus) that not even mycorrhizas are able to supply the quantities required for growth by most plants. Such soils tend to be in high-rainfall areas, where most of the soluble nutrients are leached out of the soil and carried away by the excess water (see Activity 3.4 later). These soils are often waterlogged for much of the year and oxygen availability is low. Therefore plant litter only breaks down slowly and nitrate may be lost through bacterial respiration. Where the plant productivity exceeds the rate at which dead organic matter is oxidized, peat soils develop, composed of plant remains. These soils have mineral elements locked up within complex organic molecules, but because of the lack of oxygen for microbial metabolism, they are not released. This becomes a self-perpetuating cycle, with the lack of nutrients inhibiting microbial activity further. The plants able to tolerate these conditions adopt an extreme stress-tolerant strategy, in which nutrients are very strongly conserved. The mosses classified under the genus *Sphagnum* form a major component of peat-forming communities on permanently wet, nutrient-poor soils. These plants are very long-lived and efficient at retaining minerals within their tissues from year to year. The dead organic matter produced is therefore low in mineral content and as the sites tend to be fed only by rainwater, there are very few new minerals being supplied. Very few species are able to coexist with the *Sphagnum* mosses. Some of those few species, such as the sundews (*Drosera* spp., Figure 3.73), are **insectivorous**. They trap insects on their sticky leaves and then digest them, thereby extracting the minerals contained in their tissues.

Figure 3.73 A British insectivorous species of plant, the sundew (*Drosera rotundifolia*), growing amongst *Sphagnum* mosses.

3.7.7 Nutrient cycling and soil development

Other species of plant are much less conservative with their minerals. Many species of deciduous tree shed large amounts of basic minerals with the autumn leaves. The term 'basic' is used to label a mineral element if it is capable of neutralizing acids. The major relevant basic cations include calcium (Ca) and magnesium (Mg). (You met these ions in the Block 2, Part 2, in the discussion of cation-exchange capacities.) The trees take up these cations from the soil in the spring. They use them in their leaves during summer and shed them in the autumn (Figure 3.74). The fallen leaves decompose rapidly, as fungi and other microbes can metabolize base-rich substrates quickly, and the minerals are returned to the soil (Figure 3.75). The basic cations are often relatively mobile in solution and the winter rains wash them down into the profile, where they are reabsorbed in the spring to complete the cycle. The trees are deep-rooted and able to reabsorb these cations from throughout the profile.

Figure 3.74 Deciduous woodland, illustrating the large quantity of leaf litter produced annually; this material is recycled by decomposers (such as the bonnet fungus seen fruiting here), releasing basic cations back into the soil.

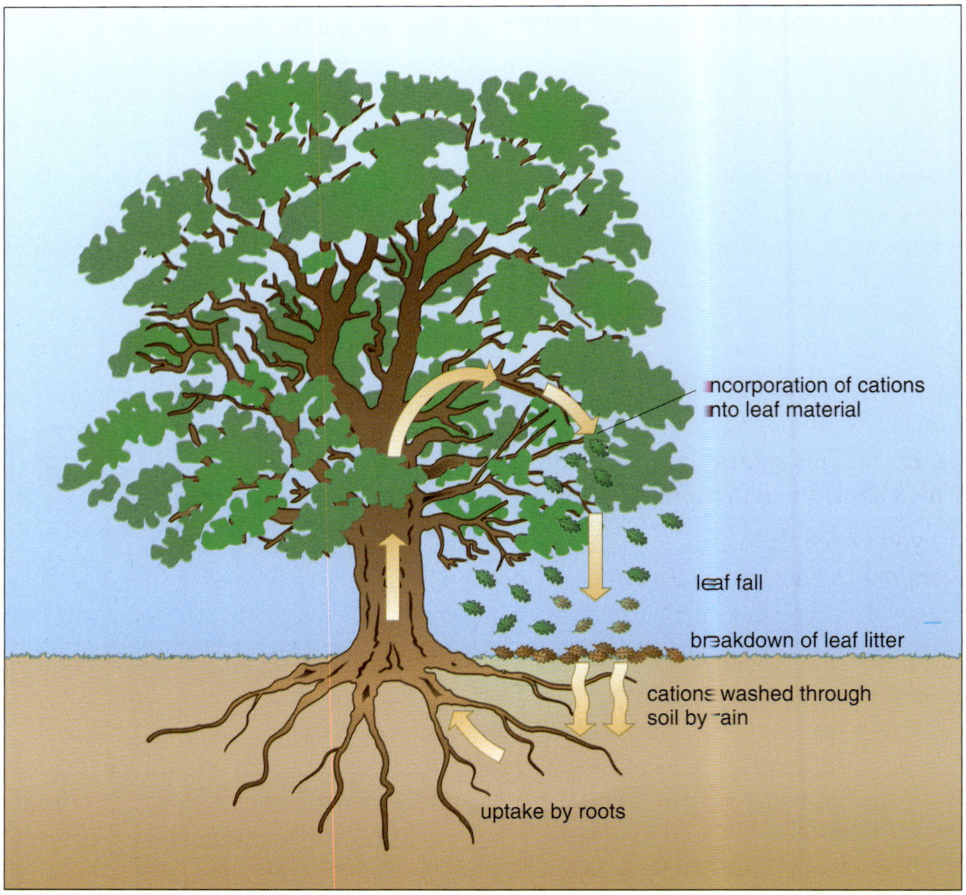

Figure 3.75 The cycling of basic cations through the soil–plant system.

Figure 3.76 The build-up of needle litter on the floor of pine forest, in this case Scots pine (*Pinus sylvestris*). The nutrient-poor needles are broken down slowly.

The constant cycling of these elements through the soil profile makes them more readily available than if they were static. This increased availability encourages microbial activity and sets up a complex food web in the soil sustaining a macrofauna of worms and moles, which serve to mix the soil. Over a period of centuries or millennia, this rapid decomposition of organic matter and constant mixing creates a deep, relatively uniform soil, often referred to in the UK as a **brown earth**. An example of a tree that cycles basic cations through the soil is the lime (*Tilia cordata,* Figure 2.26), once the most abundant tree in lowland Britain and therefore in part responsible for many of the brown earth soils of England.

Another group of trees, typified by pines (*Pinus* spp., Figure 3.76), do not recycle bases in this way. They take a more stress-tolerant strategy, rather like the *Sphagnum* mosses, though not as extreme. Their conservative use of minerals means that the bases are reabsorbed into the tree before the needles fall. (In spite of being evergreen, pines do have a turnover of needles — an individual needle remains on the tree for two or three years in most cases.) The needle litter is therefore very acidic compared to that of many broadleaved trees. Fungi and microbes decompose it less rapidly as a result. The few basic cations present are released more slowly into the soil and the soil fauna is much reduced, with little mixing of the profile occurring. The trees themselves tend to have rather shallow roots, as what few minerals are available tend to be in the surface horizons of the soil. The net result of this system, with very limited recycling of cations, is an acidic soil. In profile it is very stratified, due to the build-up of partially decomposed litter on the surface (black); beneath this there is pale sand from which the coloured components have been degraded under the acid conditions and been washed downward into a lower hard, coloured layer known as a pan. This type of soil is called a **podzol** (Figure 3.77), which is similar to the profile you met in Yarner Wood (Block 1).

Figure 3.77 A podzolized profile showing the distinct layering of the soil.

Vegetation type is therefore an important factor in soil formation. It is a two-way process: the soil type determines which plant species are successful in an area and the plants in turn influence the development of the soil. We will return to this topic again when considering ecological succession.

Activity 3.4 Plants as carnivores

You could now view the five-minute video clip *Plants as carnivores* on the DVD. This shows some of the adaptations developed by plants, which allow them to supplement their nutrition by digesting animal prey.

As you watch the video, consider why the environment on Mount Roraima in Eastern Venezuela is so poor in nutrients and where the influx of nutrients to the ecosystem comes from.

3.7.8 Summary of Section 3.4–3.7

1 Water movement through the soil–plant–atmosphere continuum is dictated by gradients in water potential. This quantity and the flux of water through the system can both be measured. Resistances to flow can be calculated and are considered greatest at the soil/root interface and at the leaf surface.

2 Plants rely on the soil as a water store. In times of shortage, plants depend on a range of adaptations to survive and water availability is often the underlying cause of patterns in vegetation.

3 Oxygen is an essential resource for all plant tissues. Underground organs can experience difficulty acquiring oxygen, especially if the soil is waterlogged. The main adaptation to this environmental stress is the presence of an internal plumbing system, known as aerenchyma, which is able to transport oxygen from the atmosphere to the roots.

4 Soil water status can therefore have two independent impacts on vegetation: it influences both the water supply for growth and the oxygen supply for root respiration.

5 Plants require a balanced supply of a suite of nutrients for growth. The one that is in shortest supply is denoted as the growth-limiting nutrient.

6 Nitrogen is the single most important mineral nutrient for plants and often the growth-limiting one. Some species have overcome this restriction by entering mutualistic relationships with bacteria that are able to utilize the vast resource of atmospheric nitrogen gas.

7 Plants can survive in nutrient-poor soils by a number of mechanisms. Mutualistic fungi known as mycorrhizas can assist with nutrient acquisition. Capture and digestion of insects is an extreme adaptation to the unavailability of soil minerals.

8 The availability of plant nutrients, especially basic cations, is a function of plant behaviour with respect to nutrient recycling. The soil pH and its development are heavily influenced by this behaviour.

Learning outcomes for Section 3

After working through this section you should be able to:

3.1 Describe in quantitative terms what happens to light energy from the Sun between reaching the Earth's atmosphere and being incorporated into the tissues of photosynthetic plants as net primary production.

3.2 Use data about energy flow to calculate energy transfer and efficiency in primary and secondary production. (*Question 3.1*)

3.3 Explain why light is a dangerous commodity to plants and describe, using appropriate examples, the ways in which plants can protect themselves from light-induced damage.

3.4 Explain the importance of scale to the study of ecosystems.

3.5 Recognize and be able to identify different types of ecosystem from energy budget diagrams. (*Question 3.2*)

3.6 List the components of water potential and combine them, using plus and minus signs appropriately. (*Activity 3.1; Question 3.3*)

3.7 Calculate water potential gradients within the soil–plant–atmosphere continuum and be able to comment on their significance. (*Activity 3.2; Question 3.3*)

3.8 Estimate the ability of soil to store water for use by plants and recognize plant adaptations to unreliable water supply, describing the effect of soil water regime on the vegetation of a habitat. (*Questions 3.4 and 3.5*)

3.9 List the methods by which plants can obtain their mineral nutrition and comment on the role played by soil nutrient availability in determining vegetation patterns. (*Questions 3.6*)

3.10 Describe the role of mycorrhizas and insectivory in facilitating nutrient capture by plants. (*Activities 3.3 and 3.4; Question 3.7*)

Ecological dynamics

4

So far, we have considered how vegetation relates to the environment in which it grows. Vegetation is not static, however. Patterns of distribution change with time. The description and study of these changes is called **ecological dynamics**. In this section we consider first how plants move in space and then how the vegetation at a particular point alters through time. Later we address how susceptible a stand of vegetation is to a change in its environment and how the diversity of the system controls this susceptibility. The final part of the section looks at how we measure and express species diversity in quantitative terms such that it can be used as an index of environmental change.

4.1 Dispersal

Our consideration of the requirements of life in Section 3 has allowed us to see how the fundamental niche of a species can be defined. For plants, a niche is defined in terms of the range of environments that can supply the resources required by a species (light, water, oxygen, CO_2, mineral nutrients), other abiotic factors (temperature regime, shelter from wind, anchorage) and biotic factors (e.g. presence of pollinators and mutualistic fungi). However, as indicated in Section 2.5.1, a species rarely fills its fundamental niche, and whilst one of the aspects that may limit its occupation of this potential niche is the presence of competitors, another important factor is dispersal, i.e. the movement of a organism within space.

4.1.1 Variation in geographical distribution of species

Most species have a limited geographic distribution. However, some species of higher plant, such as the common reed (*Phragmites australis*) and bracken (*Pteridium aquilinum,*) both of which you have already encountered in this course, are very cosmopolitan, occurring around the globe throughout the tropics and temperate regions. They are, however, restricted to particular habitat types within this broad geographical range. In the animal kingdom, some species are even more widely dispersed. Humans (*Homo sapiens*) are perhaps the most ubiquitous macroscopic, terrestrial species on a global scale, occupying a wide variety of habitats. The nearest challengers would be those species that spread with us, such as rats (*Rattus rattus*, Figure 4.1).

Figure 4.1 The black rat (*Rattus rattus*), which has spread from its original home in Asia to all parts of the world with the assistance of human activity, such as stowing-away on ships and even planes. It has a very broad fundamental niche which, with the help of humans, it has been able to fill.

At the other end of the scale, several species are confined to a single ocean island or a single mountain top. The majority of species, however, fall between these two extremes, with ranges that are restricted both geographically and in terms of habitat type.

4.1.2 Colonization of new habitats

An interesting question in ecology is whether the range of a species is limited by its ability to disperse to new areas or by its ability to establish and survive in them. In 1980, a natural catastrophe created conditions for an experiment that addressed this issue.

Figure 4.2 Mount St. Helens erupting in 1980. The ash discharged from the volcano created new sterile habitats for life to colonize.

The huge volcanic eruption of Mount St. Helens on the western side of North America buried a vast area of land in volcanic ash, which was devoid of plant life in the immediate aftermath of the eruption (Figure 4.2). Higher plants have been relatively slow to recolonize the bare ash. Scientists undertook experiments to determine whether seeds were arriving and, if so, whether they were able to establish. The results showed that many small-seeded species did arrive, mainly carried by the wind, but they struggled to establish in the inhospitable soil, because their food reserves were so small. Conversely large-seeded species, with larger food reserves, did establish when they were artificially planted, but their dispersal mechanisms were not efficient at spreading them unassisted from neighbouring undisturbed forest.

Figure 4.3 Early colonists on the ash fields of Mount St. Helens. Lupin (*Lupinus lepidus*) is in the foreground. This picture was taken just two years after the eruption.

The range over which large-seeded species occur is therefore limited by their dispersal abilities. The ash fields represent part of their potential range or, in other words, are within their fundamental niche, but the potential has not yet been realized. One example of this group is the prairie lupin (*Lupinus lepidus*, Figure 4.3), which has large seeds (> 10 mg) and is able to establish on the ash fields by its ability to send down a deep root into the drought-prone soils, enabling it to tap reserves of water lower in the profile. However, a study of the distribution of its seedlings showed the majority to be within 1 m of the parent (Figure 4.4a). In contrast small-seeded (< 1 mg) species, such as the cascade aster (*Aster ledophyllus*), although able to disperse to the ash field (17% of seedlings located at > 5 m from the parent plant: Figure 4.4b), are unable to grow in the harsh environment, primarily due to lack of moisture. One can describe the new environment as being outside the fundamental niche of the aster. Again, given a longer time frame than the 20 or so years since the eruption, the soil environment will improve as the few plants that do arrive and survive, begin to add organic material to the soil (see Section 4.2.2).

- ❍ The prairie lupin, which does colonize successfully, is a member of the legume family (Fabaceae). What does this information imply about the species impact on the newly developing soil?
- ● Legumes are able to fix atmospheric nitrogen through their mutualistic relationship with specialist bacteria. The lupins will therefore be enriching the soil with nitrogen as their tissues die and are recycled.

- ❍ Using Figure 4.4, calculate the ratio of aster to lupin seedlings that appear at distances > 5 m from the parent plant.
- ● 17% of aster seedlings are located at this distance compared to just 2% of lupin seedlings, giving a ratio of 8.5 : 1.

Figure 4.4 Distribution of seed dispersal distances for (a) lupin; and (b) aster, with photographs of the species alongside.

This contrast between the fates of large versus small-seeded plants is another example of an ecological trade-off. Small seeds favour wide dispersal, but tend to require favourable conditions at new sites, whilst large-seeded species, although often handicapped in terms of their dispersal ability, are able to tolerate a wider range of conditions during establishment. For example, they are better able to establish in dense leaf litter, because they are able to develop a tall shoot, whilst living on their stored reserves, without the need for light as an energy source. They are also able to tolerate drought-prone soils, by sending down a deep root very quickly after germination. The latter ability gave them the advantage on Mount St. Helens, making them the first successful colonists.

Species of disturbed habitats (i.e. ruderals) have the most developed adaptations to aid dispersal. Example are listed below.

- The field poppy (*Papaver rhoeas*; Figure 2.77) has tens of thousands of very small seeds shaken from a 'pepper dispenser' by the wind (Figure 4.5a).
- Seeds of cleavers (*Galium aparine*) have curved hooks to attach themselves to the fur of passing animals (Figure 4.5b).
- Corn chamomile (*Anthemis arvensis*) seeds survive the digestive tracts of birds, which then scatter them widely in their droppings (Figure 4.5c).
- The exploding pods of the tiny plant called hairy cress (*Cardamine hirsuta*) can throw their seeds more than a metre from the parent (Figure 4.5d).
- The dandelion (*Taraxacum* spp.) has a feathery parachute attached to each of its seeds, allowing them to catch the slightest breeze and often travel hundreds of metres (Figure 4.5e).

Plants of regularly disturbed habitats need either to escape through space by dispersal mechanisms such as these, or to escape through time by lying dormant in the soil seed bank as discussed earlier. Plants of more stable, predictable habitats, such as the forest floor, tend not to have such specialized dispersal mechanisms — the need for them constantly to find new habitats is much less. The contrast is only one of scale, however. All species need to disperse to some degree as no habitat is truly stable. A forest's floor may be a hospitable environment for decades, centuries or even millennia, but with the Earth's constantly changing climate, a species needs to be able to relocate if it is to avoid extinction. Examples of poorly dispersing species are given in the next section.

4.1.3 Barriers to dispersal

Dispersal of a species is often limited by the presence of barriers. On a global scale, these may take the form of oceans or mountain ranges. The reason Britain has a relatively small number of species within its native flora (around 1600), is that at the end of the last ice age, as plants spread north in the wake of the retreating ice, only a limited number reached Britain before the English Channel was formed by the rising sea-level. Other species whose fundamental niche would include habitats in Britain, were marooned on continental Europe by the Channel forming a dispersal barrier (Figure 4.6).

Ireland's flora is yet more impoverished (< 1000 species of higher plants), which is in part due to the fact that the Irish Sea formed before the English Channel, creating an even earlier barrier to the dispersal of plants from their refuges in the south. This is also the reason why there are no snakes in Ireland. The Irish Sea created a barrier before they could disperse, so the island was never recolonized by them after the last ice age. (There is a popular myth that Saint Patrick drove all the snakes out of Ireland, but the scientific evidence suggests that they simply never arrived!)

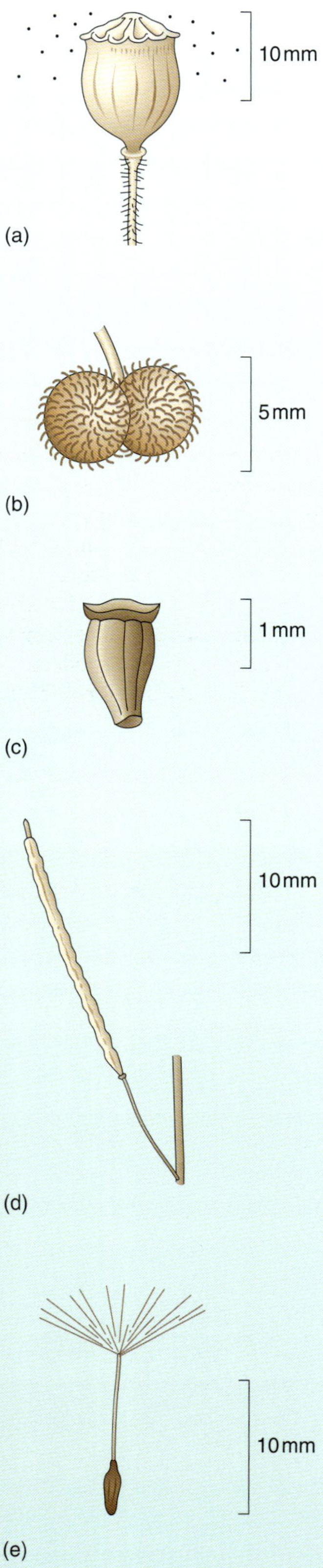

Figure 4.5 An illustration of the range of dispersal mechanisms amongst ruderal species: (a) a field poppy pod scattering its tiny seeds; (b) cleaver fruits, which readily stick to fur or clothing, due to their hooked barbs; (c) the seed of corn chamomile; (d) a pod of hairy cress waiting to explode; (e) a dandelion seed flies by using a parachute.

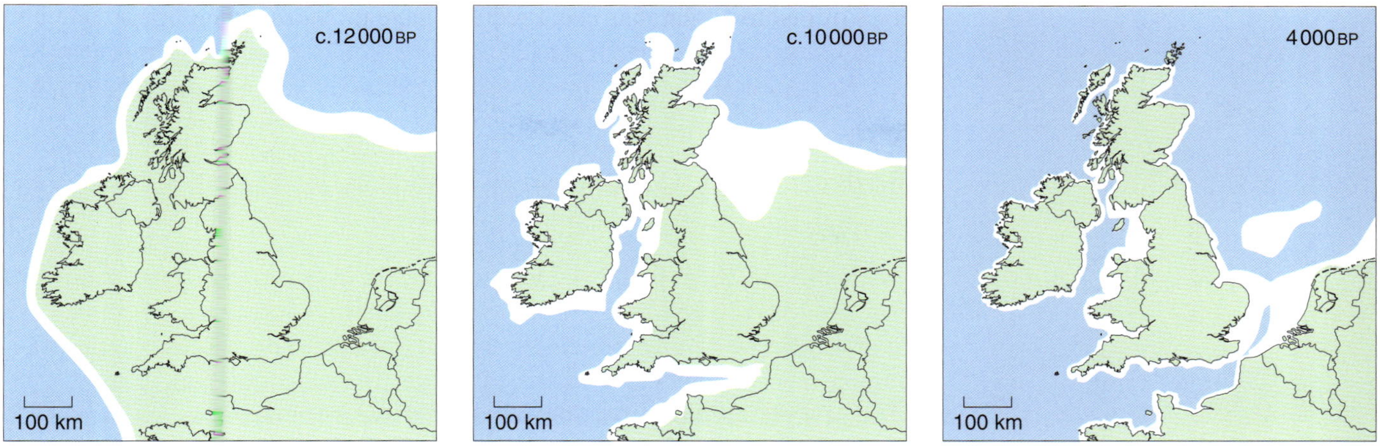

Figure 4.6 How the coastline of Britain developed. Maps showing the coastline as it appeared (a) 12 000, (b) 10 000 and (c) 4000 years ago. (BP = years before present.) Green: land or ice sheet; white: shallow sea; blue: deep sea. The black lines show modern political boundaries, for reference.

- ❍ Based on the information in Figure 4.6, for how many years was Great Britain still joined to the continent of Europe via a land bridge after its land link with Ireland was broken? How many additional species may have arrived in Britain during that period?

- ● It was joined to the continent for 2000 years after Ireland separated. In this time, up to 600 additional plant species may have crossed the bridge.

On a more local scale, barriers to dispersal are a major problem for the restoration of wildlife habitats. Within the Teign catchment in recent years, there have been a number of new woodlands established as isolated stands on former arable farmland, in response to subsidies being offered by the government. Tree species are planted and, with appropriate aftercare, they are able to thrive, creating a closed-canopy woodland after as little as 10 years. The development of a woodland ground flora, however, is often disappointing, with very few species colonizing the new habitat. Woodland species are often poor dispersers because they are adapted to a normally stable habitat. More importantly, if the woodland is isolated from other woods, the species would be unlikely to colonize the intervening arable areas and would therefore need to 'jump' from one wood to the next, because the intervening area would be outside of their realized niche due to competition from non-woodland species. This would be a very infrequent event for non-specialist dispersers.

4.1.4 Plants as indicators of a habitat's history

'Ancient woodland' is a term applied to areas in Britain that are known from documentary evidence to have been continuously wooded since 1600 AD. The probability is that they have never been ploughed and their ground flora has some continuity with the wildwood, which was at its zenith 4000 BP. Such woodland can now often be recognized on the ground by the presence of species that are stress tolerators, adapted to living in a shaded environment, and as such are unable to compete in open habitats (Figure 4.7). They therefore tend to disperse poorly to new woodland sites. Yet these same species must have spread during a relatively short period into Britain between the arrival of deciduous forest trees following the last ice age, which ended about 9000 BP, and the formation of the English Channel (*c.* 7000 BP) and then throughout the country before the beginning of large-scale

removal of forest by Neolithic man (*c.* 4000 BP). One can calculate the rate at which the species would have had to move north from continental Europe in order to reach Britain before the land bridge was broken. This suggests that populations of these woodland species, classed as poor dispersers, were once expanding at rates of up to 100 m per year. Recent studies of dispersion in one of them, oxlip (*Primula elatior*), suggest it is able to advance into new territory at the rate of just 1 m per year.

Why were the species in Figure 4.7 such competent dispersers in former times compared to today? The likely reason is that in extensive forest (i.e. when the whole of England was almost an uninterrupted expanse of deciduous woodland), the suitable niche for the species was continuous. In such a situation, chance dispersal of seeds on the feet of animals would have often been successful, because the point where the seed was subsequently dropped would have a high probability of being a suitable habitat for the species. In contrast, in the modern fragmented landscape, in which less than 1% of the land surface is likely to be a suitable habitat, such instances of 'jump' dispersal are doomed to failure and the species spread mainly by very short range seed scatter within the mother plant's habitat.

Question 4.1

What are the main factors that would limit the colonization by specialized woodland-floor species, of newly planted woodland in which only the tree species had been introduced into previously arable land?

4.1.5 Plants dispersed by human activity

In contrast to the poorly dispersing species of the woodland floor, other species have used the modern landscape to their advantage in terms of dispersal. A colourful example of this situation is the rosebay willowherb (*Chamerion angustifolium*). It was a relatively sparsely distributed plant, largely confined to the uplands in northern Britain until the middle of the 19th century. At this time, a strain that grew in lowland habitats (and may have been an introduction from abroad) appeared in several locations through southern England. Over the next 100 years, it became a common and abundant plant of both town and country, often colonizing large areas and creating sheets of purple flowers. Each plant can produce tens of thousands of wind-borne seeds, which may be carried in the slipstream of trains (Figure 4.8). The extensive new railway network, which was built during the 19th century, is believed to have been the agent of its dispersal, providing both transport for its seeds and a suitable soil (in the form of the stone chippings used as railway ballast) for its establishment. Rubble resulting from industrial dereliction or from bomb-craters during the World War II, was readily colonized by the species.

Figure 4.7 Herbaceous plants of the woodland floor that are largely restricted to 'ancient' woods in lowland England: (a) herb paris (*Paris quadrifolia*); (b) lily of the valley (*Convallaria majalis*); (c) wood sorrel (*Oxalis acetosella*); (d) oxlip (*Primula elatior*).

4.1.6 Habitat fragmentation

We have seen that many species disperse over relatively short distances. To colonize distant areas, there needs to be a corridor of suitable habitat along which to disperse. Before the influence of humans, the Earth was covered by large continuous tracts of natural vegetation, whether it be woodland, wetland or grassland. Plants therefore had the ability to migrate considerable distances uninterrupted. The process of **habitat fragmentation**, which breaks up these continuous tracts into smaller isolated habitats through the change of land use by humans, is now occurring on a global scale. It may benefit a few species (particularly those adapted to the nutrient-rich, disturbed habitats, which tend to be associated with human activity), but will constitute barriers to dispersal for the majority. This is of relevance to the current consideration of climate change and its effect on natural and semi-natural vegetation (see Section 2.5.3 on Woodward's modelling).

Figure 4.8 Rosebay willowherb (*Chamerion angustifolium*) benefiting from human disturbance of the landscape. Its rapid colonization of Britain during the 19th century was believed to be assisted by its seeds travelling in the slipstream of trains.

- ❍ Refer back to Figure 2.85 and make a rough estimate of by what distance vegetation types of northern Europe are predicted to be displaced northward by climate change in the next 50 years.
- ● The vegetation types appear to be shifted north by approximately 600 km.

Predictions suggest many species will need to relocate at distances of the order of hundreds or even thousands of kilometres in order to survive over a time period as short as 100 years. This may be achievable for those species that have evolved mechanisms for long-distance dispersal, but may prove impossible for the majority. There is a dilemma for nature conservationists, who would normally disapprove of translocating species to new habitats, because in the face of climate change it is probable that some species will need to be moved by artificial means in order to protect them from extinction.

4.2 Succession

If an area of bare soil is left undisturbed for 100 years and the range of species that utilize the area is monitored, a gradual change in growth types will be observed. This **succession** of different species at one spot over the course of time is an important ecological phenomenon. It illustrates that species alter the habitats in which they live and often bring about their own demise by rendering their habitat suitable for a competitor, which in time will displace them.

The process of succession is categorized under two broad headings depending on the state of the soil at the start of the process. If the soil lacks organic matter and has no seed bank, because no vegetation has grown on it previously, we use the term **primary succession**. If, on the other hand, the soil has recently supported

Figure 4.9 Environments that lack proper soil and which are subject to primary succession of vegetation: (a) volcanic ash; (b) dune of blown sand; (c) glacial moraine debris; (d) mining spoil; (e) exposed river sediment.

plants and contains some humus and seeds, the resultant process is termed **secondary succession**. An example of secondary succession would be in recently felled woodland in which the dense canopy of tall trees has been removed. Light reaching the woodland floor will allow new smaller species lying dormant in the soil seed bank, to thrive in a habitat previously unavailable to them. These species are sequentially replaced by taller-growing ones, which shade the preceding lower-growing types, until, if left undisturbed, the original tree species may return.

In Activity 4.1 (later) you can view a video clip, which gives an illustration of this process. However, let us first consider habitats which are formed without soil, in which primary succession proceeds.

4.2.1 Primary succession

Primary succession begins at a site when a new soil has just been formed. Such soils can arise in a number of ways, such as;

- volcanoes producing a landscape of sterile ash;
- wind-blown sand creating a new dune;
- crushed rock being exposed by the retreat of a glacier;
- mineral extraction activities generating mounds of waste material;
- sediment being deposited within shallow lakes or rivers.

These situations are illustrated in Figure 4.9. Such soil-less habitats provide very harsh growing conditions for plants. Water availability is a problem because the water-storage function of a well developed soil is missing. Essential nutrients, particularly nitrogen, are often in short supply because the plant–soil cycles that accumulate such minerals within a habitat have not begun to function.

4.2.2 Process of soil development

The habitats shown in Figure 4.9 already have the mineral component of a soil, but when first formed, they lack the organic portion. Organic matter builds up within a soil as a result of plant growth and decay, but to achieve this some plants have to colonize the barren ground surface first.

○ From what you have learned of vegetation requirements, why might plants find these environments hostile?

● Two main requirements are liable to be lacking. Firstly, without any organic matter to structure the soil, the water-holding capacity and hence the available water content will be low. Plants are likely to be exposed to drought. Secondly, the mineral nutrient that is primarily stored in the organic fraction of the soil is nitrogen. Therefore, unless plants are able to fix nitrogen directly from the atmosphere, their growth will be severely nutrient-limited.

In fact, the organisms that play a major role in the early stages of primary succession over a wide range of habitat types are not plants at all, but a group known as the **lichens**. These are exceptionally stress-tolerant in their strategy: they grow extremely slowly, are long-lived, have a very low turnover of their tissues and can survive periods of desiccation (Figure 4.10). Lichens are actually a mutualistic partnership between two organisms from entirely different

Figure 4.10 Lichens colonizing bare rock.

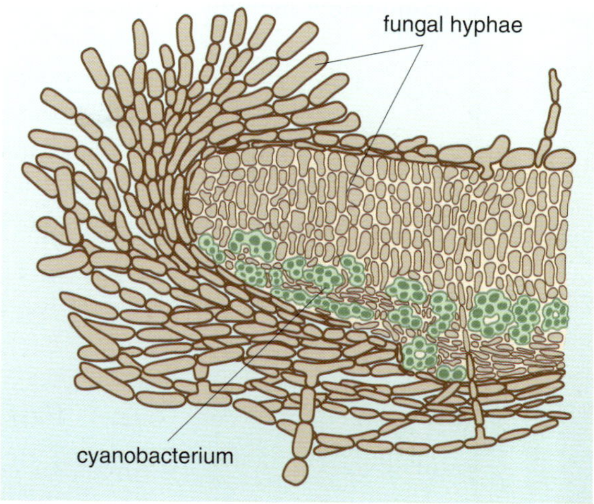

Figure 4.11 Diagram of a lichen containing a cyanobacterial partner.

kingdoms; a fungus is one of the partners and either a unicellular alga or a bacterium is the other. (One of the first people to recognize this remarkable fact was the children's writer, Beatrix Potter — see Box 4.1.) The fungal tissue provides the outer, desiccation-resistant tissues, whilst the photosynthetic alga or blue–green bacterium (cyanobacterium) provides the sugars via photosynthesis (Figure 4.11).

The blue–green bacterium has the ability to fix nitrogen. Lichens containing this partner are therefore not dependent on soil reserves of nitrogen, but can obtain all they need directly from the atmosphere. When the lichen tissue dies, it becomes incorporated into the soil as nitrogenous organic matter. Because lichens are so slow-growing, it takes a long time for the carbon and nitrogen pools to increase.

Box 4.1 The tale of Beatrix Potter

Beatrix Potter (Figure 4.12) is reported to have had a lonely childhood, having been tutored at home by a governess. To amuse herself, she became interested both in drawing and in natural history. She made a great many illustrations of plants and animals, but particularly of fungi and lichens. Gradually becoming something of an expert in the identification and classification of these groups, she drew them very meticulously and took great pains to make the illustrations true to life by dissecting specimens.

Her uncle was a chemist of some distinction and was able to introduce her to the leading botanists of the day at Kew Gardens, but they were not welcoming to a teenage girl who had never attended school. She correctly identified that lichen, although then believed to be a single entity, was in fact a dual organism composed of a fungus and an alga. She did not publish her findings and was beaten into print by a German botanist who had independently made the same observation. She did, however, write a paper on the propagation of moulds by spores and through the good offices of her uncle had her paper read aloud to the Linneaen Society, one of Britain's oldest and most learned societies. Again, however, the work of a young woman was not taken seriously and she was given little credit for her painstaking work. In response to the lack of appreciation, Beatrix put away her biological drawings and turned her talents to sketching pictures of rabbits in order to amuse the children of one of her former governesses. And so began the career for which she is now internationally renowned.

Figure 4.12 Beatrix Potter (1866–1943).

Figure 4.13 A moss (*Bryum* sp.) growing on bare rock.

Figure 4.14 Glacier Bay, Alaska, showing a glacier still in retreat.

Another group of extreme stress tolerators are the mosses (bryophytes), introduced in Section 2.2, and these also appear as early colonists in many environments undergoing primary succession (Figure 4.13). The lichens and mosses and a few small species of higher plant constitute the pioneer vegetation, the first stage of succession, in high-latitude climates.

4.2.3 Succession in action

A site that has proved to be useful for studying the process of primary succession is Glacier Bay in Alaska (Figure 4.14). During the past 200 years there, the regional climate has been warmer than the preceding period, labelled 'the little ice age,' during which glaciers advanced to the sea. The glacier that created Glacier Bay, has retreated almost 100 km back up its valley, as the ice at lower altitudes melted faster than it was replaced from above. This has been a gradual process, which has slowly exposed an area of crushed rocks along its former path (Figure 4.15). This raw soil was initially without an organic fraction, but as plants began to colonize, the soil developed. The area of glacial retreat now gives a perspective on soil development through time, because new rock debris has been constantly being exposed for two hundred years. Analysis of the soil samples from a line of points along the valley enables us to construct a time series for soil development. Figure 4.16 shows the results of one such analysis. It demonstrates the build-up of organic matter occurring over time.

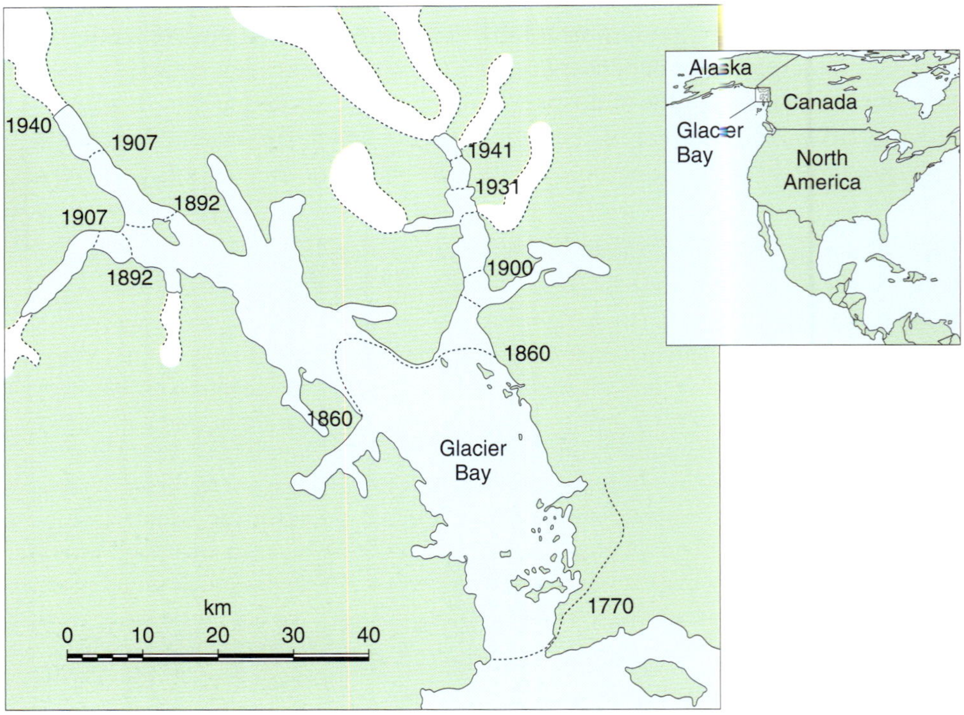

Figure 4.15 Map of the Glacier Bay area showing dates of glacial retreat.

- ❍ Study Figure 4.16a. How would you describe the shape of the curve during the first 50 years?
- ● The gradient of the curve becomes increasingly steep with time, suggesting a self-promoting process, sometimes referred to as positive feedback (a concept you met in Block 2, Part 1). It means that the higher the organic carbon content of the soil, the faster new organic material accumulates.

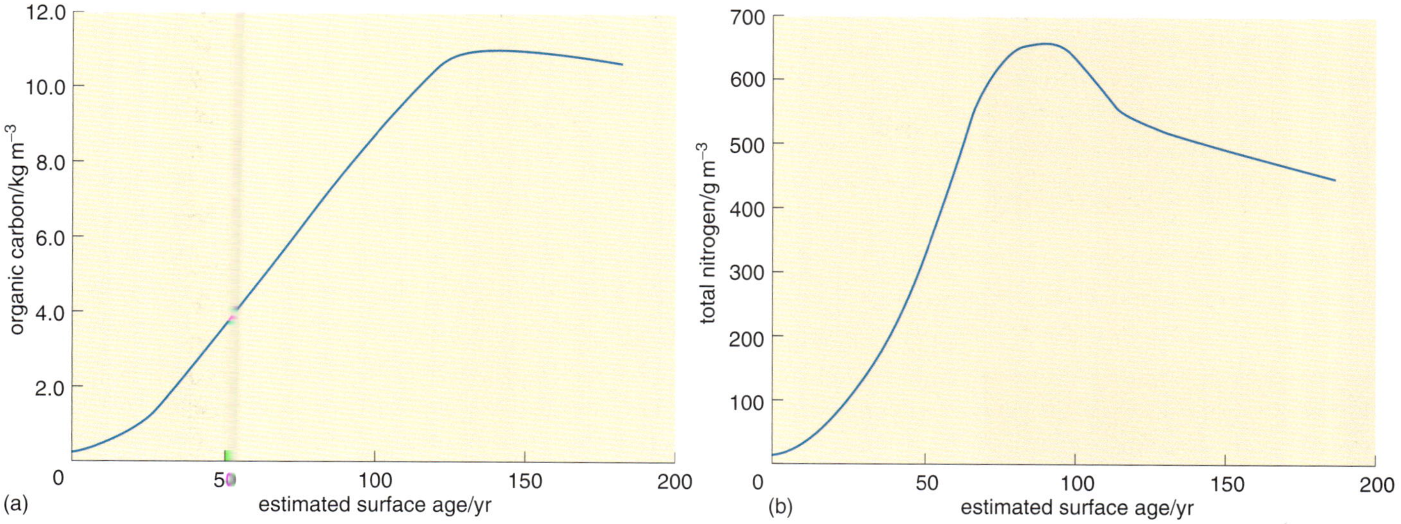

Figure 4.16 Rates of soil development following the exposure of rock fragments by a retreating glacier. Accumulation of (a) organic carbon and (b) nitrogen in the upper layers of the soil at Glacier Bay, Alaska.

This self-promoting process is an example of **facilitation**. The pioneer vegetation alters the soil composition in such a way as to allow the establishment of other, faster- and taller-growing species, which are more demanding in terms of their requirements for soil nitrogen and soil water-holding capacity. In other words, the early colonists are relieving a resource constraint that prevented some of the more productive species from establishing. Over time, increasingly productive and taller vegetation types establish themselves (Figure 4.17). In the case of Glacier Bay, the next stage in the succession of vegetation types is the dominance of shrubby willow species (*Salix* spp., Figure 4.18). These are in turn replaced by alder trees (*Alnus tenuifolia*), which permit atmospheric nitrogen to be fixed (Figure 4.19). Figure 4.16b presents the increase in soil nitrogen during the succession.

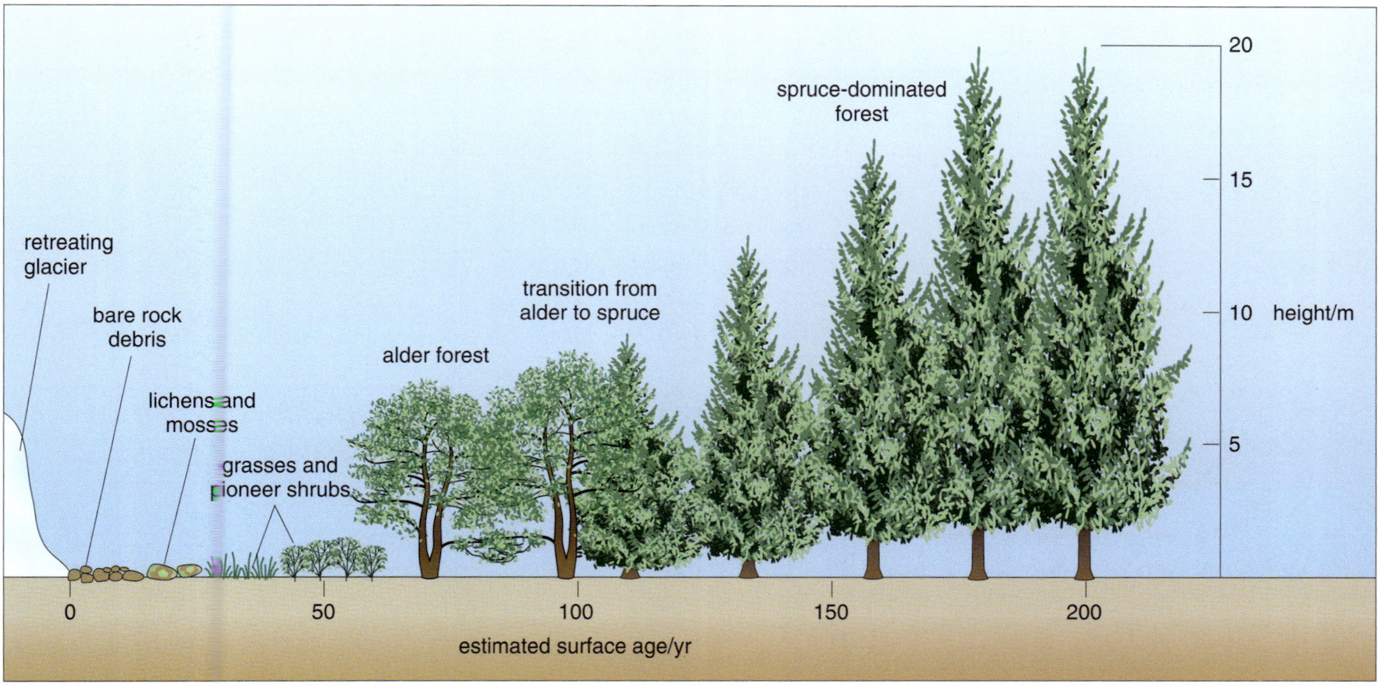

Figure 4.17 An illustration of the successional sequence of vegetation types at Glacier Bay.

Figure 4.18 Dwarf willow (*Salix* sp.), one of the pioneer shrubs.

Figure 4.19 The nitrogen-accumulating tree, American alder (*Alnus tenuifolia*).

❍ Considering Figure 4.17, what is the most noticeable trend in the vegetation types over time?

● The increasing height of the vegetation.

The alders dominate the vegetation for about 70 years, until displaced by the taller tree, Sitka spruce (*Picea sitchensis*, Figure 4.20). The spruce forest appears to form the end of the succession and is sometimes referred to as the **climax community**. It is able to maintain itself from generation to generation and appears stable. The soil development also slows down and stabilizes. Spruce is able to outcompete other species under that prevailing climate and the term **climatic climax** has been coined to describe the vegetation type that is dominant whilst the climate is stable (see Section 4.3 on ecosystem stability.)

This was once considered to be the end of the story but, like all concepts in science, the assumption that primary succession results in a stable end-point has been challenged. It is rare that vegetation reaches a true equilibrium with its environment. The environment itself is never entirely stable. Weather conditions vary from year to year and soils continue to mature. On top of which, chance events such as tornadoes or fires will introduce an element of physical disturbance. In spruce forests similar to the one at Glacier Bay, but older, species of hemlock (*Tsuga* spp., Figure 4.21) have established themselves as a major component of the vegetation able to coexist with the dominant spruce.

Figure 4.20 The Sitka spruce (*Picea sitchensis*) in its native habitat. The same species can be seen in rather uniform blocks of dense plantation across much of upland Britain.

Figure 4.21 The western hemlock (*Tsuga heterophylla*), a late-comer in the successional sequence.

These two trees tend to form patches within the wider forest, because they both regenerate from seed more effectively under the canopy of the other species than under their own. In this way, a given point in the forest continually cycles between one species and the other.

- ❍ Give a reason why a tree species may be more successful at regenerating under the canopy of another species rather than its own. (*Hint*: look at Section 2.5.3.)
- ● Under the canopy of a different species, the tree seedling will not be attacked to such a degree by leaf-eating and sap-sucking insects as it would under the canopy of its own species. This is because the herbivores tend to specialize in terms of their food plant, and therefore the ones most likely to attack the seedling will be dropping out of the canopy of the same species.

Succession not only occurs with respect to the vegetation, but the soil microbial community undergoes a parallel change with time as a new habitat is colonized (Figure 4.22). The microbial community in newly formed soils tend to be dominated by bacteria. This changes gradually over time, with fungal species becoming increasingly important.

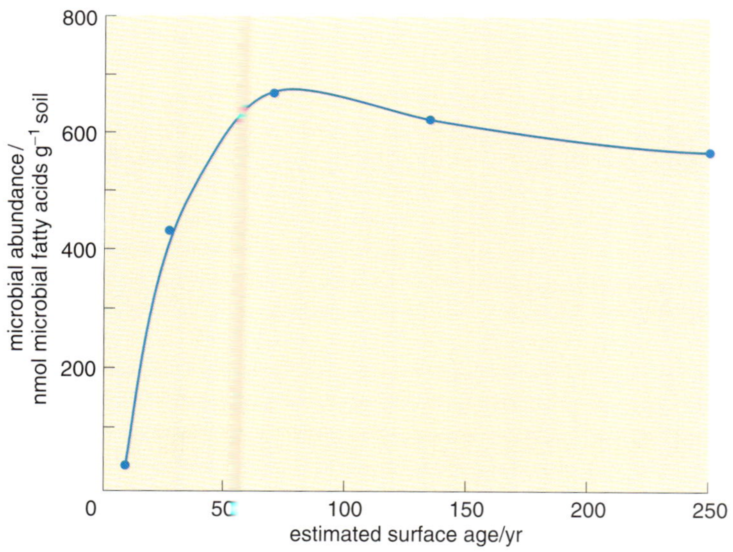

Figure 4.22 Changes in the abundance of soil microbes during primary succession. The vertical axis denotes the quantity of microbial fatty acids extracted from soil. These fatty acids are used as markers of microbial abundance because they are readily extracted from the soil and identified.

Parts of the forest, where the ground is relatively flat, tend to become wetter, due to a build-up of water-retaining organic matter on the forest floor.

- ❍ What is the effect of wet or waterlogged conditions on nutrient cycling? (*Hint*: look back at Section 3.5.1.)
- ● There are two important processes to consider. The activity of bacteria and fungi decreases in waterlogged soil due to a lack of oxygen for respiration. Their decomposer function is therefore reduced and nutrients are not released from organic matter efficiently. Available nitrogen in the form of nitrate is scavenged by bacteria that use an alternative respiratory pathway. The net effect is a lack of plant nutrients in the soil.

Figure 4.23 A moss-dominated bog typical of former areas of spruce forest in North America with poor drainage. This vegetation type is known locally as muskeg bog.

Waterlogged soils will tend to accumulate organic matter because of the restrictions placed on decomposers in an anoxic soil. The accumulating organic matter may then further retain water, creating another positive-feedback situation. At the same time, the lack of available nutrients limits the growth of plants, rendering the trees short and stunted, allowing more light through to the forest floor. The end-point of this process is the re-establishment of mosses. These are stress-tolerant species able to tolerate low nutrient availability and the anoxic soil. Under such circumstances, the once majestic spruce forest can be returned to a low-growing bog of mosses (Figure 4.23). This illustration demonstrates that succession is not necessarily a unidirectional process and is dependent on the detail of the local environment.

4.2.4 Changes in plant survival strategy during succession

The succession of vegetation types can also be considered in terms of the predominant strategy exhibited by species at each stage. The early colonists are primarily stress tolerators (e.g. lichens) able to establish in habitats very poor in nitrogen. Some will also have ruderal characteristics, as such plants tend to have good dispersal abilities, finding new sites rapidly and being able to establish before others arrive. The subsequent arrivals show increasingly competitive strategies.

Secondary succession occurs on sites after the original vegetation has been removed. This is most usually as a result of human activities such as clearance of forest for agriculture and subsequently abandoning the land, allowing natural vegetation to re-establish. The soils are already developed in these situations and may contain a seed bank and other organs such as tubers and rhizomes, from which plants can regenerate.

The early colonizers here would depend on the antecedent vegetation type as well as the vegetation of surrounding areas, which could supply seeds and other **propagules** by dispersal. In a secondary succession, the life strategy of the pioneer vegetation is typically ruderal, rather than stress-tolerant. It is the species with the best dispersal ability or the largest viable seed bank that can respond most rapidly to the availability of a new site and become the first to dominate the vegetation. Plants with other strategies may also become established, but are likely to be shaded by the initial cover of ruderals. The lack of light will inhibit their growth and delay their appearance in the canopy, but if they are able to tolerate the shade, they will be well placed for replacing the early colonists when those species come to the end of their life cycle. This is referred to as the **tolerance** mechanism of succession. In terms of life strategies, the early stages are dominated by ruderals, these are then replaced by increasingly competitive species, with the final stage being dominated by a more stress-tolerant species as the nutrient sources released by the initial disturbance once again become locked up in hard-to-decompose organic compounds. These compounds are the residue after the initial decomposition process and tend to be very stable molecules that bacteria and fungi are unable to utilize. Therefore they accumulate over the years and the nitrogen and other minerals they contain are excluded from the pool of nutrients that are being actively cycled between plants and the soil.

Again it is debatable whether truly stable climax vegetation is ever reached. Many systems show evidence of cycling between two or more different states and shifting patterns are often to be seen.

Activity 4.1 Secondary succession in woodland

You could now watch the five-minute video clip that documents the successional process that occurred in an English woodland following the great storm of 1987.

As you watch, note down which of the life strategies (competitor/stress tolerator/ruderal) is being displayed by each of the species mentioned in the sequence.

This type of extreme weather is one mechanism by which so-called climax communities retain their dynamism.

4.2.5 Halting the progress of succession

Succession, either primary or secondary, is usually a directional process, although the end-point is not always predictable. The process can be halted by management of the vegetation. Constant removal of plant material by such actions as grazing, cutting or burning will prevent the species that would normally have dominated a stage of the succession from fulfilling their potential. If the interference is extreme, e.g. clear cutting woodland or an intense fire killing all established vegetation, then the process of secondary succession will be effectively restarted (Figure 4.24). If, however, the interference is milder, but more frequent, such as mowing the lawn, then succession is not re-started but merely arrested.

(a)

(b)

Figure 4.24 Some tree species such as birch (*Betula pendula*) display ruderal traits and their seedlings can show a phase of explosive growth in newly felled woodland. (a) An area of recently felled woodland, into which new oak trees have been planted (the white tubes are tree guards to protect them from rabbits). (b) The same area after just four years of tree growth. The birch seedlings have blanketed the area, retarding the growth of the oaks that are hidden beneath them.

(a)

(b)

Figure 4.25 (a) Brambles (*Rubus fruticosa*) encroaching into an old meadow. (b) Subsequent encroachment of ash trees into the meadow following the cessation of cutting.

Figure 4.26 The Broadbalk reversion plot at Rothamsted, which was originally farmland in 1850, but has been left unmanaged for 150 years as an experiment to observe succession.

If a lawn in Britain were left uncut, the creeping grass species (*Festuca* spp., *Agrostis* spp.) would be typically replaced first by taller-growing grasses and weeds (*Arrhenatherum elatius* and *Anthriscus sylvestris*), after a couple of years by brambles (*Rubus* spp., Figure 4.25a) and shrubs (e.g. hawthorn, *Crataegus* spp.) and eventually by trees (e.g. ash, *Fraxinus excelsior*, Figure 4.25b). This example of secondary succession is not uncommon in Britain, particularly on abandoned pastures. Grassland would be very rare in Britain if human influence were removed. All the pastures, meadows, roadside verges and amenity grasslands would disappear as a result of succession if they were not cut or grazed regularly. An experiment to prove this has been conducted at Rothamsted Research Station in southern England. Part of a field was fenced off and left unmanaged for 150 years. A secondary succession took place and the site is now mature deciduous woodland (Figure 4.26).

Grassland in much of northern Europe is an example of **deflected** climax vegetation. It fulfils the criteria for a climax community insofar as it is stable in the long term. Some meadows are documented to have been continuously grassland for more than 500 years and there is no evidence for directional change, in terms of the community composition, over that period. Grassland is not regarded as a true climax, however, because it needs regular management to preserve it by preventing replacement by later successional stages. The deflected climax community, therefore, is a result of a successional process that has been halted by some external (often human) intervention.

Many habitats regarded as valuable for wildlife (e.g. meadows, reedbeds, fens) are in fact examples of either a deflected climax or a transitory successional stage. In order to maintain them, as is often the goal of conservationists, it is necessary to manage by cutting, grazing and burning to slow down, stop or even reverse the successional process (Figure 4.27).

❍ Returning to the Teign Valley in Devon, what would be the long-term impact on habitat diversity if human activities were to cease?

● The diversity of arable fields, grasslands, heaths and hedges would be lost as a relatively uniform broadleaf woodland re-established itself in the landscape.

In some parts of the world (e.g. the Australian bush), fire is part of the natural system rather than a management tool. The vegetation has adapted itself to periodic burning, resulting in a vegetation type that is dominated by grasses and shrubs rather than tall trees (Figure 4.28).

(a)

(b)

Figure 4.27 (a) One method of deflecting a successional sequence is to burn the vegetation such as in a reedbed. (b) If the area were left unmanaged, the reedbed would gradually be invaded by grey willow (*Salix cinerea*) and other shrubs.

Figure 4.28 Shrubby eucalypts of the Australian bush, a vegetation type adapted to regular burning.

Question 4.2

Think about the following British vegetation types and describe them in terms of their successional state (i.e. decide whether they can best be described as a stage in primary or secondary succession, or as a relatively stable 'climax' community or an example of a 'deflected' climax community).

(a) Grassland on a sand dune system; (b) heather on lowland acid soils; (c) ash woodland in a river valley; (d) tall grassland in a forest clearing; (e) birch trees on a drained bog.

4.2.6 Summary of Sections 4.1 and 4.2

1 Plant species do not fulfil their fundamental niche solely because of competition from other species, but also due to limits on their ability to disperse to all environments that may potentially suit them.

2 Plants have developed a wide range of adaptations to aid their dispersal.

3 Patterns of species occurrence are strongly influenced by natural barriers to dispersal, such as oceans and mountain ranges. Human activity is changing the landscape and creating new dispersal barriers for many species. These will be very significant in the event of climate change.

4 Primary succession describes the process by which vegetation colonizes a new area, altering the properties of the developing soil as it does so.

5 A climax community is regarded as the end-point of succession, when the vegetation becomes stable until a change in the climate restarts a process of change. Some ecologists now question whether any vegetation is ever truly stable and instead suggest that it may display cycling between vegetation types in which different species dominate.

6 Secondary succession occurs when vegetation has been removed from an area. It differs from primary succession in that soil already exists and contains organic matter and perhaps propagules.

7 Succession may be halted by vegetation management, creating a deflected climax community. Many vegetation types of conservation interest that may appear natural are in fact deflected climax communities, which require on-going management to conserve them.

4.3 Ecosystem stability

An ecosystem is defined as stable if the rates of key processes, e.g. primary production, biomass accumulation, nutrient cycling, supply of soluble nutrients, are more or less constant and vary within defined limits at a particular time-scale. Figure 4.29 illustrates the situation for the density of the silver-studded blue butterfly (*Plebius argus*, Figure 4.30) from a hypothetical heathland ecosystem. This butterfly needs short sparse vegetation, it lays its eggs on heathers (*Calluna vulgaris, Erica cinerea* and *E. tetralix*) and gorse (*Ulex* spp.), which are used as food plants by the larvae.

The importance of scale is clear. The system appears to fluctuate wildly on a one-year time-scale but looks much more stable when viewed over ten or a hundred years. Even over 500 years and incorporating major fires, the system is stable: the important point is that it recovers.

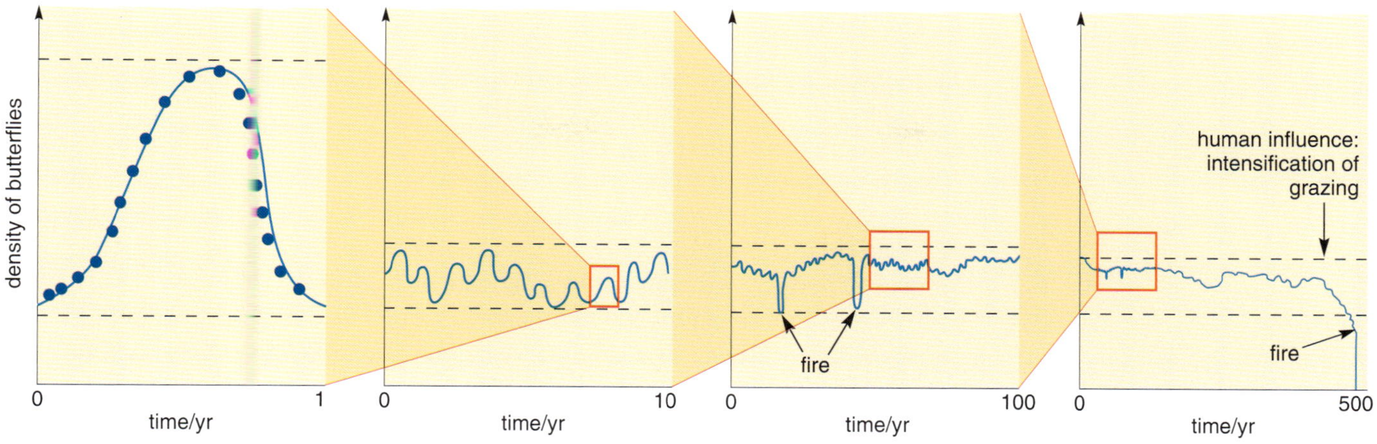

Figure 4.29 Density of silver-studded blue butterfly (*Plebius argus*) from a hypothetical heathland ecosystem viewed at different time-scales. The heath is assumed to be disturbed by fire every few decades (visible at the century scale). Dashed lines indicate the boundaries that define the limits of stability. Boxes indicate the time segment shown at the next lower time-scale and the vertical scales are arbitrary.

Figure 4.30 The silver-studded blue butterfly.

- ❍ How do you interpret the events that follow the introduction of heavy grazing?
- ● In terms of *Plebius argus* density, the heath system goes outside the boundary limits and does not recover: it can be regarded as unstable.

Heavy grazing tends to eliminate heather and heath species. Where heavy grazing occurs, mat grass (*Nardus stricta*) often comes to dominate the flora.

There are two aspects to ecosystem stability: **resistance** to change, i.e. how difficult it is to cause a change, and **resilience**, i.e. how quickly the ecosystem can recover after disturbance. In the heathland example, it is resilience that is emphasized. But for other processes it may be more appropriate to consider resistance.

One factor that appears to be important in determining ecosystem stability is **functional redundancy**, e.g. having several species performing the same function. If one species declines or disappears, another species replaces it, e.g. insect foods for the greater horseshoe bat. Functional redundancy tends to be low for ecosystems in harsh environments, such as tundra, hot springs or hot deserts. It is higher for complex systems, such as tropical rainforests. This observation may be taken to imply that functional redundancy is dependent upon species diversity. In tropical rainforests there is very high diversity, but many of the species are highly specialized, and so do not necessarily contribute to functional redundancy. Therefore, species diversity is not synonymous with functional redundancy; intermediate diversity may lead to higher levels of functional redundancy. However, species diversity can be a useful indicator of the quality of an environment and environmental change (see Block 7).

4.3.1 Indicators of habitat degradation

Throughout Sections 2 and 3, numerous references have been made, or implied, to the importance of community complexity in ecosystem function, e.g. stratification of communities, effects of pollution, numbers of trophic levels, stability. Therefore, it follows that if we have some indication of community complexity, it might enable us to make predictions about the ecological 'health' of a particular community. How do we assess community complexity? The number of species in a community is known as its **species richness**, and it follows that a more complex community has more species. However, this concept can be refined by taking into account the relative abundance of the species (**equitability**). Together, species richness and equitability are taken to give a measure of **species diversity**. It may seem unnecessary to include a consideration of equitability when assessing species diversity of a community, but intuitively a meadow containing ten plant species that are present in roughly equal amounts is more diverse than an arable field also containing ten species, but where 99.9% of the plants are of a single crop species.

The existence of high species diversity, and the presence of species with particularly exacting habitat requirements are used by ecologists and environmental scientists to identify particularly rare and valuable communities. As you know, human activities often degrade habitats through pollution and intensive management practices. Species that have the most narrow and specific requirements are the first to be eliminated by human activities, as the following case studies show.

Case study 1: Changes in the status and distribution of butterflies

Butterfly records for 1850–2000 reveal a complex pattern of change, but the dominant trend is one of widespread losses. These have sometimes been part of long-term declines, but there has been a noticeable acceleration in the last 50 years. The reasons for the decline of butterflies are complex.

Most species have declined primarily due to habitat loss, such as the destruction of semi-natural vegetation that occurs by the fertilization of semi-natural grasslands, draining of fenland and removal of hedgerows or woodland. However, many butterflies have very precise habitat requirements and have been affected as much by changes in land management, e.g. cessation of coppicing of woodland and changes in grazing intensity.

- ❍ What do you think would happen to the vegetation if land management practices such as grazing, mowing or coppicing were stopped?

- ● Succession would occur and the vegetation in each case would proceed towards mature woodland.

Many of our most precious wildlife habitats represent intermediate stages in succession. Without appropriate intervention, either by the action of abiotic factors (e.g. annual flooding or fire) or biotic factors (e.g. the action of grazers such as rabbits and sheep) all of the sites would gradually develop into woodland.

Together these changes have led to increasing fragmentation and isolation of habitats, causing further losses as populations have become less viable, and the natural pattern of extinction and recolonization has been disrupted. However, the general decline of butterflies has not affected all species equally.

Butterflies of Britain and Ireland can be divided into two categories: (1) common and widespread species, which occur throughout the farmed countryside and in urban habitats; and (2) rarer and more localized species, which are usually restricted to specific habitats. You have already met two habitat specialists. Both species are woodland butterflies, but the high brown fritillary (*Arginnis adippe*, Figure 2.60) requires the open spaces created by regular coppicing, while the white admiral (*Limenitis camilla*, Figure 2.4) needs extensive shade to do well. Consequently over the last century, in which the practice of coppicing has declined, the high brown fritillary has become increasingly rare and the white admiral has spread and flourished.

Recently published data, based on major butterfly surveys carried out in the 1970s and the late 1990s, have enabled comparisons to be made of the fates of a significant number of species of butterflies from both categories. For the purposes of the study, Britain was divided into a grid of 10 km × 10 km squares and the presence or absence of each species within each square was recorded. Figure 4.31 illustrates the winners and losers amongst the butterfly species in Britain over the last decades of the 20th century.

You can see that for the habitat-specialist butterflies (Figure 4.30a), as expected, the white admiral is above the line with the 'winners' and the high brown fritillary is below the line with the 'losers'.

- ❍ Look at the two graphs in Figure 4.31. (i) Roughly what proportions of the butterflies surveyed have made gains and losses amongst the habitat specialists? (ii) How does this pattern compare with the wider-countryside species?

- ● (ii) Amongst the habitat specialists, roughly half the species appear to have made gains, and half made losses over the period specified. (ii) In contrast, none of the wider-countryside species appears to have lost ground, and all bar one of them have increased their distribution.

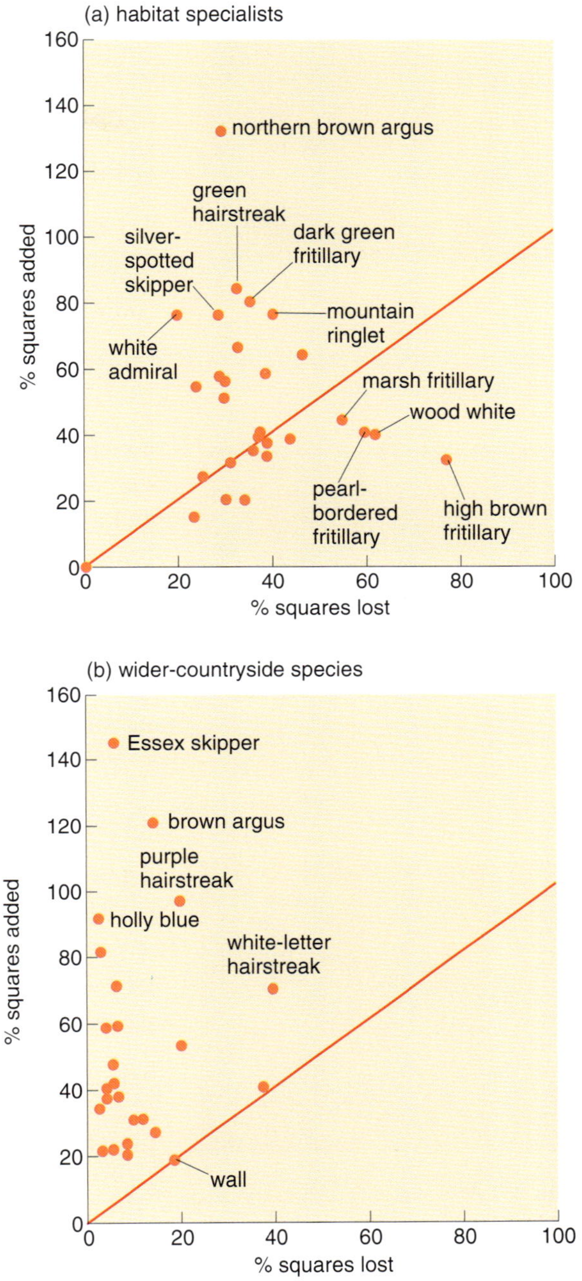

Figure 4.31 Gains and losses amongst butterfly species in Britain between 1970–82 and 1995–99, based on the number of 10 km × 10 km squares recorded in each period: (a) habitat-specialist butterflies; (b) wider-countryside species. Species lying below the diagonal line show a net decrease in recorded squares between these periods and those above the line show a net increase.

Figure 4.32 White-letter hairstreak butterfly (*Satyrium w-album*), whose populations went into decline as a result of Dutch elm disease.

Figure 4.33 Brown argus butterfly (*Aricia agestis*), a butterfly species that is increasing in frequency.

Figure 4.34 Dove's-foot crane's-bill (*Geranium molle*), a food plant of the brown argus butterfly, which has benefited from the recent agricultural practice of set-aside.

Interpreting the data is not straightforward. Some of the habitat specialists in particular are very rare and highly localized; therefore a failure to locate a single colony within a grid square has a big effect on its percentage occurrence. Similarly, the greater coverage of the 1990s survey revealed some new colonies of rare butterflies that had gone unrecorded before. Working at a 10 km square grid size can underestimate local changes. For example, numerous colonies within a locality may be lost without a change to the distribution map, provided a single colony remains within the grid square. Thus, the graphs may give a distorted view of the true nature of changes to the distribution and abundance of butterfly species. When the data are inspected more closely, it appears that while five species, including the white admiral, have genuinely expanded their range, virtually all the other habitat specialists (20 of them) have undergone a decline.

In contrast to the habitat specialists, most wider-countryside species in Britain have been re-recorded in a higher proportion of the 10 km squares that were recorded in the 1970s. The data are subject to the same sort of potential problems associated with under- and over-recording as for the habitat-specialist species, but when these are taken into account, it appears that 14 of the wider-countryside species have actually expanded in recent years while only two have declined. One of the losers is the white-letter hairstreak (*Satyrium w-album,* Figure 4.32) which breeds on elm trees. This butterfly underwent an initial decline after the Dutch elm disease epidemic in the 1970s, but it is now recovering in many areas because a substantial proportion of the root stocks of the trees destroyed by the disease were not killed. New growth, suckering up from ground level, has been incorporated into hedges, and it is these shoots that provide the egg-laying sites and larval food.

The brown argus (*Aricia agestis*, Figure 4.33), on the other hand, is a recent success story. It is a small butterfly characteristic of southern chalk and limestone grassland, but occurs in a number of other open habitats as far north as north Wales and Yorkshire. Traditionally, this species laid its eggs on the leaves of its food plant, common rock-rose (*H. nummularium*, Figure 2.38). Prior to 1990, the brown argus had declined due to the loss of its calcareous grassland habitat through intensive agriculture and abandonment. In the last decade, this loss has been more than compensated for by its spread into alternative habitats where it breeds on several species of food plant other than the common rock-rose. Several factors have influenced this spread. Warm summers in the early 1990s allowed populations to build up and spread. A new habitat, non-rotational set-aside, has been created on a large scale since the 1992 reforms of the European Common Agricultural Policy, where food plants such as crane's-bills (e.g. *Geranium molle*, Figure 4.34) can become established. It remains to be seen whether the brown argus will continue to prosper when non-rotational set-aside is eventually phased out.

Why should the habitat specialists be more prone to decline? The main reason appears to be that, by definition, the colonies of habitat specialists tend to be smaller and more isolated. The flight areas for these species tend to be small, sometimes less than a 100 metres in any direction, so their dispersal capability is low. Therefore, if chance effects such as unfavourable weather, fires or disease wipe out a particular colony, the site may remain unoccupied because it is too isolated to be recolonized. The more fragmented the colonies become, the more difficult it is for individuals with limited dispersal abilities to move into the site to recolonize it and so the more likely it is that local extinctions become permanent.

Question 4.3

Complete the following passage, selecting appropriate words or phrases from the list below it. (*Note*: not all the words and phrases will be used.)

Plant distribution is determined primarily by ————, ———— and land use. The distribution of ———— invertebrates is determined mainly by the availability of food plants, but ———— invertebrates are greatly affected by abiotic factors such as ———— and ————. Bird distributions are mainly determined by either the availability of food plants or the ———— that live upon them. Therefore, when land use changes, as a result of human activities such as ————, ———— or construction, the distribution and abundance of plant species and their associated invertebrates and birds also changes. The species that have the most ———— requirements in terms of resources and conditions, or are the most pollution ———— die out, leaving a more ————, ———— but ———— diverse ———— behind.

soil type	water flow rate	most	tolerant
herbivores	oxygen availability	least	parasites
temperature	agricultural intensification	less	sensitive
community	industrial pollution	more	terrestrial
population	volcanic activity	species	generalist
storm damage	invertebrates	pH	aquatic
climate	specialist		

Case study 2: The fall and rise of the otter

The otter (*Lutra lutra*, Figure 4.35) suffered a sudden and severe population crash in southern Britain, starting in 1957. This was followed by a recovery in the 1970s and 1980s.

The evidence supports an hypothesis that the immediate cause of the crash was very high mortality amongst breeding-age adults following the introduction of the highly toxic and persistent organochlorine insecticides, dieldrin and aldrin, as seed dressings and sheep-dips in 1956. The crash occurred against a background of a steady decline in otter numbers resulting from loss of habitat caused by the intensification of agriculture and engineering work to river-courses throughout the late 19th and the 20th centuries. These changes were most extreme in southeastern England. The situation was exacerbated by the fact that the otter population suffered persecution in the decades preceding dieldrin use and this had already reduced its numbers to a low level. Persecution took the form of hunting by otter hound packs, and 'predator control' by gamekeepers and water bailiffs in order to protect shooting and sport-fishery interests. The effects of the army of gamekeepers were probably more significant than that of the otter hounds in depleting an already stressed otter population.

Figure 4.35 The otter (*Lutra lutra*), a top predator in many British rivers.

Two major actions have been taken to bring about the recovery: (1) bans on all agricultural uses of the organochlorine insecticides by 1981–82; and (2) legal protection of the otter in 1978 and 1982.

There were regional differences in the extent of the decline in otter numbers and the rate of recovery. Southeastern England had the most rapid decline in numbers and was slowest to recover. The southwest saw the smallest decline and recovered most quickly. The main reasons for this were threefold. Firstly, the southeast, and especially East Anglia, had a less suitable and more fragmented habitat for otters because of extensive arable farming with associated pollution of rivers, and large-scale engineering work on the watercourses to bring about improved drainage and water management in the Fens. Secondly, farming in the west was mostly pastoral, with cattle-raising predominating. Legislation to limit the use of organochlorine pesticides was introduced earlier in the southwest. Thirdly, the southeast had more intensive game preservation and otter hound activity than the southwest.

In an effort to assist the recovery in the east, captive-bred otters were released, and have been shown to have bred in the wild. Figure 4.36 summarizes the changes in otter populations across England between 1950 and 1996. Almost certainly, the population of East Anglia would have continued to decline to extinction without the release of captive-bred otters starting in 1983 (see lowest line). The horizontal bars indicate the span of dates at which the decline 'bottomed out' (reached its nadir).

It is notable that the difference in the severity of the decline in the east and the west was not very great at the period 1966–71, but the gap widened as the population of the west recovered and that of the east declined further. There is a noticeable correlation between the status of the western and eastern otter populations and the use of organochlorines in agriculture in these two halves of southern Britain.

In contrast to the case of the water vole (*Arvicola terrestris*), research has shown that the feral American mink (*Mustela vison*) was not a causal factor in the otter's decline, nor did its presence deter the otter's recolonization of mink-occupied areas. Indeed, otters appear to show strong antagonism towards mink and to cause a decrease in their numbers and site occupation (presumably by predation) everywhere that the two species occur together and where otter density has shown a marked recovery, e.g. in the southwest of England, where there has been a 50% reduction in mink occupation since 1984–86.

Currently, the main threats to the continued recovery of the otter population appear to be: high mortality on the road, the low age at death, and the potential problem of contagious lethal diseases contracted from captive relatives of the otter (such as ferrets and mink).

In conclusion, one of the most important causative factors underlying the decline of British wildlife would appear to be the loss or degradation of habitat, most usually associated with the intensification of agriculture, building development and industrial activity. The loss and fragmentation of habitat reduces the carrying capacity of the landscape for affected species, sometimes to the point at which viable breeding populations cannot be sustained. These changes increase the likelihood of local extinctions resulting from chance events, such as pollution, disease, fires and freak weather conditions, and inhibit recolonization, particularly if the species concerned have very specific habitat requirements and limited dispersal ability. Top-down controls such as hunting by otter hound packs and gamekeepers are of secondary importance. These factors only become significant in extinctions if the hunting is very intensive and/or the populations of prey are already much reduced and fragmented.

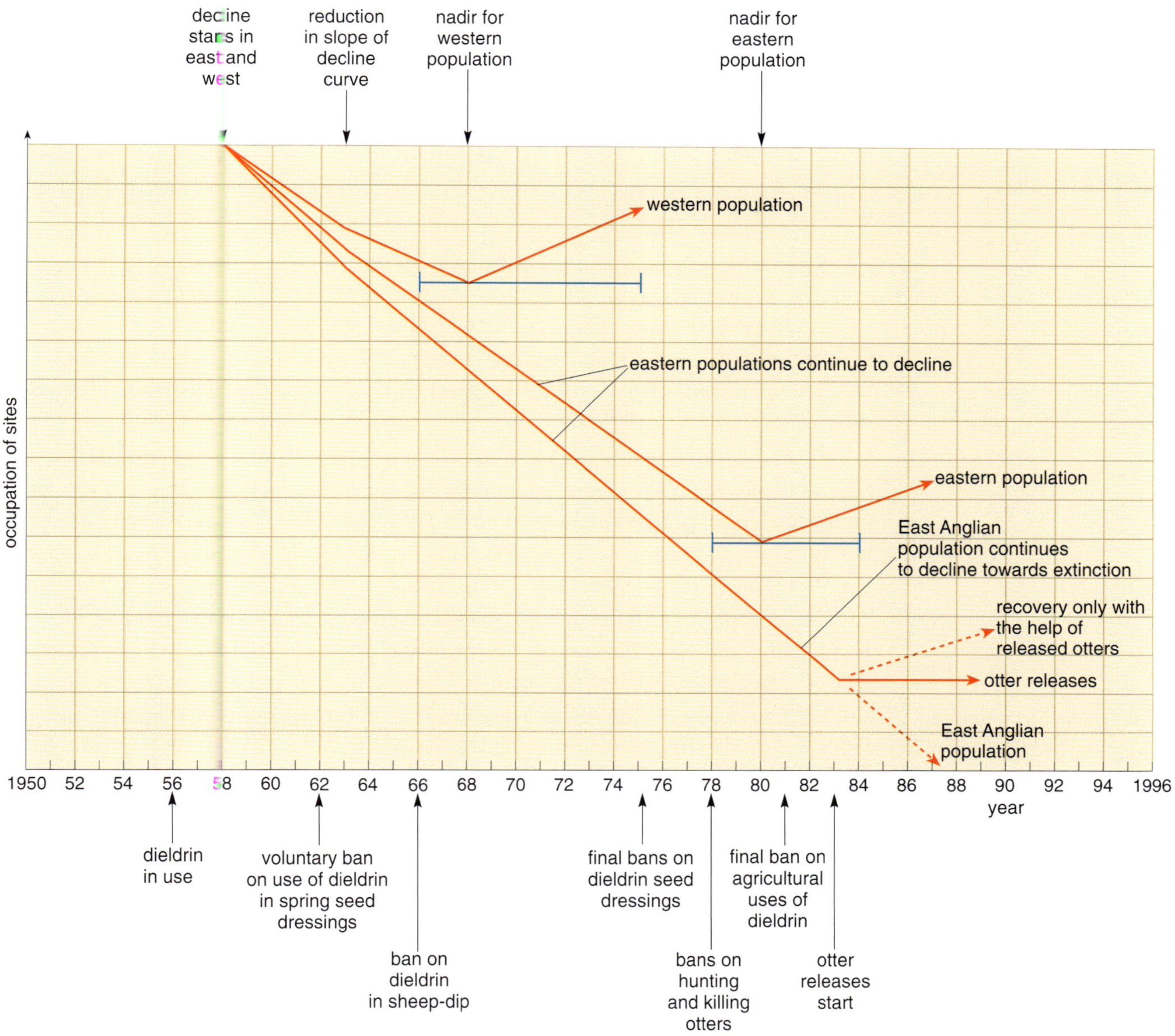

Figure 4.36 Diagrammatic representation showing the timing of the start of the decline in English otter numbers, and its nadir. The blue horizontal bars indicate the span of dates at which the nadir occurred in different areas. Along the horizontal axis are shown the dates of introduction of important bans on otter hunting and the use of dieldrin in sheep-dip (in the west) and in seed dressings (in the east).

It follows from these conclusions that the very sensitive and demanding species can be used by environmental scientists as early indicators of trouble, very much in the manner of the canaries that were taken by miners into coal mines. The death of the canary alerted the miners to a build-up of toxic gases in the pits while there was still time for them to escape. The elimination of such 'canary' species, and the associated reduction in species diversity in the environment at large, gives us an early warning of undesirable and perhaps potentially lethal changes to the environment while there is still time to reverse the damage. Furthermore, these findings suggest that manipulation of bottom-up controls, i.e. habitat creation and management, are likely to be of prime importance in the battle to conserve British wildlife, but that manipulating top-down controls, i.e. predation, is also likely to have a role to play (see Block 7).

4.3.2 Summary of Section 4.3

1. An ecosystem is defined as stable if the rates of key processes, e.g. biomass accumulation, are more or less constant and vary within defined limits at a particular time-scale.
2. Ecosystem stability is affected by resistance to change, i.e. how difficult it is to cause a change, and resilience, i.e. how quickly the ecosystem can recover after disturbance.
3. Functional redundancy, e.g. having several species performing the same function, so that if one species declines or disappears another species replaces it, is an important determinant of ecosystem stability.
4. Species richness (the number of species in a community) and equitability (the relative abundance of the species) combine to give a measure of species diversity.
5. Species that have the most narrow and specific requirements are the first to be eliminated by human activities.
6. Over the last 150 years, Britain's butterflies have undergone a general decline. There have been a few cases of individual species increasing their numbers and distribution, e.g. the brown argus and white admiral, but many species, especially habitat specialists, have become extinct or suffered a contraction in their range. The principal causes of decline are loss of habitat and changes in land management practices.
7. The otter suffered a major population crash in the late 1950s and 1960s, from which it is still recovering. The principal cause of the crash was the introduction of the use of organochlorine pesticides, but the crash was exacerbated by persecution of an otter population already diminished by habitat loss, especially in the eastern counties. A recovery has been observed following bans on the use of the pesticides and otter hunting.
8. Reduction and fragmentation of habitat makes more difficult the recolonization of otherwise suitable sites after local extinctions brought about by chance events such as fire, disease or bad weather, especially when the organisms concerned have limited means of dispersal.
9. Organisms with particularly demanding environmental requirements are the most sensitive to environmental change and their local extinction, along with an associated drop in species diversity, can be used by environmental scientists as early indicators of habitat degradation.

Learning outcomes for Section 4

After working through this section you should be able to:

4.1 Explain the importance of dispersal ability in defining the realized niche of a species and list a number of common dispersal mechanisms. (*Question 4.1*)

4.2 Use examples from different time-scales to show how habitat fragmentation limits the dispersal of species and comment on the implication of this for future changes. (*Question 4.1*)

4.3 Describe the principle of vegetation succession and use examples to illustrate the difference between primary and secondary succession. Comment on the influence of soil development and microbial communities on the process of succession. (*Activity 4.1*; *Question 4.2*)

4.4 Define the concept of a deflected climax, give examples, and comment on the implications for the conservation of such vegetation. (*Question 4.2*)

4.5 Explain what is meant by a stable equilibrium with respect to an ecosystem stability and describe the factors that contribute to it.

4.6 Explain, using appropriate examples, how changes in species diversity can be used by environmental scientists to indicate a deterioration in environmental conditions. (*Question 4.3*)

4.7 Use examples to illustrate how human activity has influenced the distribution and abundance of members of the British fauna through its effects on environmental factors that represent bottom-up and top-down controls.

Comments on activities

Activity 2.3

(a) Your calculations should look something like the following:

Table 2.14 The calculation of a Spearman rank correlation coefficient.

Site number	Rank of mink abundance	Rank of vole abundance	Difference in ranks	Difference squared
1	11	1	−10	100
2	6	2	−4	16
3	9	3	−6	36
4	8	5	−3	9
5	7	4	−3	9
6	10	8	−2	4
7	5	7	+2	4
8	2.5	6	+3.5	12.25
9	2.5	9	+6.5	42.25
10	1	10.5	+9.5	90.25
11	4	10.5	+6.5	42.25
				total: 365

$n = 11$, sum of differences squared = 365. Therefore

$$r_s = 1 - \frac{6 \times 365}{11(121 - 1)} = -0.659$$

(b) From the tables in *Using the Spearman Rank Correlation Coefficient*, $p < 5\%$; therefore the null hypothesis of no correlation between mink abundance and water vole abundance is rejected, and we conclude that there is a statistically significant correlation between these two variables, and because $r_s < 0$, the correlation is deemed to be negative.

As with the heather hypothesis in Block 1, a significant relationship has been found, but conclusions must be drawn with care. The evidence from these data is consistent with the view that the presence of mink at a site causes a decline in the winter vole population, but it does not prove that the mink are the cause of the vole's decline. It is possible that another environmental factor may be causing the two species to use different habitats

Activity 3.1

If your osmometer has not worked, then check that your potato did not split when you bored the hole. If it did, try again with another potato. If it is still not working, try adding more salt directly to the well in the potato.

Your responses to the questions at the end of the activity should have been along the following lines:

1 The salt solution. Osmotic pressure increases as the solute concentration increases.

2 The tapwater. Its osmotic pressure is lower than that of the the salt solution and therefore its water potential is less negative. (Remember $\Psi = -\pi$, if hydrostatic and matric pressures are insignificant.)

3 Water moves from the tapwater into the salt solution, i.e. down its water potential gradient, as we would predict.

Activity 3.3

The huge amount of excess rainfall received by Mount Roraima, which flows so spectacularly over its cliffs, carries away most of the soluble nutrients. Weathering of the rock will supply some new ones, but this is a very slow process compared to the rate of loss. Plants therefore depend on the importation of nutrients form the relatively rich rainforest below. They are transported as small packages of flying nutrients, also called insects.

Activity 4.1

The video on woodland succession introduced you to four species in an English wood. Your interpretation of their life strategies, as defined by Grime, should have been along the following lines:

- Foxglove (*Digitalis purpurea*) is a ruderal species. It grows rapidly from long-lived seeds, when conditions are favourable. It invests all its energy into a single flowering, to maximize seed production.
- Rosebay willowherb (*Chamerion angustifolium*) also shows ruderal characteristics: rapid growth and production of large numbers of seeds which are readily dispersed. However, it is taller-growing and longer-lived than the foxglove — both competitive traits. It is therefore best classed as competitive ruderal.
- Birch (*Betula pendula*) is a fast-growing tree, attaining a height of 20 m, and lives for up to 60 years. It is a competitor that relies on outgrowing its neighbours in order to survive.
- Oak (*Quercus robur*) is much slower-growing, but is able to tolerate shade cast by the birch. It is very long-lived (> 500 years) and may grow to 30 m. It is classed as a stress-tolerant competitor, because although its ultimate strategy is to outgrow and shade its neighbours, it can tolerate the stress of low light and can survive on nutrient-poor soils.

Answers to questions

Question 2.1

(i) Swamp. The Letter 'S' within the code S2 refers to the broad habitat type (see Table 2.3).

(ii) *Cladium mariscus* (saw-sedge) is the only species listed above the first horizontal rule and therefore is the only constant species for that community (you would expect to find it within the vast majority of samples from a stand of S2).

(iii) The species that best distinguish between the two subcomunities are: *Calliergon cuspidatum, Solanum dulcamara, Phragmites australis, Menyanthes trifoliata, Potentilla palustris, Utricularia vulgaris* and *Mentha aquatica*. Their frequency varies between the two subtypes by at least two classes. They are listed between the horizontal rules and are known as the preferentials. The other 14 preferentials listed in the table are either less frequent or less discriminating.

(iv) *Juncus subnodulosus*, although more frequent within stands of S2 than *Carex rostrata*, is not considered diagnostic of either subcommunity and is therefore classed as an associate. Associates often occur in a number of other communities and may not be strongly linked to the particular community at all, in spite of occurring in up to 60% of samples.

Question 2.2

(a) Bracken was encountered in a number of distinct habitats (oak woodland, coniferous woodland, open moorland) and so can be regarded as non-specific.

(b) Swimming mayflies are habitat-specific, being restricted to moving freshwater of good chemical quality and intermediate flow rate.

(c) Wheatears move between habitat types during their seasonal migrations and are therefore non-specific.

(d) Brown hairstreak butterflies are habitat-specific, being found only in deciduous woodland with an understory of blackthorn.

(e) Humans are non-specific, insofar as they are found almost throughout the catchment, yet they often modify the environment to such an extent (urbanization) that they could be regarded as partially habitat-specific, if the urban environment were viewed as a distinct habitat type.

(f) Sea lavender is specific to stabilized saltmarshes, inundated by high tides. Indeed, it is a good indicator species for this type of environment.

Question 2.3

The BMWP index for January 1999 is 54 (obtained by summing the scores for the individual invertebrate families present in the samples at that time). Thus we can see from the data in Table 2.7 that there is a sudden sharp decline in the BMWP indices in January 1999, followed by a steady recovery. Low BMWP indices are associated with high BOD readings, which typically result from organic pollution. Therefore, the slurry spillage probably occurred between the November 1998 and January 1999 sampling times.

Question 2.4

(a) Decline of the high brown fritillary butterfly: bottom-up, resulting from a decline in coppicing.

(b) Increase of the white admiral butterfly: bottom-up, a result of the decline in coppicing and the spread of conifer plantations.

(c) Decline of the water vole: top-down, a result of spread of the feral mink; also decline of habitat (bottom-up), but mink considered to be more important.

(d) Decline of the brown argus butterfly: bottom-up, a result of the decline of the unimproved chalk grasslands.

(e) Decline of the otter: bottom-up, in the form of poisoning by dieldrin and loss and fragmentation of riverine habitat; top-down, in the form of otter hunts with hounds and persecution by gamekeepers and water-bailiffs, but top-down controls are only really significant once bottom-up controls have reduced population levels to a low level.

(f) Decline of the wolf: top-down, persecution by humans, although hunting to extinction followed a long period in which the human population increased and put pressure on the wolf through loss of habitat by deforestation.

Question 2.5

(i) The wild daffodil (*Narcissus pseudonarcissus*) is not found growing wild in environments denoted by position B in Figure 2.65, because although these are within its fundamental niche, i.e. its physiology would allow it to grow there, it is outside its realized niche, i.e. competition from other species under conditions of high light availability will exclude it.

(ii) The species is not found in environments denoted C, because the combination of prolonged waterlogging and low light availability is beyond its physiological tolerance range. A daffodil transplanted into that environment would probably die, even in the absence of competition from other species.

Question 2.6

You may have selected any three of the following possible answers.

1. In spite of its deliberate manipulation to favour wheat, variation within the field may result in some patches representing environments beyond the fundamental niche of wheat (too waterlogged, too shaded due to a high hedge, etc.) in which more stress-tolerant species are able to grow.
2. Species with very rapid germination and early growth may get themselves established in the field before the wheat plants have a chance to outgrow and suppress them.
3. Where disturbance has occurred (for example a tractor rut created by a fertilizer application), the wheat crop may take some time to fill the gap by lateral growth, giving species with good dispersal mechanisms and rapid growth rates the chance to gain a foothold.
4. A monoculture of wheat is susceptible to pests and diseases such as fungal parasites. If patches of wheat succumb to such a disease, other species, resistant to that particular fungus, may have the opportunity to invade.
5. Following a period of torrential rain, part of the field may be submerged under a pool of water for several days. Wheat plants are intolerant of flooding and their growth will be halted, giving flood-tolerant species the chance to replace them.

Question 2.7

Heather is a stress tolerator. It is long-lived (30 years). It is able to survive on soils that are very deficient in the principal mineral nutrient, nitrogen. It is relatively low-growing and does not flower in its first year of growth.

Bracken, whilst having a similar lifespan and inhabiting similar habitats, has a higher growth rate and tends to grow taller than the heather. In relative terms, it is a more competitive species.

Question 2.8

Deciduous broadleaf woodland. The minimum temperature is too cold for tropical rainforest or evergreen broadleaf woodland to survive (the latter can tolerate temperatures no lower than −15 °C). Boreal coniferous forest and tundra would both be able to survive this temperature, but the table suggests they are only found in more extreme conditions (i.e. where the minimum temperature drops below −40 °C on occasions). The higher threshold temperature for leaf growth of deciduous broadleaf woodland suggests that the vegetation type would be able to outgrow conifers if the mean growing season temperature is suitable (as it is in England).

Question 3.1

(a) The herbivores ate 8% of *NPP*, and 3% of that went into secondary production. Therefore the efficiency of conversion of *NPP* into herbivore flesh is 3% of 8% = 0.24%.

To calculate the lions' efficiency of conversion, we first need to calculate the herbivores' production in terms of energy. This is done by multiplying *NPP* by the conversion factor just calculated:

$$12\,700\ \text{kJ m}^{-2}\,\text{yr}^{-1} \times 0.24\% = 30.5\ \text{kJ m}^{-2}\,\text{yr}^{-1}$$

The efficiency of conversion from this herbivore flesh into lion tissue is

$$\frac{\text{lions' production}}{\text{herbivores' production}} = \frac{0.004}{30.5} = 0.00013 = 0.013\%$$

(b) The amount of energy assimilated by the lions is 80% of 0.28 kJ m^{-2} yr^{-1} = 0.224 kJ m^{-2} yr^{-1}, so the proportion stored in their tissues is

$$\frac{0.004}{0.224} = 0.018 = 1.8\%$$

This is a tiny value. The vast bulk of the lions' energy intake must have been used in respiration for warmth, movement and repair.

Question 3.2

(a) This energy budget is typical of a detritus ecosystem, such as a mature temperate forest.

Reason: A high proportion of the primary production goes to form detritus and there is only one consumer level, so the grazing chain is weak. An insignificant amount of the primary production goes into storage, and the system is in balance because the energy entering the system is equal to the energy leaving it.

(b) This energy budget is typical of a storage ecosystem, such as immature temperate forest.

Reason: A high proportion of the primary production goes into storage, so that the energy entering the system exceeds the energy leaving it. Little of primary production is eaten because the grazing chain is weak.

(c) This energy budget is typical of a grazing ecosystem, such as a freshwater stream.

Reason: A high proportion of the primary production is eaten by herbivores, therefore the grazing chain is strong, and it can support a number of additional consumer levels. The primary production not consumed goes into detritus; little of it is stored. Therefore, the system is in balance because the energy entering it is equal to the energy leaving it.

Question 3.3

The pressure on the sponge is force/area.

force = 10 N

area = 0.2 m × 0.1 m = 0.02 m^2

Therefore the pressure applied is

$$\frac{10\ \text{N}}{0.02\ \text{m}^2} = 500\ \text{N m}^{-2} = 500\ \text{Pa}$$

The tension at which the water was being held within the sponge is therefore equal and opposite to this quantity and so we give it a negative value:

$$\psi = -500\ \text{Pa}$$

Question 3.4

For the CAM plant:

$$\text{WUE} = \frac{4.3\times10^3}{2.9\times10^5} = 1.5\times10^{-2} = 0.015$$

For the C3 plant:

$$\text{WUE} = \frac{6.2\times10^3}{5.0\times10^5} = 1.2\times10^{-2} = 0.012$$

The CAM plant has the higher WUE value, so can be said to use its water more efficiently.

Question 3.5

The sand has an available water capacity of 0.05. In a profile of 0.8 m, therefore, it can retain 0.8 m × 0.05 = 0.04 m of water. At a rate of 0.002 m d^{-1}, this will be exhausted in 0.04/0.002 = 20 days. The silt will supply water for 40 days, as its AWC is twice that of the sand. Similarly, the clay will supply water for 60 days and the loam for 80 days.

Question 3.6

(i) The potential productivities per plant are 65 g for Ca, 21 g for Al and 80 g for P. Therefore Al availability would be limiting growth as it has the lowest potential productivity. Aluminium would be limiting growth through a toxicity effect.

(ii) Following liming, the relevant productivities become 120 g for Ca, 80 g for Al and 45 g for P. Phosphorus would therefore be a limiting nutrient as it supports the lowest potential productivity. Adding calcium to a soil can restrict P availability by combining with phosphate ions to form insoluble calcium phosphate.

(iii) Liming would therefore increase the productivity of *Anthoxanthum* from 21 g to 45 g per plant.

Question 3.7

For one gram of tissue, we know the volume (V) is 1000 mm^3. We can use the relationship $V = \pi r^2 h$, and solve the equation for h.

For the hypha:

$$V = 1000\ \text{mm}^3 = 3.14 \times (0.002\ \text{mm})^2 \times h$$

Therefore

$$h = \frac{1000\ \text{mm}^3}{3.14 \times 0.000\,004\ \text{mm}^2} = 79\,600\,000\ \text{mm}$$

which we can round up to 80 000 m.

For the root:

$$V = 1000\ \text{mm}^3 = 3.14 \times (0.2\ \text{mm})^2 \times h$$

Therefore

$$h = \frac{1000\ \text{mm}^3}{3.14 \times 0.04\ \text{mm}^2} = 7960\ \text{mm} \approx 8\ \text{m}$$

The ratio of hyphal length to root length = 80 000 m : 8 m = 10 000 : 1.

Alternatively, you could have rearranged $V = \pi r^2 h$ to give $h = V/(\pi r^2)$, which shows that length (h) changes in proportion to the inverse square of the radius ($1/r^2$). Therefore, if the hypha has $\frac{1}{100}$ of the root's radius, it will have 100 × 100 = 10 000 times its length.

Question 4.1

The species' dispersal abilities would be a major factor. The isolation of the new woodland from other established woodlands is important in this respect. There needs to be a source of seeds for the ground flora that is close enough to enable dispersal mechanisms to transfer them to the site. Other factors, such as the former arable soil lacking organic matter and having an over-abundance of available nutrients, would favour competitive species such as bramble (*Rubus* spp.) over more typical woodland species such as the bluebell (*Hyacinthoides non-scriptus*).

Question 4.2

(a) Grassland on a sand dune is likely to be a stage in primary succession. Grasses colonize stabilized sand, but are gradually replaced by shrubs and trees as the soil develops.

(b) Heather on lowland soils is likely to be a deflected climax. It would tend to be replaced by shrubs and trees if it were not maintained by grazing and/or burning.

(c) Ash woodland on wet soils can be regarded as a stable climax. Ash is a tall forest tree tolerant of periodic waterlogging. In a natural river valley, the erosive power of water would keep the vegetation dynamic, so even this vegetation is not entirely stable.

(d) Tall grassland within an area of cleared woodland is likely to be a stage in a secondary succession. Following the removal of trees, grasses and herbs would exploit the gap in the canopy until tree seedlings were able to re-establish themselves, providing there was no continued management or disturbance.

(e) Birch trees on a drained bog is most likely to be a stage in a secondary succession. Draining the bog would have caused the original stable community dominated by bryophytes (e.g. *Sphagnum* spp.) to be lost and birch trees, being pioneer species with good dispersal ability able to colonize low-nutrient soils, would be one of the first species to establish. Birch rarely forms a climax community, being short-lived and intolerant of shade. If the site remained well drained, oaks would be likely to invade; if the drainage failed and the site became waterlogged, *Sphagnum* might re-establish.

Question 4.3

The completed passage is given below (with alternative correct answers shown where appropriate):

Plant distribution is determined primarily by *soil type*, *climate* and land use. The distribution of *terrestrial* invertebrates is determined mainly by the availability of food plants, but *aquatic* invertebrates are greatly affected by abiotic factors such as *water flow rate* and *oxygen availability/pH*, and bird distributions are mainly determined by either the availability of food plants or the *herbivores/invertebrates* that live upon them. Therefore, when land use changes, as a result of human activities such as *agricultural intensification*, *industrial pollution* or construction, the distribution and abundance of plant species and their associated invertebrates and birds also changes. The species that have the most *specialist* requirements in terms of resources and conditions, or are the most pollution-*sensitive* die out, leaving a more *generalist*, *tolerant* but *less* diverse *community* behind.

Acknowledgements for Part 2 *Life*

Grateful acknowledgement is made to the following sources for permission to reproduce material in this book:

Figure 2.1: Oxford Scientific Films/Colin Milkins; *Figures 2.2, 2.3, 2.4, 2.5, 2.6, 2.8, 2.11, 2.12, 2.16, 2.18a, 2.22, 2.23, 2.24a, 2.25, 2.27, 2.29, 2.31, 2.34, 2.36, 2.38, 2.45, 2.47b, 2.49, 2.50, 2.60, 2.61, 2.63, 2.66, 2.71a, 2.73, 2.75, 2.77, 2.80, 2.81, 2.82, 2.84, 3.6, 3.7, 3.9, 3.25, 3.46, 3.48b, 3.50, 3.55, 3.56, 3.57, 3.61, 3.63, 3.71, 3.72, 3.73, 3.74, 3.76, 4.7, 4.8, 4.9, 4.10, 4.13, 4.18, 4.21, 4.25a, 4.26, 4.27b, 4.30, 4.34, 4.35*: © Mike Dodd/Open University; *Figure 2.13*: Oxford Scientific Films/Sophie Evans; *Figures 2.15, 2.71b, 4.28*: Owen Mountford/Centre for Ecology and Hydrology; *Figure 2.18b*: Science Photo Library/Eric Grave; *Figure 2.18c*: Science Photo Library/Astrid & Hanns Frieder Michler; *Figure 2.24b*: Oxford Scientific Films/Kathie Atkinson; *Figures 2.26, 3.42, 3.54, 3.60*: David Gowing/Open University; *Figure 2.30*: Data Courtesy of Professor Mick Crawley/Imperial College; *Figure 2.32*: Oxford Scientific Films/Paulo de Oliveira; *Figure 2.34*: Oxford Scientific Films/Deni Brown; *Figure 2.35*: Oxford Scientific Films/Peter Parks; *Figure 2.37*: Oxford Scientific Films/Richard Ringland; *Figure 2.39*: Oxford Scientific Films/Stephen Dalton; *Figure 2.42*: Oxford Scientific Films/TC Nature; *Figures 2.43, 2.59c*: Oxford Scientific Films; *Figure 2.47a*: Oxford Scientific Films/Tim Shepherd; *Figure 2.56*: © Dave Hollis/Wildscape; *Figures 2.57, 2.59a, 2.59b*: Oxford Scientific Films/Robin Redfern; *Figures 2.62, 4.33*: *The Millennium Atlas of Butterflies in Britian and Ireland*, © Courtesy of Butterfly Conservation/Centre for Ecology and Hydrology; *Figure 2.70*: © Clare Lawson; *Figure 2.72*: Jill Rapson/Massey University; *Figure 2.83*: Simone Pitman/Open University; *Figure 3.3*: Taiz, L. and Zeiger, E. (1998) *Plant Physiology*, 2nd edn, Sinauer Associates, Inc.; *Figure 3.11*: Eric Grave/Science Photo Library; *Figure 3.12*: Oxford Scientific Films/Michael Leach; *Figure 3.15*: Data Courtesy of Tim Hess/Cranfield University; *Figure 3.32*: Courtesy of the Desert Research Foundation of Namibia; *Figure 3.36*: © Eye of Science/Science Photo Library; *Figure 3.48a*: © From *Biology of Plants* by P. Raven, R. F. Evert and S. E. Eichhorn, © 1999 by W. H. Freeman and Company/Worth Publishers. Used with permission; *Figures 3.49, 3.66*: © Courtesy of National Soil Resources Institute, Cranfield University; *Figure 3.51*: Oxford Scientific Films/Tobias Bernhard; *Figure 3.58*: Rob Nunnington/Oxford Scientific Films; *Figure 3.59*: Science Photo Library/Dr Jeremy Burgess; *Figure 3.77*: Geoscience Features Picture Library; *Figure 4.1*: Science Photo Library/Carole A. McKeone; *Figure 4.3*: Bob Spicer/Open University; *Figures 4.4a, 4.27a*: Courtesy of Glenn and Barbara Halliday, NPSO Salem, Or., USA; *Figure 4.4b*: Courtesy of Dr Iza Gogoff; *Figures 4.14, 4.19, 4.20*: Richard Bardgett/Lancaster University; *Figure 4.22*: Data Courtesy of Richard Bardgett/Lancaster University; *Figure 4.23*: Oxford Scientific Films/Duncan Murrett; *Figure 4.27a*: Oxford Scientific Films/Chris Knights; *Figure 4.32*: Oxford Scientific Films/Terry Button; *Figure 4.36*: by permission of the Vincent Wildlife Trust.

The figures listed below have been adapted from the following sources:

Figure 2.14: Figure adapted from Beckett, B. S. (1982) *Biology — A Modern Introduction*, 1st edn, Oxford University Press; *Figure 2.67*: Data adapted from Ellenberg, H. (1952) Physiologisches und okologisches verhalten derselben Planzenarten, *Berliner Deutsch Botanische Gesturn*, **65**, pp. 350–361; *Figure 2.69*: Adapted from Miles, J. (1979) *Vegetation Dynamics*, Chapman and Hall Publishers, pp. 38–39; *Figure 2.85*: Adapted from Woodward I., 'Plants in the Greenhouse World', Inside Science No. 21, *New Scientist*, 6 May 1989; *Figure 3.39*: Adapted from Salisbury F. B. and Ross C. W. (1978) *Plant Physiology*, 2nd edn, Wadsworth Publishing Company; *Figure 3.43*: Data adapted from Smith, L. P. and Trafford, B. D. (1976) *Climate and Drainage*, MAFF Technical Bulletin No. 34, HMSO; *Figure 3.44*: Data adapted from Smith, L. P. and Trafford, B. D. (1976) *Climate and Drainage*, MAFF Technical Bulletin No. 34, HMSO; *Figure 3.64*: Data adapted from Davies, M. S. and Snaydon, R. W. (1973) 'Physiological differences among populations of *Anthoxanthum odoratum* L. collected from the Park Grass Experiment, Rothamsted: Part I Response to calcium', *Journal of Applied Ecology*, **10**, pp.33–45; Davies, M. S. and Snaydon, R. W. (1973) 'Physiological differences among populations of *Anthoxanthum odoratum* L. collected from the Park Grass Experiment, Rothamsted: Part II Response to aluminium', *Journal of Applied Ecology* **10**, pp. 47–55; Davies, M. S. and Snaydon, R. W. (1974) 'Physiological differences among populations of *Anthoxanthum odoratum* L. collected from the Park Grass Experiment, Rothamsted: Part III Response to phosphate', *Journal of Applied Ecology* **11**, pp. 699–707; *Figure 3.68*: Data adapted from Howeler, R. H., Asher, C. J. and Edwards, D. G. (1982) 'Establishment of an effective endomycorrhizal association on cassava in flowing solution culture and its effects on phosphorus-nutrition', *New Phytologist*, **90**(2), pp. 229–238; *Figure 3.70*: Data adapted from Bethlenfalvay, G. J., Bayne, H. G. and Pacovsky, R. S. (1983) 'Parasitic and mutualistic associations between a mycorrhizal fungus and soybean: the effect of phosphorus on host plant–endophyte interactions', *Physiologia Plantarum*, **57**(4), pp. 543–548; *Figure 4.15*: Figure adapted from Crocker, R. L. and Major, J. (1955) 'Soil development in relation to vegetation and surface age at Glacier Bay, Alaska', *Journal of Ecology*, **43** (2), pp. 427–448; *Figure 4.16*: Data adapted from Crocker, R. L. and Major, J. (1955) 'Soil development in relation to vegetation and surface age at Glacier Bay, Alaska', *Journal of Ecology*, **43**(2), pp. 427–448.

Every effort has been made to trace all the copyright owners, but if any has been inadvertently overlooked, the publishers will be pleased to make the necessary arrangements at the first opportunity.

Index

Note: Entries in **bold** are key terms. Page numbers referring to information that is given only in a figure or caption are printed in *italics*.

A

B

T

U

V

W

X

Y

Z